Thin Films

Modeling of Film Deposition for Microelectronic Applications

VOLUME 22

Serial Editors

Honorary Editors

* Passed away in 1995.

Recent volumes in this serial appear at the end of this volume.

Thin Films

Modeling of Film Deposition for Microelectronic Applications

Edited by

Stephen Rossnagel
IBM Corporation
T. J. Watson Research Center
Yorktown Heights, New York

VOLUME 22

ACADEMIC PRESS
San Diego London Boston New York Sydney Tokyo Toronto

This book is printed on acid-free paper. ∞

Academic Press, Inc.
525 B Street, Suite 1900, San Diego, California 92101-4495, USA
http://www.apnet.com

Academic Press Limited
24-28 Oval Road, London NW1 7DX, UK
http://www.hbuk.co.uk/ap/

International Standard Serial Number: 1079-4050

International Standard Book Number: 0-12-533022-7

PRINTED IN THE UNITED STATES OF AMERICA
96 97 98 99 00 01 QW 9 8 7 6 5 4 3 2 1

Contents

A Process Model for Sputter Deposition of Thin Films Using Molecular Dynamics

C.-C. Fang, V. Prasad, R. V. Joshi, F. Jones, and J. J. Hsieh

Feature Scale Transport and Reaction during Low-Pressure Deposition Processes

Timothy S. Cale and Vadali Mahadev

Contributors

Numbers in parentheses indicate the pages on which the authors' contribution begins.

MICHAEL J. BRETT (1), Department of Electrical Engineering, University of Alberta, Edmonton, Alberta T6G 2G7, Canada

TIMOTHY S. CALE (175), Department of Chemical, Biological, and Materials Engineering, Center for Solid State Electronics, Arizona State University, Tempe, Arizona 85287

STEVEN K. DEW (1), Department of Electrical Engineering, University of Alberta, Edmonton, Alberta T6G 2G7, Canada

C.-C. FANG (117), Process Modeling Laboratory, State University of New York, Stony Brook, New York 11794

S. HAMAGUCHI (81), T. J. Watson Research Center, IBM Corporation, Yorktown Heights, New York 10598

J. J. HSIEH (117), T. J. Watson Research Center, IBM Corporation, Yorktown Heights, New York 10598

F. JONES (117), T. J. Watson Research Center, IBM Corporation, Yorktown Heights, New York 10598

R. V. JOSHI (117), T. J. Watson Research Center, IBM Corporation, Yorktown Heights, New York 10598

VADALI MAHADEV (175), Department of Chemical, Biological, and Materials Engineering, Center for Solid State Electronics, Arizona State University, Tempe, Arizona 85287

V. PRASAD (117), Process Modeling Laboratory, State University of New York, Stony Brook, New York 11794

TOM J. SMY (1), Department of Electronics, Carleton University, Ottawa, Ontario KIS 5B6, Canada

Preface

Thin films in various forms have been deposited under vacuum conditions for over a hundred years. The applications touch many areas of everyday life, including the semiconductors and displays in computers, window and optical coatings, hard or decorative films, films on plastic for food packaging, gold-colored films on car trim, and so on. In many of these cases, the films are initially deposited by trial and error, with a minimum understanding of the physics of the film or the deposition process.

In the past 30 years, much progress has been made in understanding the growth and microstructure of flat films. The nucleation processes for thin films have been well understood and documented. The interplay between microstructure and the deposition process has also been the subject of much research. In many cases it has been possible to understand the phenomonological relations between deposition conditions (kinetic energy, temperature, and directionality) and the resulting film microstructure. The classic example of this is the well-known Thornton Zone Diagram, which described the physical film microstructure as a function of deposition pressure (relating to various kinetic energy effects in sputtering) and sample temperature (relating to surface mobility and diffusion for the depositing film atoms). This visual model can be used as a diagnostic to predict how changing a deposition parameter (e.g., pressure, temperature, or even film species) might result in a change in the microstructure and the appearance of the film.

One of the longstanding challenges of thin film technology has been to obtain an understanding of the deposition process deep enough to explain how a certain film ended up with the properties it has and, even more importantly, to *predict* how a change in some aspect of the deposition would result in a change in the film. This challenge has resulted in the development of a wide range of computer models, which use various mathematical models and the physical properties of the system to describe the film deposition process.

It is this topic, the computer modeling of film deposition processes, which is the subject of this book. However, even this topic is much too detailed and wide ranging for coverage in a few hundred pages. A subset of this field is the deposition and modification of films used for semiconductor applications, and coverage of this subset is the primary goal of this volume. Semiconductors are used not only in computers worldwide, but also in the automotive industry, in communications, in credit cards and other identification applications, and even in such routine applications as home appliances.

At the cutting edge of semiconductors, the complexity and expense of trial and error can be enormous. Simply analyzing a film which has a thickness of a hundred or so angstroms is very difficult, requiring about 1 worker-day of work for TEM preparation and analysis. Also, as the dimensions of the films on the wafer shrink, the wafer size is increasing. At present, 200-mm wafers are the standard, but the expectation is a transition to 300 mm in the next few years. Considering a wafer with perhaps 200 chips, each with 256–1000 million circuits on each chip with five to six levels of metallurgy and a few dozen vias and/or lines for each bit, one comes up with a number in the trillions of potential samples to analyze for each wafer. Obviously, any physical analysis will be a token measure of what has occurred in the processing of these films.

Computer modeling, as it is currently practiced, is still an approximation of reality. There are two different general approaches to modeling, and this volume examines examples of both. The models fall into the general classifications of analytic or discrete. In the analytic approach, the film surface is described by an equation, usually in the form of a sequence of line segments, and this equation is modified by the various processes (deposition, diffusion, etc.) that occur during deposition. In the discrete approach, the substrate surface is viewed as an array of distinct points or disks. Typically, these points or disks are representative of a large number of atoms, but are treated physically as a single atom. Physical processes may occur on these disks, such as adsorption, diffusion, and incorporation, which will be consistent with single atom physics. However, the use of a disk to represent a large number of atoms (typically 1000–100,000) drastically reduces the computational requirements and makes solutions more practical. This discrete approach is often described by the term "molecular dynamics" (MD).

In this volume, two chapters are included on the analytic-or-continuum approach, and two chapters are included describing discrete or MD-like models. In each case, the authors have included a description of the thin film deposition model and real, experimental results to which the model can be compared.

Each of the modeling approaches described in this volume is predictive as well as explanatory. Actually, this is the key goal of modeling technology, the so-called "virtual experiment," where the computer model takes the place of the physical experiment. In this way, new technologies, new sample geometries, and even chemical effects can be simulated and predicted without the need for experimental confirmation. While this approach is routinely used in the circuit design of chips, it is only now becoming possible for actual, physical film deposition through the use of these modeling approaches.

Chapter 1, by the group of Mike Brett and Steve Dew at the University of Alberta and Tom Smy at Carleton University, describes an approach that examines both the film deposition process and the transport process of atoms from the source (e.g., a magnetron cathode) to the sample to make the film. The transport part of the model (SIMSPUD) describes the energy and angular distributions of the species as they arrive at the substrate. This can take an analytic form for high pressures, which can be characterized by diffusive flow, or can take a discrete or Monte Carlo approach for cases, such as sputtering, where the mean free path of the particles is on the same order as the system dimensions (centimeters) so that transport is neither diffusive nor purely ballistic.

The film deposition process simulation known as SIMBAD was first developed in 1987 and uses a molecular dynamics-like approach where the film is represented by an aggregation of 10,000 to 100,000 2-dimensional disks, each representing the statistical average of several thousand atoms with similar trajectories. This approach is chosen so that the dimensions of the disks are much smaller than the surface topography, but large enough to facilitate practical computer operating time on a workstation. Disks arrive at the film surface based on the transport model and are then subject to the various physical effects (shadowing, diffusion, desorption, etc.) which occur at the surface in the particular experiment. The film is then the sum of these many thousands of disks which have arrived and been modified as they became part of the film.

Chapter 2, by Satoshi Hamaguchi of IBM, falls mostly into the analytic class of models. In this model, the surface boundary is represented by a piecewise linear function. The normal velocity of the surface at each point is then calculated by using a shock-tracking algorithm. This approach most correctly handles the physics of intersecting line segments in a way similar to shock waves in gas dynamics. The model is pointed toward sputter deposition with the additional capability of ionized deposition. In the latter case, the depositing particles may also etch the surface, resulting in local redeposition and topography modification.

Chapter 3, by C.-C. Fang and V. Prasad at SUNY–Stony Brook and R. Joshi, F. Jones, and J. Hsieh of IBM, uses a molecular dynamics approach to film deposition. The authors also use a Monte Carlo model of atom transport to describe how atoms move from a local sputtering target to the sample. This allows for gas scattering of the atoms in-flight and angular filtering, which might occur as the atoms pass through a collimator, or for long target-to-sample distances. This work focuses primarily on the relationships between the film microstructure, the adatom energy, ion bombardment, gas entrapment, and the stresses of the films.

Chapter 4, by Tim Cale and Vadali Mahadev at Arizona State University, describes a model known as EVOLVE. In this model, the transport is handled in three dimensions and the physical surface is represented in two dimensions, typically as a trench (line) or a circular via (hole). The surface is represented by line segments, and the various physical processes are determined for each segment. This approach allows the introduction of a variety of physical and chemical effects, including diffuse and specular re-emission from a surface, chemical reactions (deposition or etching), and thermal effects, such as surface diffusion, which may also alter the profile of the deposited feature.

This modeling approach has been applied to a wide variety of systems, and computational and experimental data are developed for both physical vapor deposition (PVD) and a variety of chemical vapor deposition (CVD) and plasma-enhanced CVD applications. Examples include CVD of silicon dioxide from TEOS, CVD of W, PECVD of silicon dioxide, PVD of Ti/TiN with and without collimation, and more complicated PVD systems which use deposition from ions and neutrals (I-PVD).

Even with the interesting and significant results shown in these chapters, the field of thin film modeling is still rapidly developing. This is due to the inexorable increase in computing capability. The models in this volume have two general characteristics: they approximate the surface in units which are significantly larger than a single atom, and they are linear in that a surface profile is examined a single point or line segment at a time, rather than at multiple points simultaneously.

The advent of readily available parallel computing in the mid-1990s finally opened the door to a dramatic evolution in computer modeling. Since thin film deposition involves extraordinary numbers of atoms and yet fairly low energies, it is easily possible to process a large number of points on the surface concurrently, as they are simply too many atoms apart to interact. Thus, the evolution of a complicated surface can be modeled or built from single atoms within reasonable computer–process–time limitations or expense. This opens the door for the true virtual experiment building films from first principles in single atom increments. The early work on this approach is already being done, and it is expected that within a few years some truly interesting results and predictions will become possible.

Stephen Rossnagel

Thin Film Microstructure and Process Simulation Using SIMBAD

MICHAEL J. BRETT AND STEVEN K. DEW

Department of Electrical Engineering, University of Alberta, Edmonton, Alberta, Canada T6G 2G7

AND

TOM J. SMY

Department of Electronics, Carleton University, Ottowa, Ontario, Canada K1S 5B6

I. Introduction

A. Thin Film Microstructure and Its Implications

Both sputtered and CVD-deposited films are known to have a columnar microstructure that is generally considered to be a result of nonequilibrium deposition conditions. A general classification scheme for thin film microstructure was developed by Movchan and Demchishin (*1*) and by Thornton (*2*) that charts the transition from a highly porous columnar film (zone I) to a dense columnar grain structure (zone II) and then to a recrystallized dense grain structure (zone III).

If the film is deposited at temperatures that are low with respect to the melting temperature, this microstructure is generated by the processes of low adatom mobility associated with rapid condensation and self-shadowing by the film. The microstructure consists of individual columns separated by voids and grain boundaries. An example of a zone I structure is shown in Fig. 1, a cross-sectional micrograph of a thick W film sputtered over trenches in oxide. The size of the columns is determined primarily by the surface diffusion length of the depositing flux. Columnar orientation is determined by the average angle of incidence of the sputtered flux relative to the local substrate normal and is discussed in more detail in Sec. III.B.

At high substrate temperatures corresponding to zones II and III of the structure zone model, the internal structure of the film is characterized by tightly packed grains separated by grain boundaries. The grain size is determined by the surface diffusion length and bulk diffusion processes, and can vary greatly depending on the material, deposition rate, system cleanliness, and substrate temperature. A proper simulation of film growth in this regime (i.e., "hot" Al deposition at temperatures greater than 700 K) requires physically accurate modeling of the processes of wetting, grain boundary grooving, minimization of surface and interfacial energies, and bulk diffusion.

The microstructure of the thin metal films used in VLSI interconnects has a significant effect on the properties of these films and the resulting quality of the metallization. Refractory metals are commonly used as contact, barrier, and adhesion layers and also as a material for contact and via hole filling. The very high melting point of these materials results in a zone I microstructure for both sputtered and CVD films. The nature of this microstructure affects the resistivity, etching characteristics, oxidation, index of refraction, permittivity, diffusion properties, and surface roughness of the deposited film. These properties are obviously crucial in

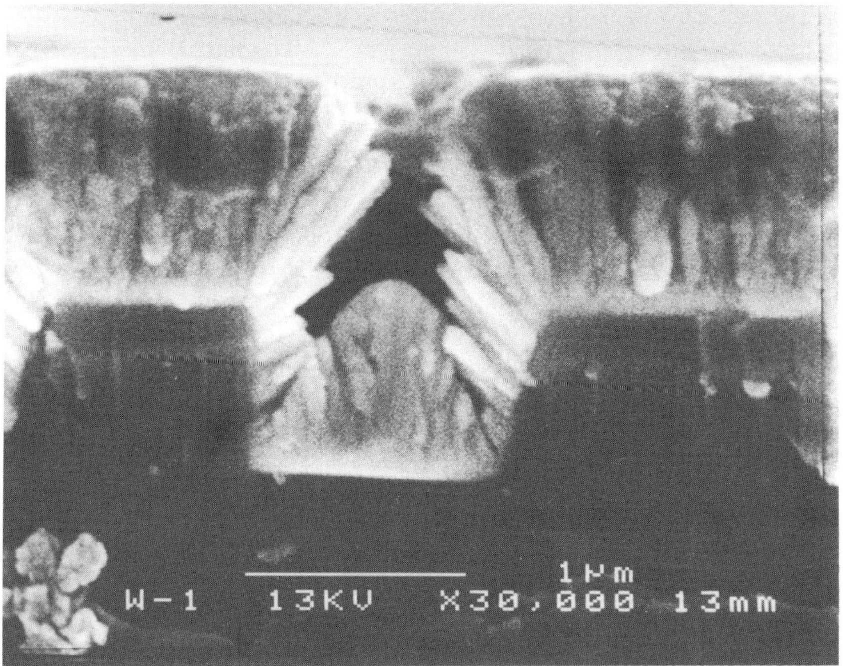

FIG. 1. SEM micrograph of a W film sputter deposited over a 1-μm-wide trench (from Ref. 63).

determining the usefulness of the layer. In addition, the microstructure of the film is, in part, a consequence of the angle between the average incident flux direction and the local surface normal. As such the microstructure of a film deposited over topography is a strong function of position and the film properties will not be homogeneous. Also, the microstructural properties of the low-resistance metal film that forms the bulk of the interconnect (usually Al) are very important. In interconnect failure by electromigration, both the rate of failure and the failure mode are strongly dependent on the microstructure of the metal lines.

B. Modeling Thin Film Microstructure

To model correctly the deposition of a thin film by physical or chemical methods, two distinct physical processes must be addressed. These are the transport of the depositing species from the source to the substrate, and the incorporation of the material into the growing film.

The modeling of the transport of flux to the film surface must take into account the process involved (sputtering, CVD), the geometry of the system (spacings, wafer size, etc.) and process parameters (power levels, pressure, etc.). The transport model should predict energy and angular distributions at the wafer surface for the various fluxes important to the deposition process. The growth of the film and the resulting film microstructure are very sensitive to these distributions. The complexity of the transport model used is determined in part by the method of deposition. For example, CVD is usually performed at high enough pressures to enable characterization of transport as a diffusive flow. In this case it is possible to derive an analytic description of the angular and energy distributions. Sputter deposition is, however, more problematic because the pressures used lead to a mean free path of the order of 1 to 10 cm. At these pressures the flux transport can be characterized as neither a pure diffusive flow nor simple line-of-sight ballistic transport. A solution to this problem is to use a Monte Carlo model of the transport which determines the flux distributions by following a large number of transport events from the target, through a number of collisions with the background gas, and to the substrate.

When modeling the growth of the film produced by incident flux, the two basic processes of shadowing and diffusion need to be considered. The primary and critical process is shadowing of incident flux. If the film is deposited over topography (such as a via or trench), certain regions of the film will receive less flux than others, resulting in a nonuniform film thickness. This effect is strongly dependent on the angular distribution of the incoming adatom flux. Thus the model must correctly predict the magnitude and direction of the flux striking each point of the film surface and also incorporate shadowing of the flux by both the initial topography and the growing film. For the case of a nonunity sticking coefficient or film resputtering, the situation becomes complex. One approach is to treat the entire film surface as a source of re-emitted material and to find a self-consistent solution to the flux striking each surface point (*3*). Alternatively, as described in this chapter, a modeling approach based on collecting an aggregation of units of film material can be used.

Film microstructure evolution is also strongly determined by the surface mobility or diffusion of the depositing species after they strike the growing film. The primary driving force behind this diffusion is a reduction in the surface and interfacial energies. Thus, for long diffusion lengths, a model must include the effect of a chemical potential created by the surface curvature of the film. At short diffusion lengths, local relaxation processes should be permitted.

Although the physical processes just described are essential to modeling the deposition of a thin film, the prediction of the development of the internal film microstructure requires additional processes to be incorporated. As atoms initially deposit, they will agglomerate into islands, each with a specific crystal orientation. The average size of the islands is determined by the diffusion length of the adatoms, whereas the shape of the islands is determined by crystal faceting processes and the wetting angle between the depositing film and the substrate. As islands coalesce to form a continuous film, grain boundary grooving processes will further influence microstructure development. The resultant wetting angles and grain boundary grooving angles are consequences of minimization of the surface energy of the film. As the film continues to grow, microstructure is further influenced by local shadowing of the incident adatom flux by neighboring grains, and by the resulting growth "competition" between grains. The more oblique and narrower the incident adatom distribution, the more porous and lower density microstructure produced.

The SIMBAD process simulator was conceived and developed in 1987 in response to a need for microstructure information about films deposited over topography. Popular commercial profile simulators at that time such as SAMPLE (*4*), SPEEDIE (*5*), and DEPICT (*6*) each used 2-D string segments to represent the film surface, and incorporated string advancement algorithms to model deposition or etching of films. Unfortunately, these could not represent the internal information needed to depict microstructure. Alternatively, research on ballistic aggregation models developed by Meakin and others (*7–9*), revealed the fundamental processes of shadowing and accretion that led to the formation of a columnar microstructure in vapor-deposited thin films. Molecular dynamics simulations, incorporating interatomic potentials and atomic hopping processes (*10*), were capable of evaluating microstructure formation only in very small (i.e., thousands of atoms) portions of film. SIMBAD, an acronym for SIMulation by BAllistic Deposition, incorporated features from each of the existing models at that time, by utilizing a ballistic aggregation of disks and physically correct surface diffusion algorithms to generate both surface profile and film microstructure information. The features and predictive power of SIMBAD have been extended considerably since the late 1980s, and are described in detail in this chapter.

The remainder of this chapter is structured as follows. Section II describes the transport and deposition algorithms used in the two primary programs, SIMSPUD and SIMBAD. Section III describes key experiments that were used to verify these models. In Sec. IV, the utility of SIMBAD is demonstrated through application to various deposition and etch processes of current interest. Section V describes a new algorithm for modeling grain structure evolution and film growth involving kinetic and surface energy considerations, and Sec. VI concludes this chapter.

II. Description of the Transport and Deposition Algorithms

Simulation of film growth requires two stages: simulation of macroscopic flux transport to the substrate to generate the correct spatial, angular, and energetic distributions of the incoming species, and the microscopic simulation of growth of the film incorporating nucleation, shadowing, and diffusion. In SIMBAD, the transport stage is known as SIMSPUD, an acronym formed from SIMulation of SPUtter Distributions. SIMSPUD was written in 1989 specifically for deposition by sputtering in order to improve the predictive power of the simulations. The latter stage of the simulation or condensation algorithm for CVD or PVD films is performed by the core SIMBAD model.

For the case of a sputter system, the two simulation regimes are illustrated in Fig. 2. The flux distribution at the substrate is a function of deposition geometry, sputter source design, target material properties, sputter gas pressure, and presence or absence of a collimator. Only after the incident sputter flux distribution has been obtained can an accurate simulation of film growth over topography be performed.

A. The SIMSPUD Vapor Transport Model

SIMSPUD is a three-dimensional Monte Carlo vapor transport simulation. In this model, individual sputtered particles are tracked from the point of emission at the target, during collisions in the sputter gas, and up to deposition on the substrate. The position, angle, and energy of each particle are tracked to collect statistics on the corresponding sputter distribution.

Particles are emitted from the target at random positions which account for the erosion profile of the sputter target. This profile is generally nonuniform for magnetron sources and can be directly measured from an eroded target. Similarly, the emission angular distribution from the target can be input from experimental results (*11*), or a simple cosine distribution can be used. The cosine is justified by a simple theory although nonideal behavior is often observed at the energies typical of sputtering (*12*). The energy of particles emitted from the target is presumed to follow a Thompson distribution (*13*). As a parameter of this distribution, an aver-

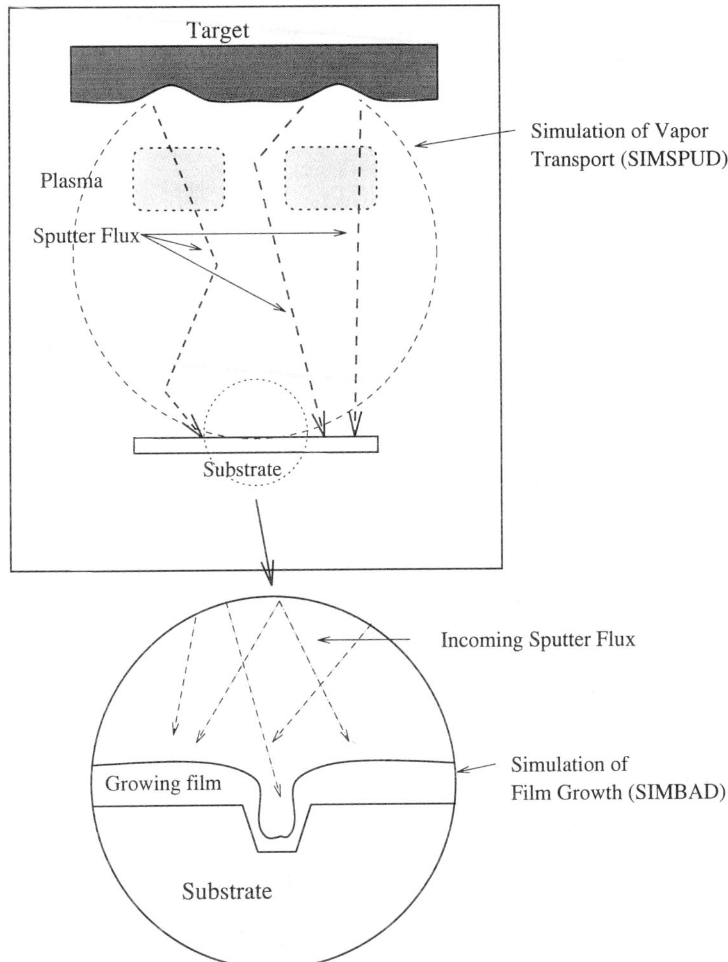

FIG. 2. Schematic drawing of a sputter system showing the transport and deposition regimes to be simulated (Reprinted from *Can. Metall. Quart.*, **34**, T. Smy, S. K. Dew, and M. J. Brett, "Simulation and experimental analysis of refractory metal and aluminum deposition over high aspect ratio VLSI topography," pp. 195–202, 1995, with kind permission from Elsevier Science Ltd., The Boulevard, Langford Lane, Kidlington, 0X5 1GB, UK).

aged sputtering ion energy is assumed with a value of 0.73 for the target voltage (*14*).

Due to the working gas pressure, sputtering particles suffer collisions *en route* to the substrate which modify the emission distributions. An elastic interaction is assumed, following a hard sphere model (*15*). The collision frequency is governed by the collision cross-section σ, which determines a mean free path λ_m given by (*16*):

$$\lambda_m = \frac{kT}{p\sigma}, \tag{1}$$

where kT is the thermal energy of the gas and p is the gas pressure. For sputtering, σ can be a strong function of particle energy. Unfortunately, the detailed energy dependence of the collision cross-section is not well established. In SIMSPUD, the relative dependence reported by Robinson for Ar (*17*) is extrapolated to other materials by assuming a suitable scale factor. In the regime of interest, this dependence fits very well to a power law described by:

$$\sigma(E) = \sigma_0\left(\frac{E}{E_0}\right)^{-0.29} \quad \text{for } E > E_0, \tag{2}$$

where $E_0 = 1\,\text{eV}$ and $\sigma_0 = \sigma(E_0)$. Below E_0, the particle is presumed to be nearly thermalized, and σ can be approximated using (*18*):

$$\sigma_t = \pi(1 + m/m_g)^{0.5}(r + r_g)^2, \tag{3}$$

where m and m_g are the masses and r and r_g are the radii of the thermalized sputtered atoms and gas atoms, respectively. Thus a description of the cross-section energy dependence requires two parameters, σ_0 and σ_t, which may be found in the literature or determined by experiment (as detailed in Sec. III.A).

Using these relationships, the energy, position, and direction of a large number of particles can be tracked. Macroscopic quantities such as deposition efficiency or blanket uniformity over a wafer are easily calculated. By considering only those striking a selected portion of the substrate, averages, distributions, and other statistical information can be estimated for the flux angular, energy, and spatial probabilities. This information can be directly input into the SIMBAD simulator for film growth depictions. The effects on the flux transport process of utilizing a physical collimator (hexagonal, square, or circular grid) are accounted for in SIMSPUD by removing those particles whose trajectories *en route* to the substrate intersect with the walls of the collimator. Thus the SIMSPUD model incorporates the effects of nonuniform target erosion, variable target-substrate and collimator geometries, material and energy-dependent scattering behavior, and working gas pressure.

B. The SIMBAD Deposition Model

The SIMBAD deposition model consists of the core growth algorithm and a number of extensions specifically focused on different film technologies. A description of the basic algorithm follows in this section, with extensions to processes such as bulk diffusion, bias sputtering, and film composition covered in Sec. IV. In the basic SIMBAD model, the film is represented in two dimensions by an aggregation of 10^4 and 10^5 2-D disks. Each disk represents the statistical average of a large number of physical atoms (usually several thousand) with similar trajectories. Thus the model tracks the statistical movement and deposition of atoms within the simulation region. The size of the disk is chosen to be much smaller than the microstructural features to be modeled, but large enough that the total film can be represented using the resources (memory and execution speed) of a typical computer workstation. Within these constraints, disk size can be varied by about a factor of about four without significantly changing the film representation. Typical widths of the simulation region are 0.5 to 5 μm.

The substrate is initialized to the device topography by a series of straight line segments, or a previously deposited film. Disks are launched from just above the film surface with an angular distribution determined by SIMSPUD or input by the user from measurements or knowledge of the deposition process. Since this is a Monte Carlo model, launch trajectories are determined randomly in a manner consistent with the angular distribution. Because the mean free path for adatoms in the gas phase is very large with respect to the simulation region, these disks move ballistically until they contact the growing film or the initial substrate. The ballistic approach intrinsically incorporates the effects of self-shadowing by the film which gives rise to microstructure evolution, and of shadowing by topography which can create poor coverage over nonplanar features. Disks may desorb (dependent on the sticking coefficient) or will be incorporated in the film after undergoing diffusion (utilizing a user-input diffusion length) to minimize the surface chemical potential and reduce local concentration gradients.

The surface potential in two dimensions can be written as (*19*):

$$\mu = \mu_0 + \frac{\gamma}{R}\Omega, \tag{4}$$

where μ_0 is the chemical potential of a flat surface, γ the surface tension of the film material, Ω the atomic volume, and R the radius of curvature. It is the gradient of this curvature which will cause the preferential

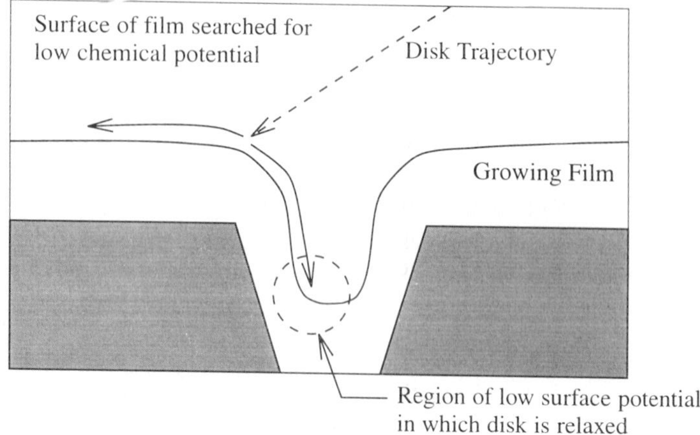

FIG. 3. In the SIMBAD model, once the disk strikes the film, the surface is scanned for areas of low chemical potential where the disk may be locally relaxed (S. K. Dew, T. Smy, and M. J. Brett, "Simulation of Elevated Temperature Aluminum Metallization Using SIMBAD," *IEEE Trans. Electron Dev.* **39**. © 1992 IEEE).

diffusion of adatoms into areas, such as inside vias, where the surface is concave. Using this expression, a relative surface potential for each surface disk can be calculated. The surface of the growing film is searched within the specified diffusion length to determine the point of lowest surface energy, as illustrated in Fig. 3, where the incident disk will be incorporated into the growing film.

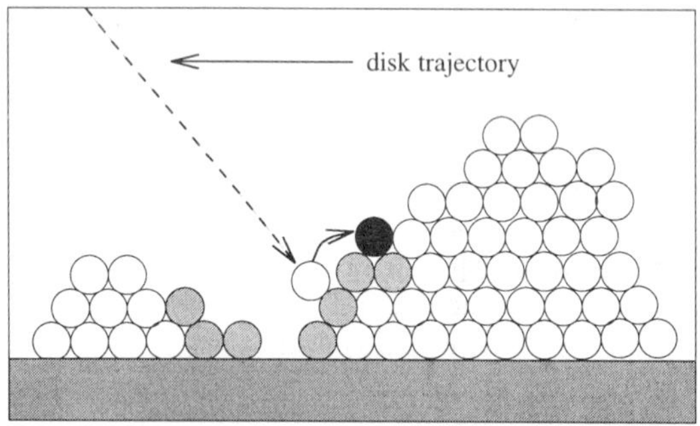

FIG. 4. Local relaxation of an incident disk in the basic SIMBAD model. Gray disks indicate potential nearest neighbors within a radius equal to four disk diameters (S. K. Dew, T. Smy, and M. J. Brett, "Simulation of Elevated Temperature Aluminum Metallization Using SIMBAD," *IEEE Trans. Electron Dev.* **39**. © 1992 IEEE).

A final relaxation step over a small region, shown in Fig. 4, is used to achieve optimum disk packing. Within a range of about four disk diameters, the disk will come to rest at a position which cradles against the largest number of previously deposited disks. The disks are packed hexagonally with each grain/column having its own unique identity, and the disks' data structures contain information on material type, interface information, and surface curvature. Thus this two-stage algorithm correctly models surface diffusion at high substrate temperatures driven by minimization of surface energy, and implements short-range disk diffusion to ensure dense disk packing.

One of the advantages of microstructure depiction by SIMBAD is the ability to calculate the relative local density of the film. A clear and simple example of local density variation due to intercolumn voiding is shown in Fig. 5, for MgF_2 evaporated over oxide lines at an incident angle of 5° from the substrate normal (*20*). Extreme self-shadowing on the left sidewall of the oxide has created a voided microstructure and corresponding lower density region. On the right, shadowing is complete and a large void has formed. The simulation of Fig. 6a depicts this microstructure. By changing random number seeds, statistical variations of the film representation can be produced, each having a unique (yet nominally correct) microstructure. Relative local density is calculated by determining the number of disks within a certain radius, and averaging this number over a series of simulations with different random number seeds. The density values are normalized to a hexagonal closest packed arrangement of disks, since this is the densest possible arrangement. The corresponding density plot shown in Fig. 6b is particularly well suited to identifying regions, such as the left sidewall film, in which the substrate topography has resulted in a porous and less dense film. Darker shades represent regions of high density on this figure. The surface profile of the film can be defined as a surface of constant (near-zero) density or taken directly as the surface disks from the disk representation.

Although SIMSPUD generates a 3-D incident flux distribution, the growth model of SIMBAD is two dimensional. As such, it is directly comparable only to very long features such as trenches. The immense memory and computation requirements of a full 3-D ballistic simulation using spheres currently precludes such an approach. However, since most ULSI structures are 3-D with corresponding 3-D issues involving shadowing and out-of-plane surface curvature, SIMBAD has written to enable "quasi–3-D" film growth. In this algorithm (*21*), shadowing in the third dimension is taken into account, but the disks are accreted in a 2-D slice of the structure. The limitation of the quasi–3-D algorithm is that it assumes radially symmetric incident flux. This restriction was removed in

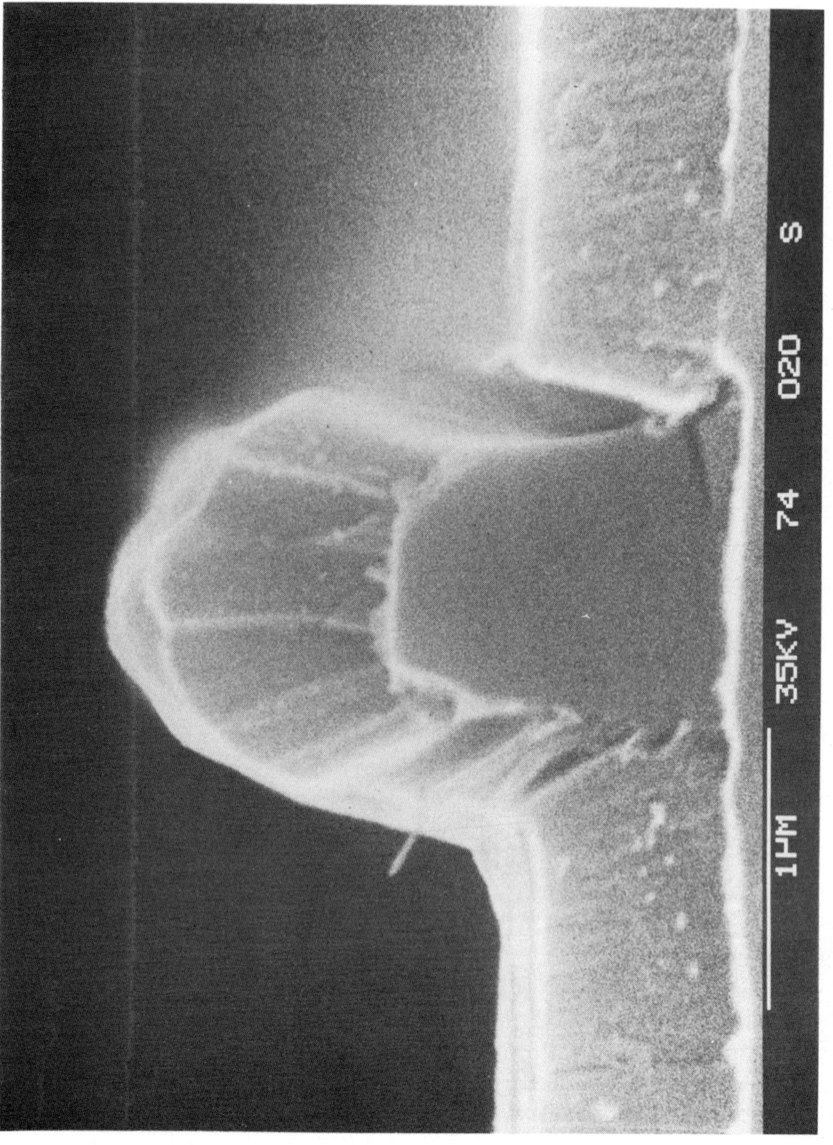

FIG. 5. SEM micrograph of an MgF_2 film evaporated over an oxide line, at an angle of incidence 5° to the left of the vertical (Reprinted from *Thin Solid Films*, **187**, R. N. Tait, T. Smy, and M. J. Brett, "A ballistic deposition model for films evaporated over topography," pp. 375–384, 1990, with kind permission from Elsevier Science S. A., Lausanne, Switzerland).

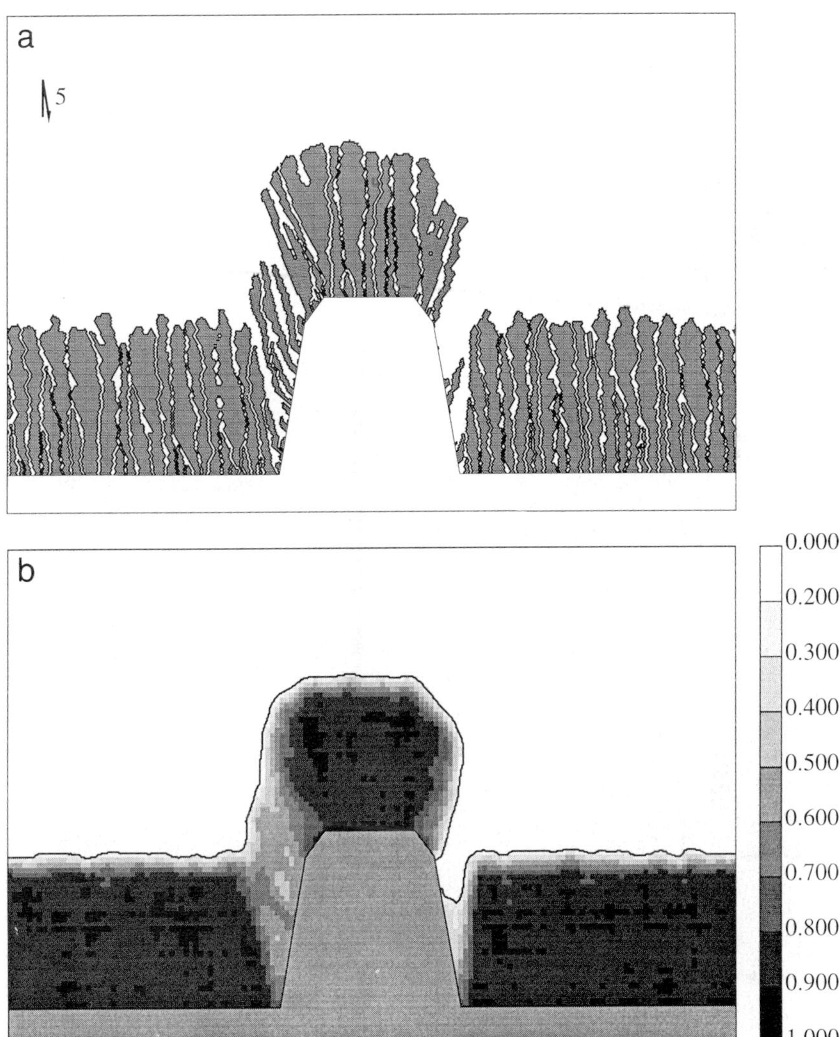

FIG. 6. (a) Microstructure depiction and (b) film local density simulation for the MgF_2 film of Fig. 5 (Reprinted from *Thin Solid Films*, **187**, R. N. Tait, T. Smy, and M. J. Brett, "A ballistic deposition model for films evaporated over topography," pp. 375–384, 1990, with kind permission from Elsevier Science S. A., Lausanne, Switzerland).

later development of an interpolated 3-D model, which is addressed in Sec. IV.F and applied to a study of the effects of diffusion in vias driven by out-of-plane curvature. Other extensions to the basic SIMBAD algorithm include the capability to simulate ion bombardment, etching processes, chemical vapor deposition, bulk (vacancy) diffusion, and alloy or reactive deposition. These enhancements are discussed in Sec. IV.

C. SIMBAD PROGRAM STRUCTURE AND USER INTERFACE

The current SIMBAD and SIMSPUD code is written in the C programming language. Its organizational structure is illustrated in Fig. 7. Both

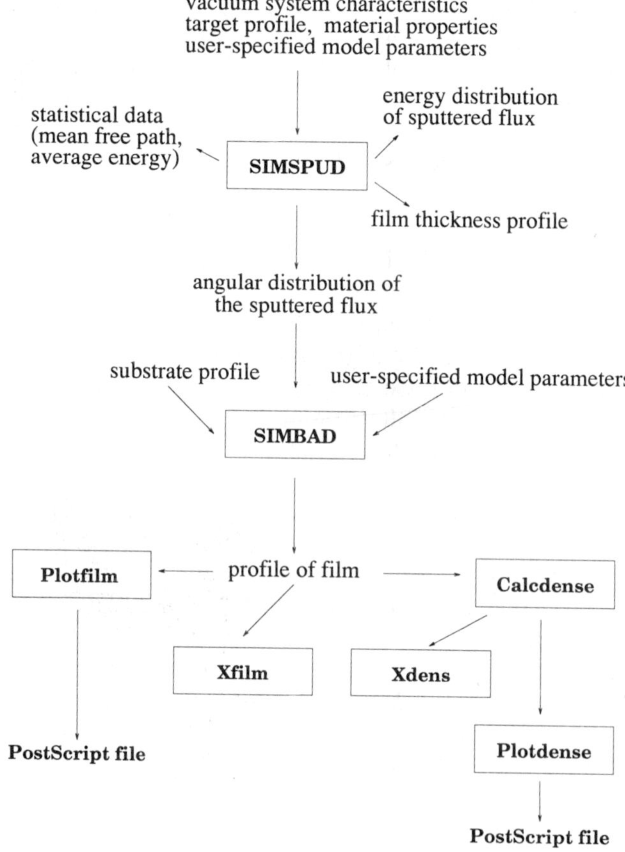

FIG. 7. Flow chart for the SIMSPUD and SIMBAD transport and deposition simulators.

simulators read input files to acquire parameter and geometry information such as substrate profile, collimator geometry, sputter target erosion profile, sticking coefficients, and adatom surface diffusion length. For the case of deposition by sputtering, SIMSPUD can provide a thickness profile across a wafer, produce flux energy distribution files, and pass angular distribution files to SIMBAD. SIMBAD generates PostScript™ output files for either density plots or (disk-based) microstructure depictions.

SIMBAD simulations can be run either in a command line mode, or by using the X Window user interface compatible with UNIX™-based operating systems. The graphical interface allows the user to draw or edit topography, set simulation parameters, monitor simulation status and view SIMBAD and SIMSPUD output, without requiring extensive UNIX experience. Figure 8 shows the appearance of some of the viewer windows encountered in the interface. They include the top-level application window, an "Edit Parameters" window, a "Film Simulation" control window, and the "xfilm" film viewer window. Run-time memory requirements are typically several hundred kilobytes to 10 MB, depending on the simulation type and resolution. Execution time is also quite variable, depending on the conditions and process being modeled, but typically is several minutes or tens of minutes on a Sun SPARCStation 10/30.

III. Model Verification

During the course of development of SIMBAD, more than 45 journal papers have been written by a group comprising some 24 students, postdoctoral fellows, and professors. A significant portion of this research has focused on fundamental verification of the simulator. This verification has ranged from studying properties and formation of columnar microstructure on planar substrates, to relative deposition rates and bottom fill in vias for sputtering with and without collimators.

A. Verification of SIMSPUD

A pinhole experiment was used to verify the validity of predictions from the SIMSPUD model (*22,23*). The configuration for this experiment is shown in Fig. 9, with film deposition occurring through a small hole (1.6-mm diameter) in a mask placed just above (13.4 mm) the substrate such that negligible scattering would occur in transport between the mask and the substrate. The thickness profile of the resulting film is highly sensitive to the incident angular flux distribution at the pinhole. After

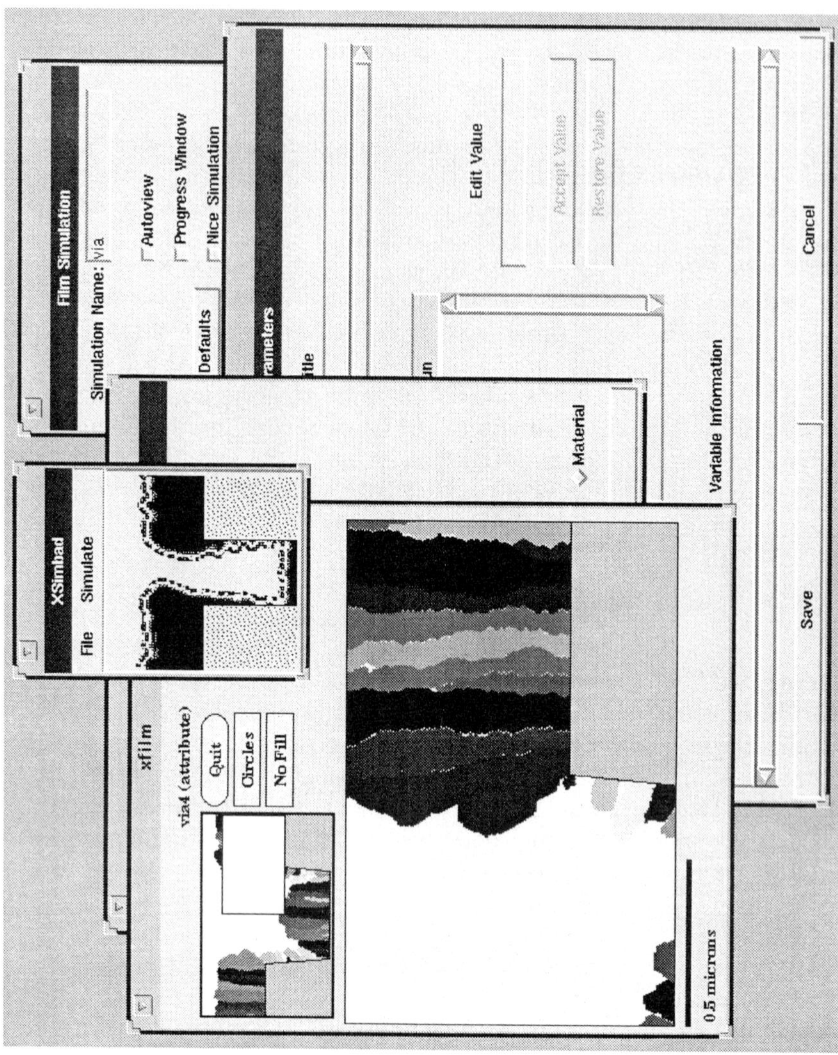

FIG. 8. Illustration of some of the viewer windows in the SIMBAD graphical user interface.

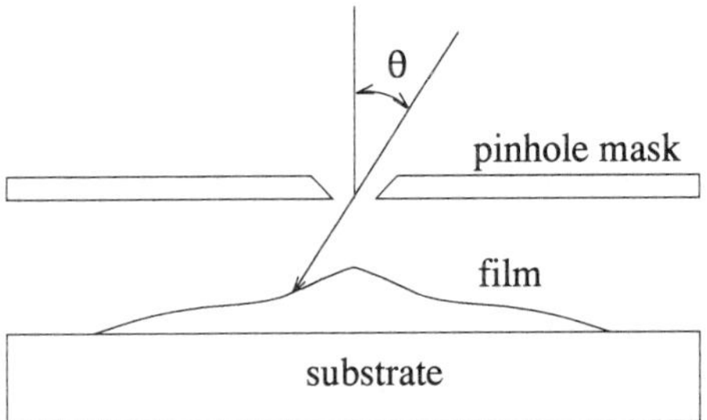

FIG. 9. Schematic of configuration used in pinhole experiments to study angular flux distributions (after Ref. 25).

deposition, the thickness profile can be directly measured using a profilometer over etched lines in the film and compared with the simulation results. For these experiments, a 5-cm-diameter sputter gun with Al or Cu targets was used as the source, placed 4.8 cm distant from the pinhole mask.

The corresponding SIMSPUD simulations took these dimensions as inputs, as well as the target erosion profile, target voltage, and material parameters such as mass and binding energy. Thermal radii for free atoms are not always available, so estimating values of 2.0 and 2.1 Å were used for Al and Cu, respectively (*24*). Emission distributions from the target surface measured by Tsuge (*11*) were used for Al, whereas for heavier Cu atoms a cosine emission was assumed. Unfortunately, values have not been established for σ_0, the cross-section at 1 eV, so this quantity was treated as an adjustable parameter for the initial comparison with experiment. The results for an Ar working gas pressure of 5.1 m Torr are shown in Fig. 10a, where a value of $\sigma_0 = 45\,\text{Å}^2$ gave an excellent fit to the experimental data. Unless otherwise stated, a unity sticking coefficient was assumed for the pinhole simulations. To make a fair test of the model, the experiment was repeated at 20.1 m Torr (where significant flux scattering will occur) using the value of σ_0 determined at 5.1 m Torr. Figure 10b confirms that the predictions are in excellent agreement with the data at this higher pressure. Notice that the curves at two pressures show qualitatively different behavior since at 5.1 m Torr there is relatively little scattering and the peak in the pinhole image is off axis at a position corresponding to the peak in target erosion profile of the 5-cm magnetron sputter source. At higher pressures however, numerous randomizing collisions of sputtered Al with the Ar gas have erased any memory of the sputter distribution originating at the target.

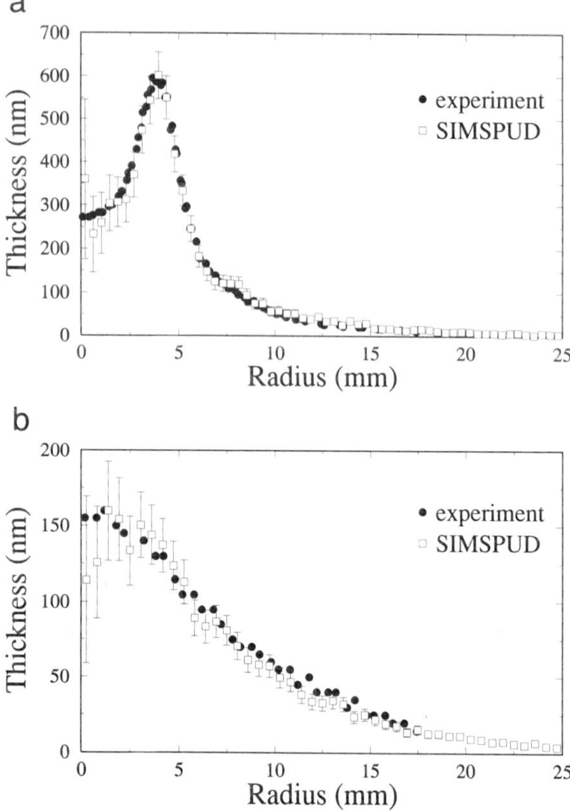

FIG. 10. Radial thickness profiles from the pinhole experiment and model results for aluminum sputtered at (a) 5.1 m Torr and (b) 20.1 m Torr. Error bars on the simulation values correspond to twice the standard deviation (from Ref. 23).

Pinhole experiments and simulations have been repeated for other metals, including Cu. Figures 11a and 11b demonstrate the excellent agreement for Cu, using a cross-section of $\sigma_0 = 27.9\,\text{Å}^2$. In Figure 11b and 20.1 m Torr, there remains an off-axis peak due to the erosion profile of the target. This occurs because Cu atoms are more massive than Al atoms and retains a larger proportion of their forward momentum after scattering off an Ar gas atom. A summary of results from pinhole experiments for Cu, Al, and other commonly used metals is given Table I (*25,26*). Since W does not scatter readily due to its high mass, the σ values for W are difficult to measure and should be used with caution.

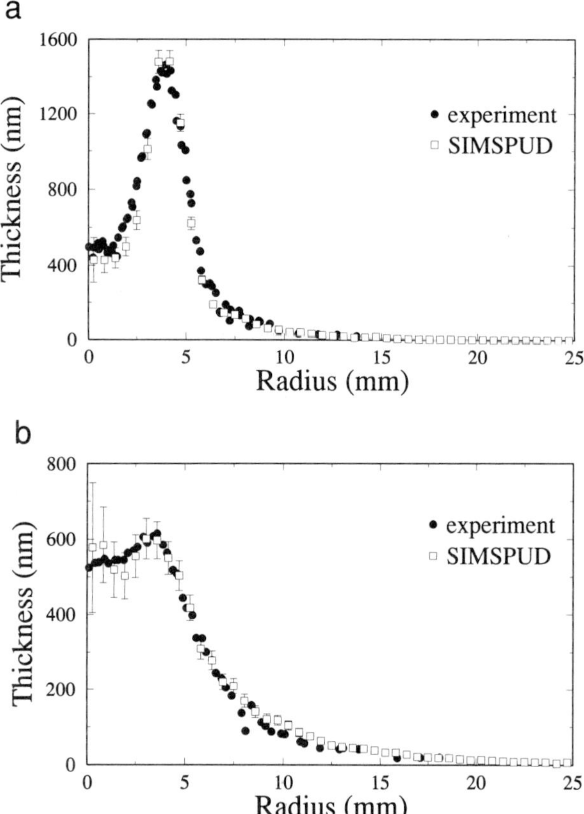

FIG. 11. Radial thickness profiles from the pinhole experiment and model results for copper sputtered at (a) 5.1 m Torr and (b) 20.1 m Torr. Error bars on the simulation values correspond to twice the standard deviation.

TABLE I

SCATTERING CROSS-SECTIONS OF METALS, DETERMINED BY EXPERIMENT / SIMSPUD COMPARISON

Cross-Section	Al	Cu	Ti	W
σ_t (Å^2)	60	54	68	40
σ_0 (Å^2)	45	28	48	25

Further demonstration and verification of the utility of SIMSPUD is given in Sec. IV.C, through analysis of flux and spatial nonuniformities created by collimated sputtering.

B. Verification of SIMBAD

Considerable effort has been expended to study and verify the microstructure predictions of SIMBAD film depictions. In the following pages, columnar structure, columnar orientation, and film density are examined. The parameter having the largest effect on film structure is the adatom diffusion length, inasmuch as it determines in which zone of the structure zone model the film is classified. Substrate temperature is related to surface diffusivity through the relation

$$D = D_0 e^{-Q_s/kT}, \tag{5}$$

where D is the intrinsic diffusivity and Q_s is the activation energy of the dominant mechanism for surface diffusion (*19*). The average adatom diffusion length L may then be expressed as

$$L = \sqrt{D_0 \tau}\, e^{-Q_s/2kT}, \tag{6}$$

where τ is the average length of time that an adatom is mobile on the surface. Unfortunately, the factor $D_0\tau$ is not well known and is difficult to measure because of system dependencies. Although for metals of melting point T_m it has been empirically determined that $Q_s \cong 5kT_m$ (*27*), in general $D_0\tau$ will be a strong function of material and deposition conditions. For instance, surface diffusion lengths for Al are observed to be reduced significantly by oxygen contamination during deposition (*28*). For this reason, in SIMBAD simulations a calibration of L need be made by comparison of surface profiles and film column sizes between simulations and films. This procedure yields a diffusion length for the process that can be scaled to other temperatures using the relationship in Eq. (6).

A well-known result of increased substrate temperature, and thus diffusion length, is an improved density of the film and larger diameter columnar grains (*28*). The SIMBAD simulations of Fig. 12 show this behavior clearly, as the diffusion length is increased from 0.01 to 0.06 μm (*29*). Note the more porous, open, zone 1 columnar structure in Fig. 12a, relative to the dense zone 2 structure in Fig. 12c. Simulation of larger diffusion lengths (i.e., temperatures near the melting point) and bulk diffusion processes is described in Sec. IV.E.

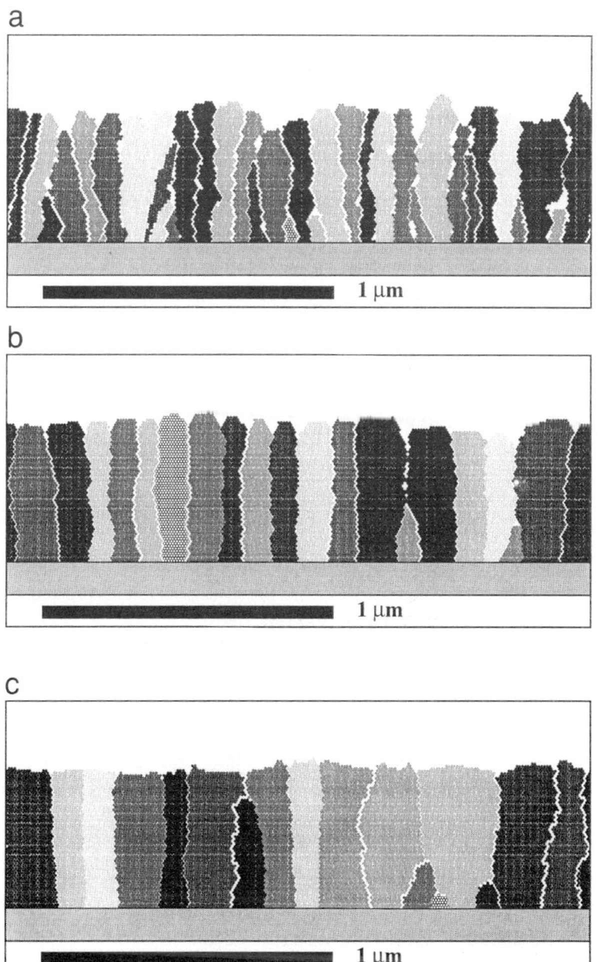

FIG. 12. SIMBAD simulations of the columnar grain size of films sputtered at increasing adatom diffusion lengths of (a) 0.02 μm, (b) 0.05 μm, and (c) 0.125 μm.

For film deposition onto a substrate at an average angle of incidence that is off normal, an empirical "tangent rule" has been used to describe the resulting orientation of the film microstructure (*30*). It relates the angle β, between the column growth direction and the substrate normal, to the angle α, between the vapor flux direction and the substrate normal,

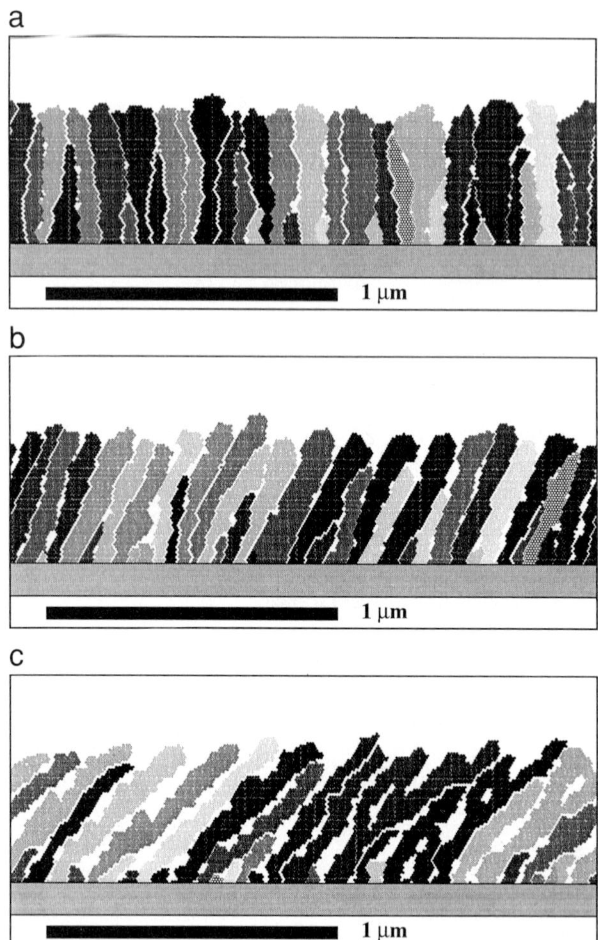

FIG. 13. SIMBAD simulations of film growth at diffusion length 0.02 μm and at flux angles of incidence of (a) 0°, (b) 30°, and (c) 60°.

as follows:

$$2 \tan \beta = \tan \alpha . \tag{7}$$

Thus the average column is oriented at an angle closer to the substrate normal than is the vapor flux. Figures 13a, 13b, and 13c show simulations of film growth for $\alpha = 0$, 30, and 60°, respectively. The orientation of the columns with increasing angle of incidence is summarized in Fig. 14 (*31*),

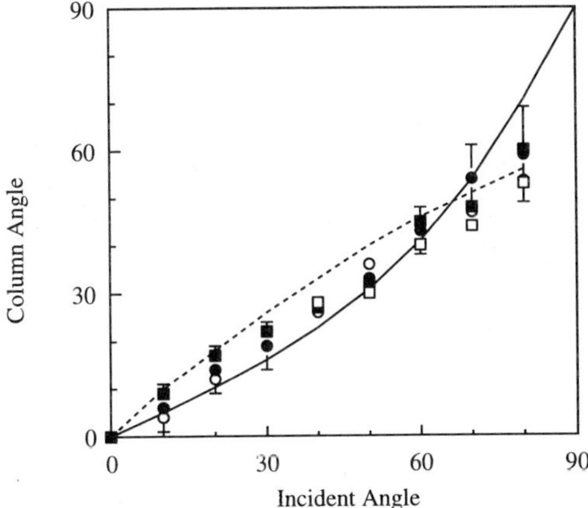

FIG. 14. Simulated column angles for films with diffusion lengths of zero disk diameter (●) and four disk diameters (■). Experimental results are for thermally evaporated MgF_2 (○) and thermally evaporated Fe (□). Theoretical column angles predicted by the tangent rule (——) and Tait's rule (- - -) are also shown (Reprinted from Thin Solid Films, **226**, R. N. Tait, T. Smy, and M. J. Brett, "Modelling and characterization of columnar growth in evaporated films," pp. 196–201, 1993, with kind permission from Elsevier Science S. A., Lausanne, Switzerland).

showing the close agreement of the SIMBAD model to the tangent rule and to experimental results of evaporated MgF_2 and Fe films. A new rule or description of column angle has been derived from geometrical principles by Tait (*31*) that improves the prediction at highly oblique angles of incidence above 60°.

A deleterious effect of oblique angle deposition onto substrates at low temperatures is increased film porosity caused by greater shadowing, exacerbated by the inability of the short diffusion length to fill shadowed regions effectively. Deposition sources for microelectronics processing may be oriented such that the flux arrives perpendicular to the wafer, but for small regions on topographical features, such as via sidewalls, the flux may arrive highly obliquely. Since this effect may lead to localized low density and potential film failure regions, it is critical to model such density changes accurately. As a preliminary verification of SIMBAD density predictions, measurements of density of films evaporated onto planar substances are shown in Fig. 15 to agree very well with the model predictions (*32*). In this figure, densities are normalized to 1 for an incident angle of $\alpha = 0°$. Similar results for measurement and simulation of W sputtered onto Si wafers are shown in Fig. 16 (*33*). The agreement for sputtered W is

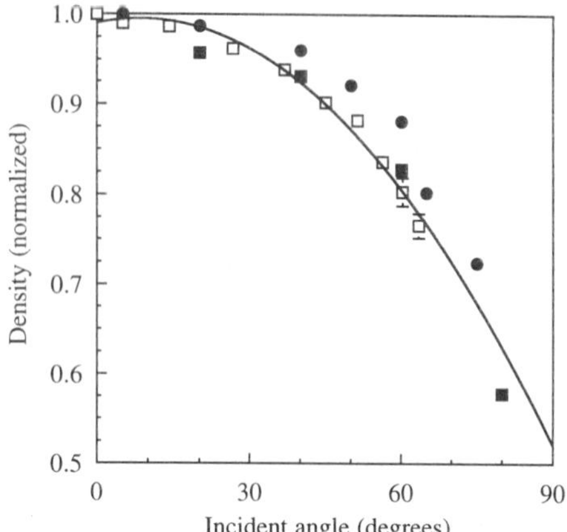

FIG. 15. Film density as a function of vapor incidence for ■, evaporated Ge films [77]; ●, evaporated Ge films [78]; □, SIMBAD simulations. The line is a fit to the SIMBAD results. Data are normalized to a density of 1.0 at normal incidence (Reprinted from *Thin Solid Films*, **187**, R. N. Tait, T. Smy, and M. J. Brett, "A ballistic deposition model for films evaporated over topography," pp. 375–384, 1990, with kind permission from Elsevier Science S. A., Lausanne, Switzerland).

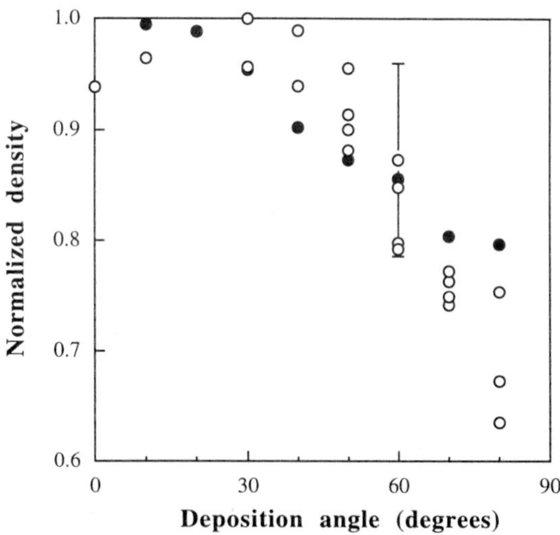

FIG. 16. Density of real tungsten films ○ and simulated tungsten films ● as a function of flux angle of incidence. Densities are normalized to peak values, and a typical uncertainty in experimental results is shown (from Ref. 33).

also reasonable in spite of large uncertainties in the experimental values because of the difficulty in making these measurements.

A further qualitative verification of density predictions was made for the more relevant case of deposition over topographical features (*33*). Films of W were chosen for this work because they will deposit onto low-temperature substrates with a very evident zone 1 columnar structure. Initially, a wet etch of $K_3Fe(CN)_6$ and NaOH was used on various W films deposited onto planar substrates to determine the dependence of etch rate on density. This study confirmed that etching proceeds rapidly when the film is porous and the etchant has access to a large surface area. Because the etching proceeds along grain boundaries and pores, rather than from the top surface only, film thickness stayed relatively constant during etching whereas film density decreased until the film was removed from the substrate. To study films on topographical features, W was sputter deposited to a thickness of about 1 μm over trenches of aspect ratio approximately 1 patterned in oxide. These films were etched for varying times and are shown viewed in cross-section in Fig. 17. The columnar zone 1 microstructure of the film in Fig. 17a is clearly visible, with the film on the sidewall showing a structure with definite voids between coarse columns. The depositing flux arrives obliquely on the sidewalls and the films at this region are thus more porous and expected to etch faster. Figures 17b and 17c confirm this expectation, because the film on the sidewall is seen to be etched preferentially. Only a small amount of material beyond the depth of focus of the microscope remains on the sidewall.

A SIMBAD simulation of the density profile of the unetched W film is shown in Fig. 18a, depicting the low-density sidewall film. Furthermore, the effects of wet etching were simulated by decreasing the local density of the simulated film with an increased etching time according to the data from etch measurements on planar films. Figures 18b and 18c show the results of this simulated wet etch. Again, a preferential removal of the low-density film regions, and a good agreement with experiment are evident. Of particular note in this experiment and simulation is that although the film showed good step coverage over the trench feature, it is likely a poor interconnect due to the high resistivity expected in the porous sidewall region.

Because the previous example dealt with deposition onto a 2-D trench structure, this section is concluded with a demonstration of the accuracy of SIMBAD in modeling films deposited onto 3-D topography (*34*). Figures 19a and 19b show films of sputtered Al and W, respectively, deposited over 1-μm × 1-μm contact holes. Notice the more evident film microstructure in W, which has a lower T/T_m for this deposition and thus a lower adatom diffusion length. The SIMBAD simulations in Fig. 20 were performed using the quasi-3-D model described earlier and show a good agreement for both the step coverage and microstructures of the deposited films.

A further verification of both SIMBAD and SIMSPUD was performed by experiments that are commonly used to determine sticking coefficients of sputtered metal atoms. Figure 21 shows a cross-section of an overhang structure fabricated by wet etching of Si to undercut an SiO_2 film. A Ti film has been sputtered over this structure. As a refractory metal, Ti film is expected to have a very small surface diffusion length (corroborated by the narrow columnar structure). The coating of film on Si below the SiO_2 overhang is a sensitive indicator of the angular distribution of the arriving flux. Presence of any Ti film directly on SiO_2 underneath the overhang would indicate a nonunity sticking coefficient or resputtering from previously deposited film. In Fig. 21, the Ti film was sputtered at a relatively high pressure of 10 m Torr to minimize resputtering created by energetic

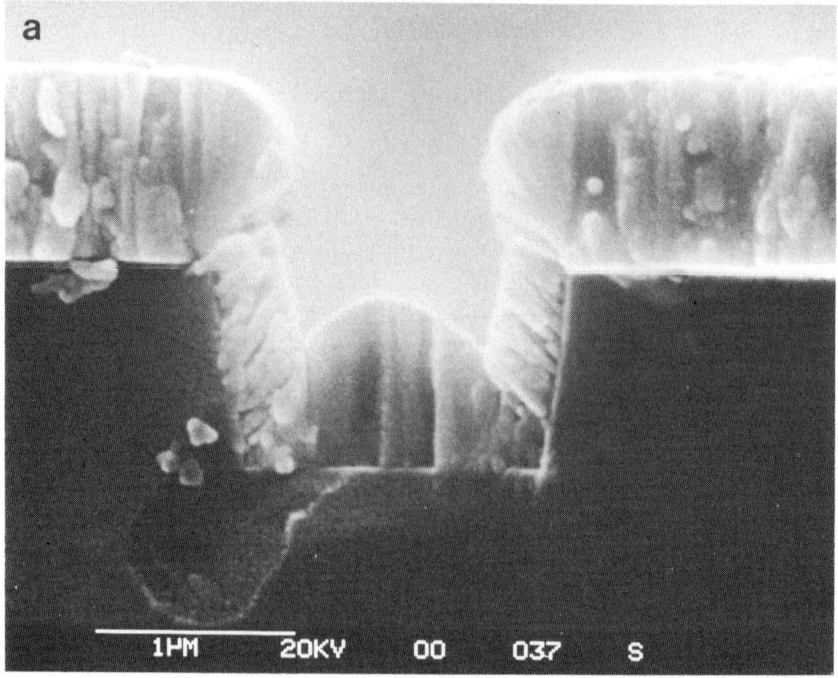

FIG. 17. SEM micrographs of cross-sections of W films deposited over a trench: (a) unetched film, (b) film after 2 s of etching, and (c) film after 10 s of etching (from Ref. 33).

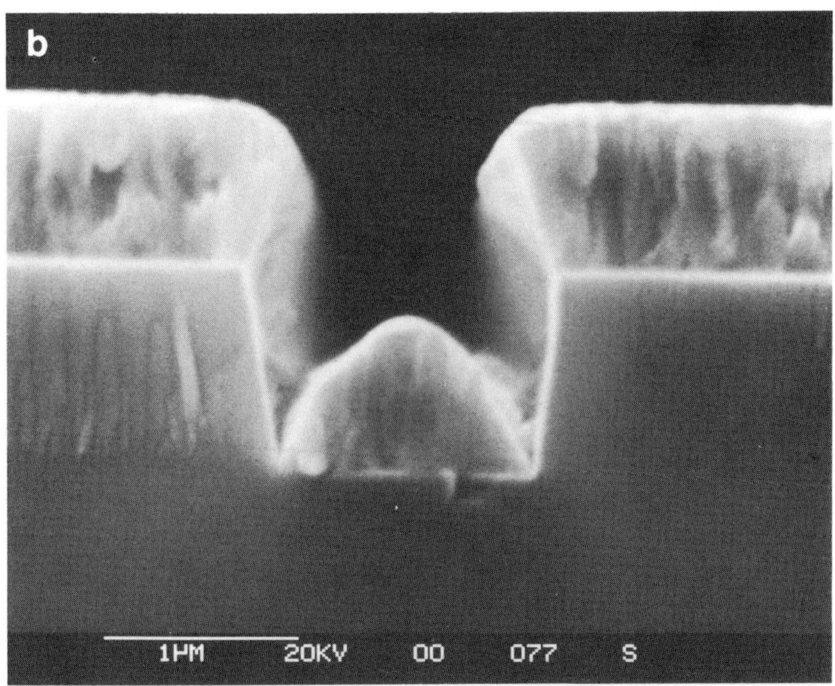

FIG. 17. Continued

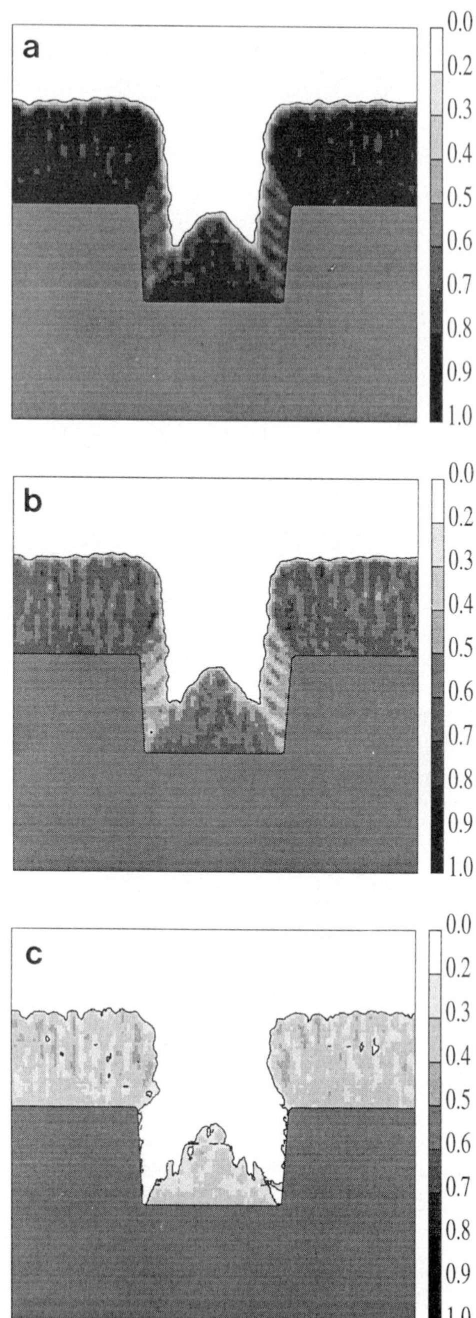

FIG. 18. SIMBAD density depictions of the films in Fig. 17: (a) unetched film, (b) films after 2 s of etching, and (c) film after 10 s of etching (from Ref. 33).

species incident on the substrate, and no Ti film was visible beneath the overhang, indicating a unity sticking coefficient.

Using the scattering cross-sections for Ti shown in Table I to generate the incident flux distribution, Fig. 22a shows a SIMBAD simulation of film deposition over this structure. A very close agreement of film coverage at all points on the substrate confirm both the incident flux distribution at the substrate generated by SIMSPUD and the SIMBAD microstructure depiction. A useful feature of this type of experiment is that the sensitivity to nonunity sticking coefficients can be evaluated by simulation. For instance, Fig. 22b shows an identical simulation, but with a sticking coefficient for Ti of 0.9. Through comparison with high-magnification SEM images, such simulations helped establish a lower limit of the sticking coefficient (consistent with experimental uncertainty) of 0.97.

IV. Examples of Application of the SIMBAD / SIMSPUD Models

In this section, applications of SIMBAD and SIMSPUD to a variety of processes and problems in microelectronics manufacturing are discussed. These include CVD, a variety of sputter deposition processes, modeling of film composition, and etching processes. For each of the examples, limitations of space prevent a full description of the experiments and simulations, thus the interested reader should consult the references given therein.

A. Refractory Metal Chemical Vapor Deposition

CVD-deposited refractory metal films have a very prominent columnar microstructure that is the result of shadowing, precursor diffusion length, and nonunity sticking coefficient. These features are readily implemented by a ballistic model such as SIMBAD, and thus it provides a relatively simple CVD model that depicts the film microstructure. The linear trajectories used by SIMBAD are valid for films deposited with micron (or smaller) features and for LPCVD pressures of less than several torr. Ideally, a cosine angular distribution is appropriate for a thermalized gas source, but minor variations from this are possible due to concentration gradients caused by high sticking coefficients. In general, the sticking coefficient is dependent on the precursor desorption rate, the surface reaction rate, and the impingement rate (all of which may vary across the film) (*35,36*), but it is often appropriate to use a single averaged value. In SIMBAD, disks that do not stick are re-emitted and may stick elsewhere on the film, or may eventually "bounce" their way out of the simulation region. The SIMBAD model provides simple modifications that allow for multiple sticking coefficients (several possible precursors) or sticking coefficients that vary with the underlying material (i.e., selective deposition).

The SIMBAD CVD model has been used to depict successfully the microstructure and step coverage of tungsten films deposited in very high aspect ratio trenches (*37*), and to model selective depositions (*38*). Here, we demonstrate the utility of SIMBAD to study general microstructure behavior at different sticking coefficients and surface diffusivities. Figure 23 shows a study of CVD deposition into a trench at two different values of surface diffusion length and three values of sticking coefficient. Increased diffusivity in Fig. 23d creates a more uniform film structure but does not noticeably improve the coverage. However, as expected, there is near-conformal film coverage at small sticking coefficients in Fig. 23c. Particularly telling of a low sticking coefficient is that the columnar orientation on the trench sidewalls is not upward toward the source but normal to the local substrate surface. This behavior is created by multiple re-emission processes that create an effective incident distribution that is symmetric about the local surface normal. A new ballistic model, GROFILMS, capable of handling more detailed CVD kinetics and spatially varying sticking coefficients, is described in its current state of development in Sec. V.

B. Bias Sputtering

Bias sputtering is a process wherein a secondary discharge is created at the substrate during sputter deposition in order to create energetic bombardment of the growing film by working gas ions (*39*). Bias sputtering will affect almost any measurable film parameter, and its popularity has fluctuated during the last decade due to the advantageous effects of planarization and film densification offset by often unwanted inert gas incorporation and increased intrinsic stress. Currently, a process involving

---

FIG. 19. SEM micrographs of (a) Al and (b) W sputtered over 1-μm $\times$ 1-μm contact holes in oxide (Reprinted from *Can. Metall. Quart.*, **34**, T. Smy, S. K. Dew, and M. J. Brett, "Simulation and experimental analysis of refractory metal and aluminum deposition over high aspect ratio VLSI topography," pp. 195–202, 1995, with kind permission from Elsevier Science Ltd., The Boulevard, Langford Lane, Kidlington, 0X5 1GB, UK).

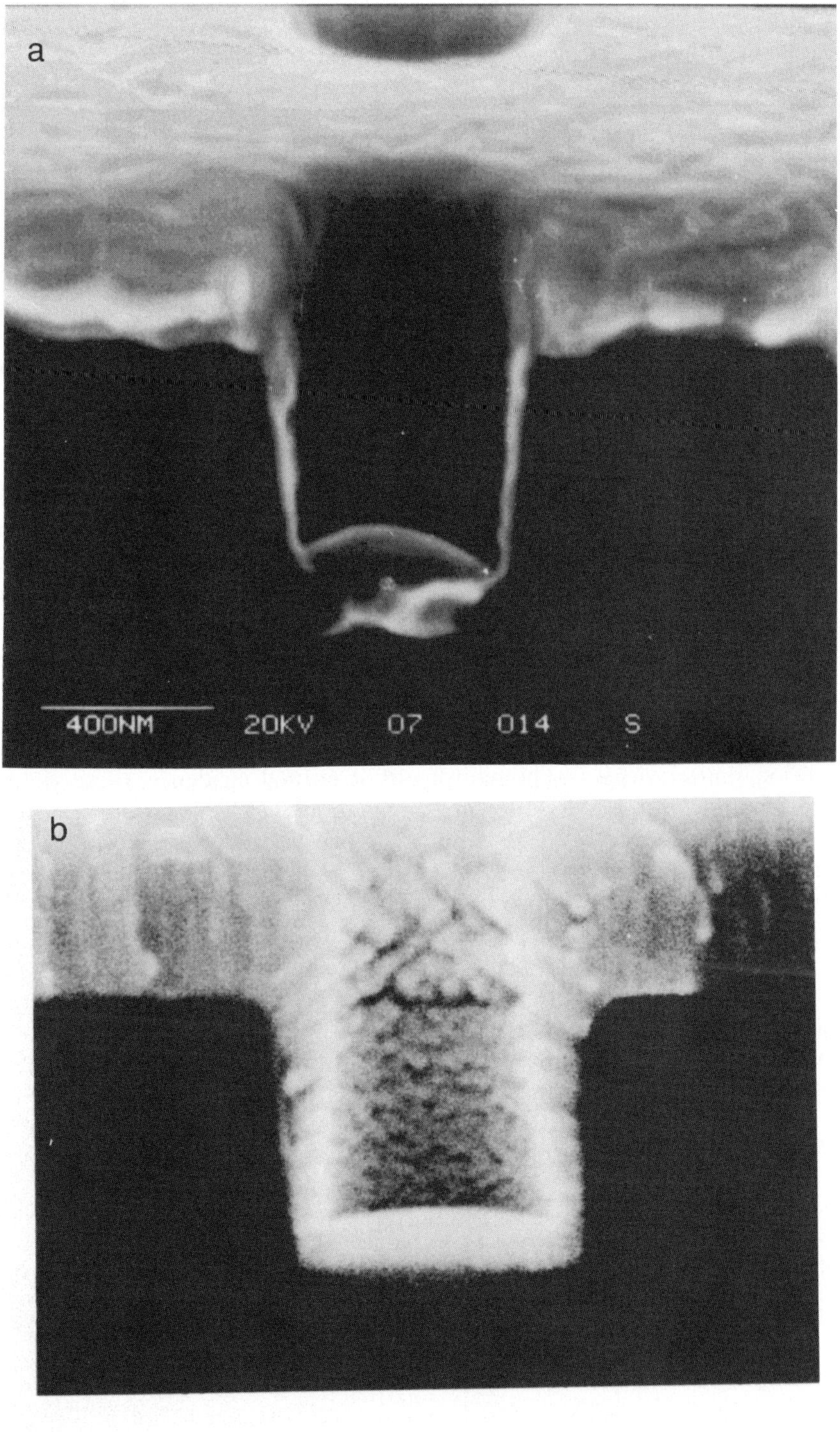

a

400NM 20KV 07 014 S

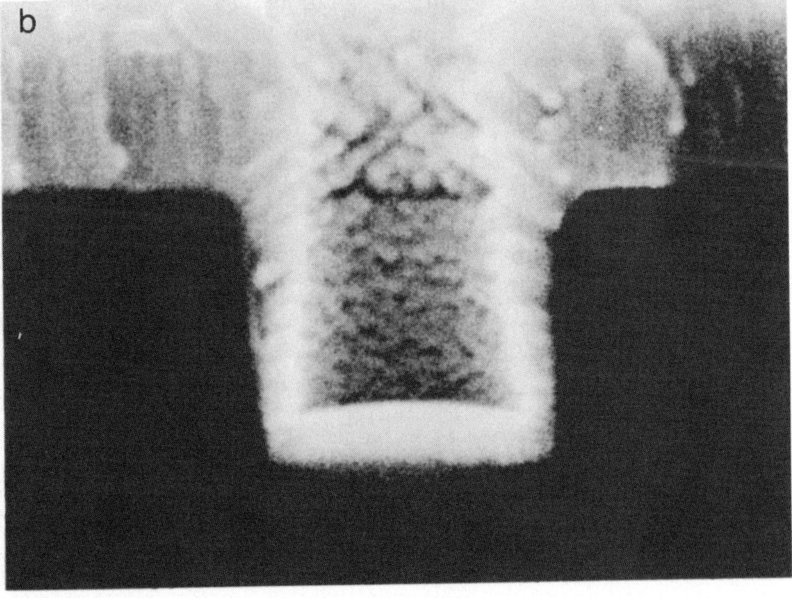

ion bombardment of the substrate, ionized enhanced magnetron sputtering (*40,41*), is being investigated as a technique to better fill the deep topography of VLSI structures.

SIMBAD was extended to model ion bombardment by generating ion disks as well as film disks. The behavior of ion disks is intended to represent the average behavior of many ions in the same manner that film disks represent film atoms. Unlike the film disks, the ion disks are not incorporated into the film. To account fully for the effects of ion bombardment, an ion disk impingement on the film may reflect off the film surface, cause resputtering of the struck film disk, and/or cause local film surface relaxation. Each of these processes is characterized by physical or empirical models.

Ion reflection probability is determined by ion energy, angle of incidence, and relative ion to atom mass. Because the details of this dependence are not well known, the reflection probabilities used in SIMBAD are determined from calculations made by Hou (*42*) with the program MARLOWE. These results, which are in qualitative agreement with the program TRIMSP (*43*), show a very high probability of reflection above a critical angle. SIMBAD calculates the local surface normal of the film at the point of impact, and if reflection occurs, a specular reflection angle is assumed. The trajectory of a reflected ion is tracked to determine if it hits the film at another location.

The probability of the ion disk causing resputtering is proportional to the ion sputter yield. For bombardment at normal incidence, these yields are tabulated for many gas/film combinations (*39*). However, a strong angular dependence of yield exists, typically with a peak in yield at an incidence angle near 45°. In SIMBAD, this angular dependence may be input explicitly, if known, otherwise an empirical dependence is used (*44*). The emission distribution of the resputtered disk follows the results of Tsuge (*11,45*), which show preferential emission in the specular direction. The trajectories of resputtered disks are tracked in case they are incident elsewhere on the film. The final process of ion bombardment-induced surface diffusion is included in SIMBAD through a local relaxation of the film in the immediate vicinity of ion impact.

Studies of planarization by bias sputtering for films in trenches were undertaken for tungsten films (*46*). The tungsten source was an rf planar

---

FIG. 20. SIMBAD microstructure depictions of the sputtered films in Fig. 19: (a) Al film and (b) W film (Reprinted from *Can. Metall. Quart.*, **34**, T. Smy, S. K. Dew, and M. J. Brett, "Simulation and experimental analysis of refractory metal and aluminum deposition over high aspect ratio VLSI topography," pp. 195–202, 1995, with kind permission from Elsevier Science Ltd., The Boulevard, Langford Lane, Kidlington, 0X5 1GB, UK).

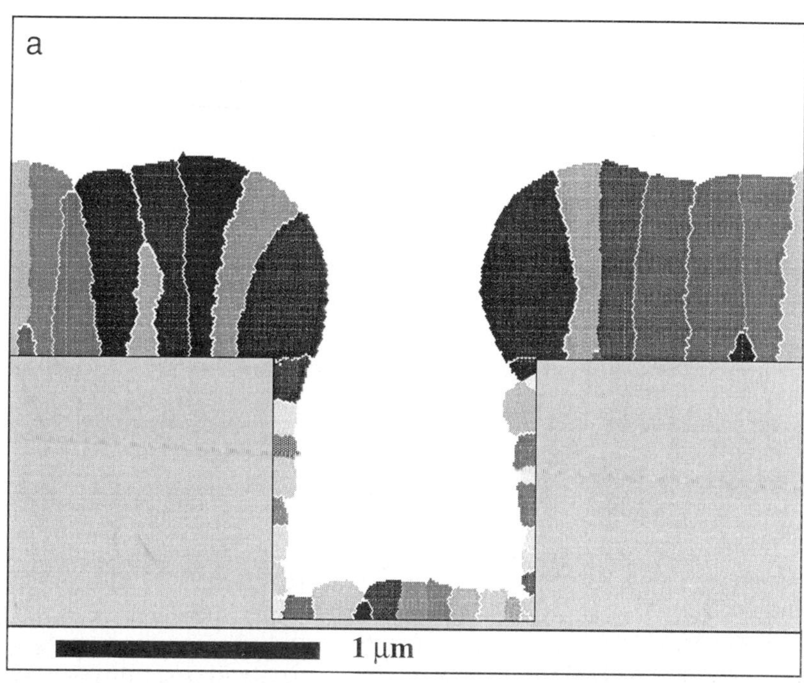
a
1 μm

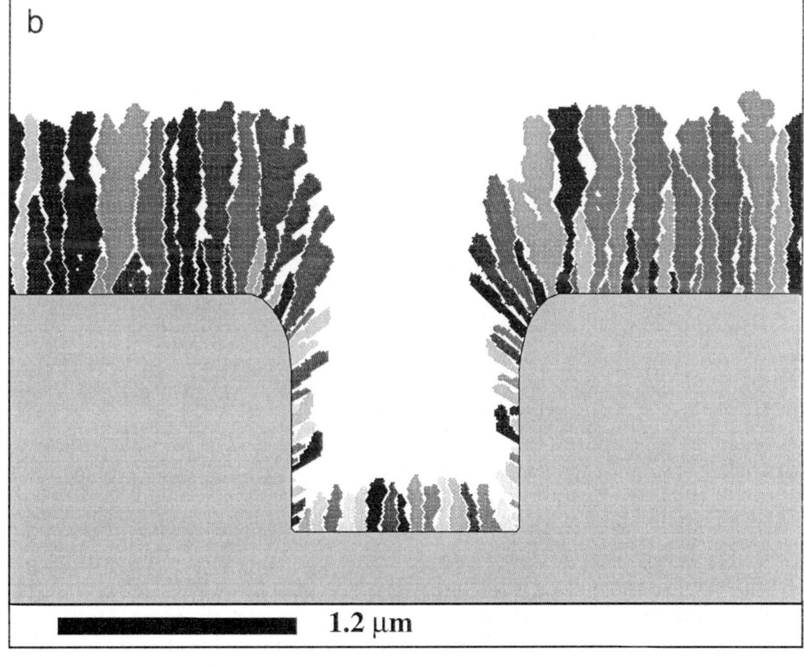
b
1.2 μm

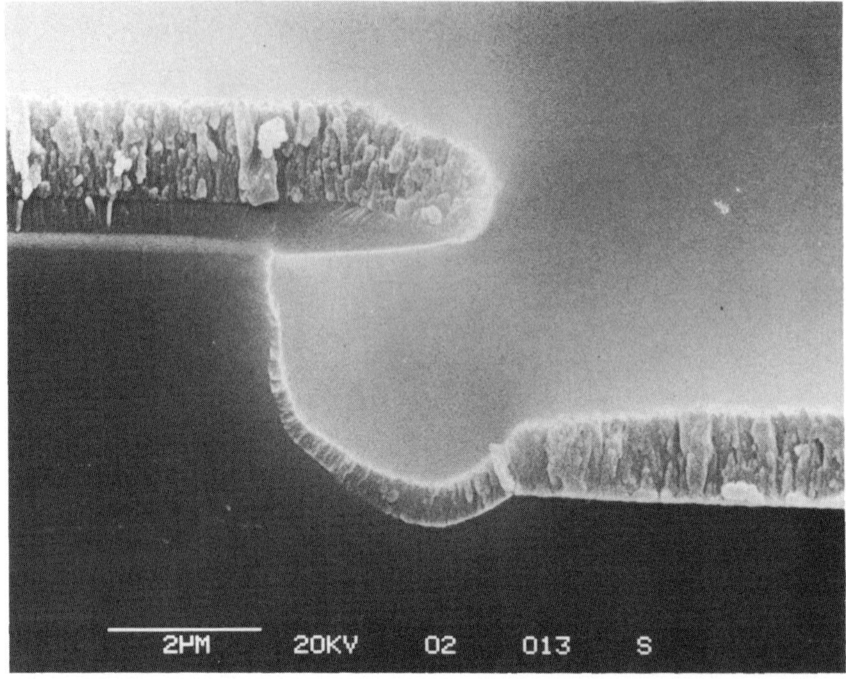

FIG. 21. SEM micrograph of a Ti film sputtered over an overhang structure. Notice the lack of film on the underside of the overhand (from Ref. 26).

diode operating at 8 m Torr of Ar working gas, and with various values of negative dc bias induced at the substrate by a secondary rf discharge. Figure 24a shows a cross-section of the W film deposited without bias, whereas Fig. 25a shows a slightly thicker film deposited at −330-V bias. Corresponding SIMBAD microstructure depictions are shown in Fig. 24b and 25b. Notice the greatly improved planarity of the bias sputtering film, and the resulting central "keyhole" void. This void is somewhat off center because the photo was taken of a slightly tilted sample. An undesirable effect of high bias is just visible in Fig. 25a, where ion reflection off the trench sidewalls has etched the substrate near the corners of the trench. Although the film simulation of Fig. 25b shows the correct film profile, SIMBAD does not currently allow erosion of the substrate. These examples demonstrate how use of SIMBAD for simulation of films undergoing ion bombardment can provide information about microstructure and the surface profile in order to assist in optimizing the deposition process.

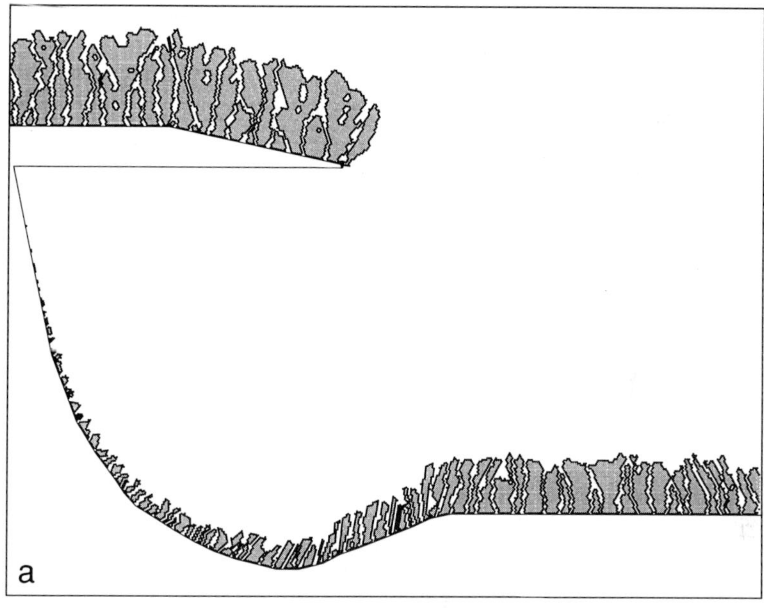

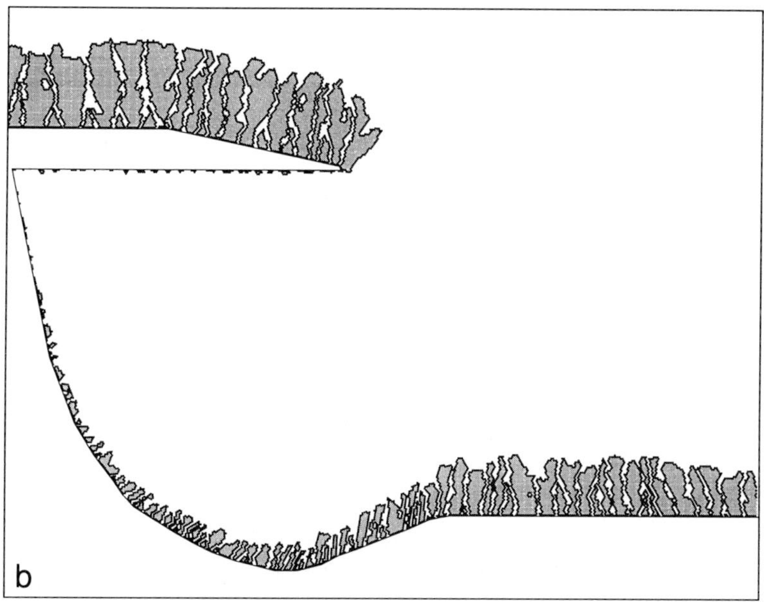

FIG. 22. SIMBAD simulations of the overhang structure of Fig. 21, using sticking coefficients of (a) 1.0 and (b) 0.9 (from Ref. 26).

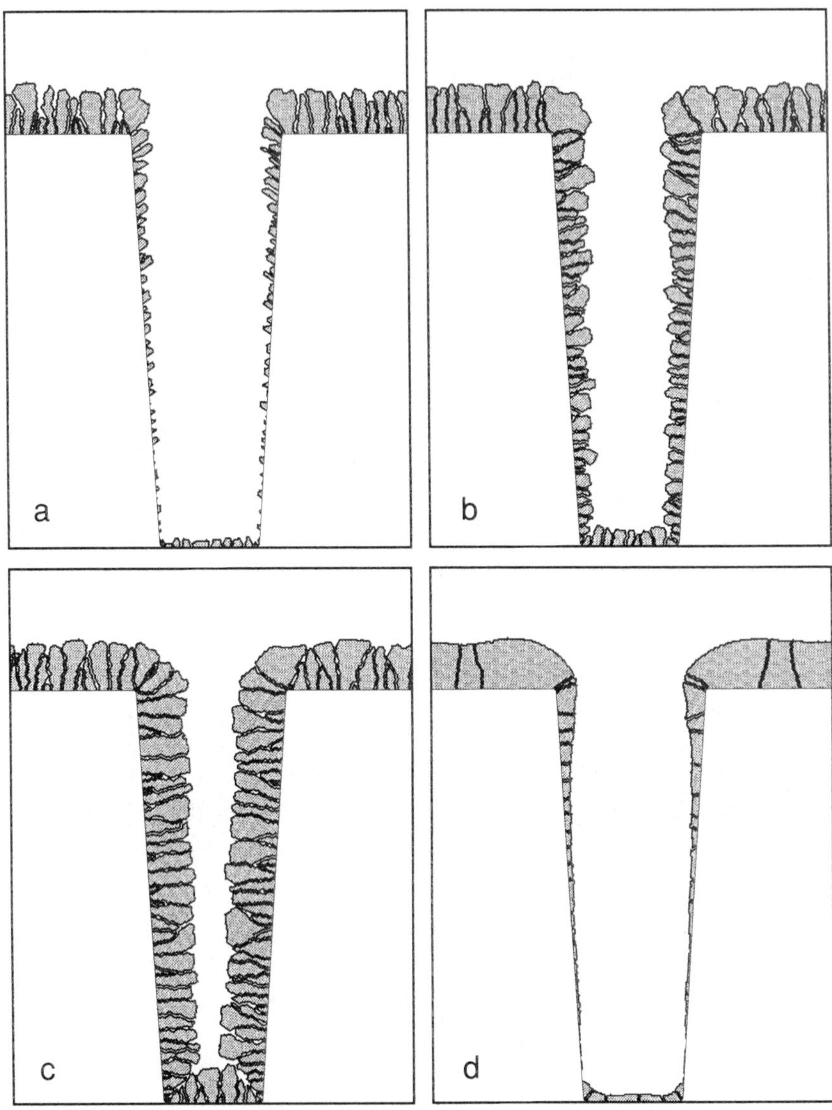

FIG. 23. SIMBAD simulation of chemical vapor deposition into trenches for various values of sticking coefficient s and surface diffusion length d: (a) $s = 1.0$, $d = 20\,\text{nm}$; (b) $s = 0.1$, $d = 20\,\text{nm}$; (c) $s = 0.01$, $d = 20\,\text{nm}$; and (d) $s = 1.0$, $d = 100\,\text{nm}$ (from Ref. 37).

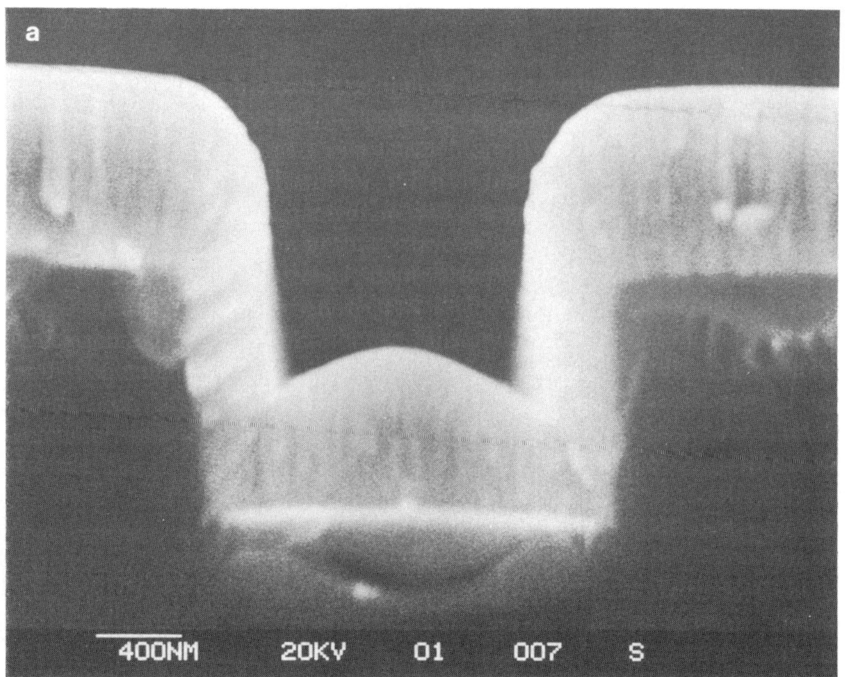

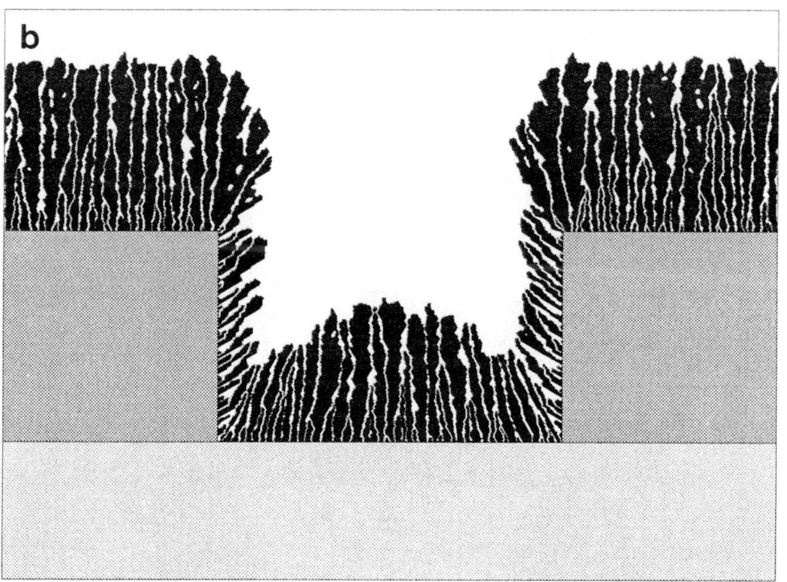

FIG. 24. (a) SEM micrograph and (b) SIMBAD microstructure depiction of a tungsten film sputter deposited without substrate bias (from Ref. 46).

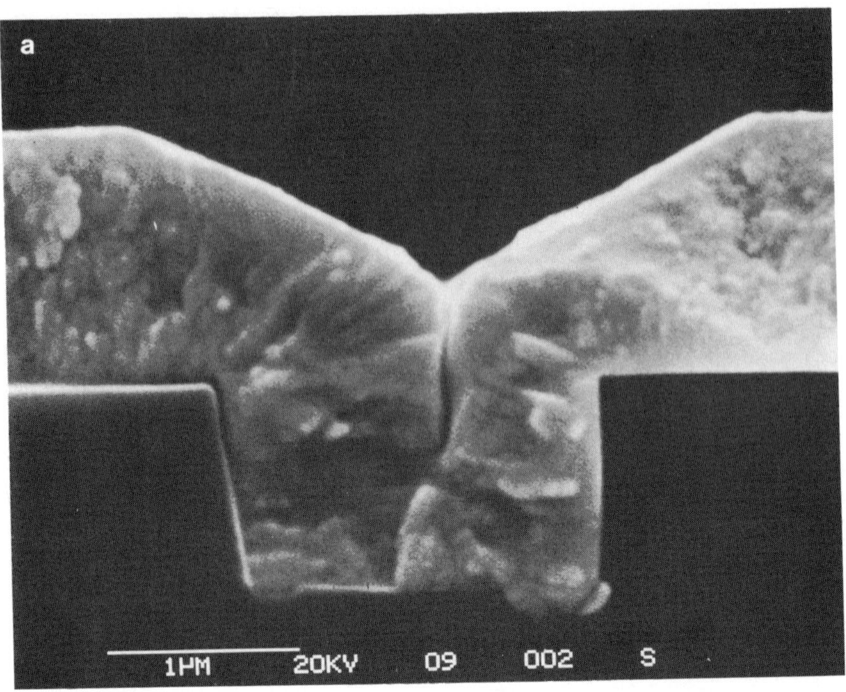

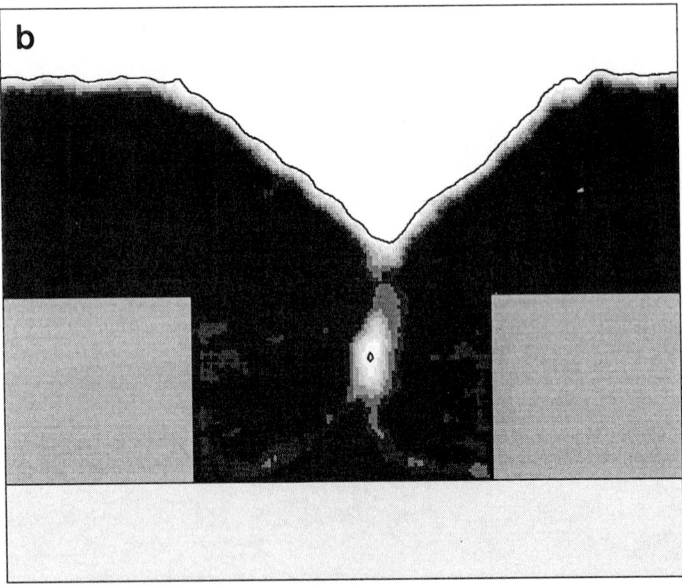

FIG. 25. (a) SEM micrograph and (b) SIMBAD density depiction of a tungsten film sputter deposited with a −330-V rf substrate bias (from Ref. 46).

C. Collimated Sputtering

Collimated sputtering has been of recent interest because of its ability to improve filling of deep topographic features (*47*). It differs from conventional sputter deposition by the introduction of a collimator between the target and the substrate. The collimator consists of an array of holes or cells which preferentially transmit flux traveling nearly normal to the substrate. A square grid collimator is shown schematically in Fig. 26. Atoms which arrive at the collimator with oblique incidence are more likely to be intercepted and deposited onto the collimator walls. The higher the aspect ratio (depth : width) of the collimator cells, the more selective it becomes, and the narrower the angular distribution ultimately incident at the substrate.

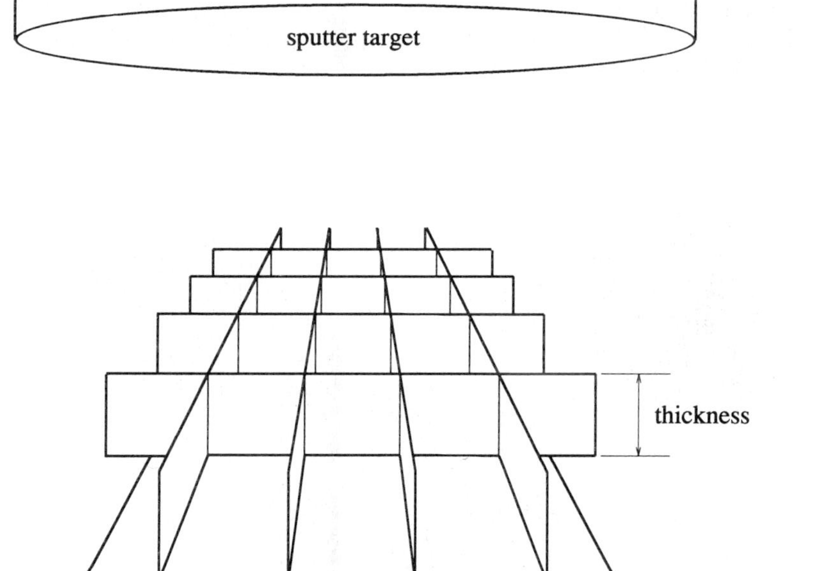

FIG. 26. Diagram of a square grid vane collimator. The collimator aspect ratio is defined by height : pitch (from Ref 49).

Collimation introduces a great deal of complexity to the sputtering process, due to the numerous new variables which will affect the deposited film. These include collimator aspect ratio and pitch, collimator-substrate separation, cell design, vane width, etc. Using SIMSPUD one can specify the geometry and position of either hexagonal grid, square grid, or round cell collimators, and thereby assist in optimizing the collimation process through simulation. For example, Fig. 27 shows the results of SIMSPUD simulation of flux angular distributions incident at the substrate for cases without and with varying degrees of collimation (*48*). The narrowing of the distribution evident for the 3 : 1 collimator of Fig. 27 comes at a cost of reduced deposition rate at the substrate. Figure 28 shows both experimental and simulated results for the relative deposition rate, and also includes the simulated distribution FWHM width (*49*). SIMSPUD results show clearly the trade-off that must be made between deposition rate and collimation.

Once the collimator geometry has been specified in SIMSPUD, the generated angular distributions may be utilized by SIMBAD to study film coverage over topography. In Fig. 29, the growth of a 0.2-μm-thick Ti barrier layer in a (3-D) contact hole of aspect ratio 3 : 1 was simulated for cases with and without collimation (*48*). The bottom coverage of 15% for the case with no collimator increases to 30% for the 3 : 1 aspect ratio collimator.

Simulation of collimated sputtering has revealed more subtle dependencies that must be addressed in process design. A collimator in close

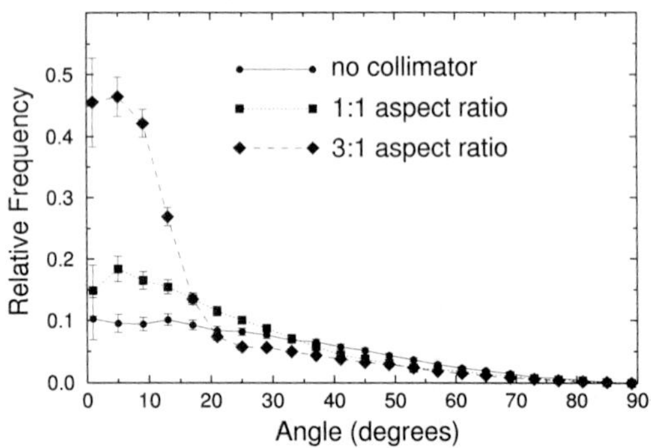

FIG. 27. Angular distribution for Ti sputtered at 5 m Torr with no collimator and with collimators having aspect ratios of 1 : 1 and 3 : 1 (Reprinted from *Can. Metall. Quart.*, **34**, T. Smy, S. K. Dew, and M. J. Brett, "Simulation and experimental analysis of refractory metal and aluminum deposition over high aspect ratio VLSI topography," pp. 195–202, 1995, with kind permission from Elsevier Science Ltd., The Boulevard, Langford Lane, Kidlington, 0X5 1GB, UK).

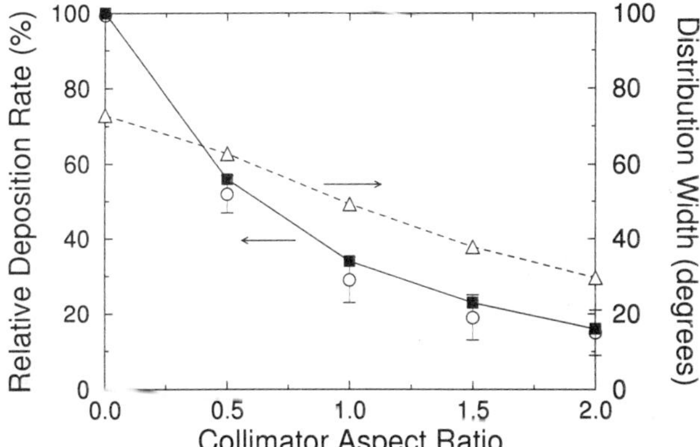

FIG. 28. The effect of collimator aspect ratio on the simulated deposition rate (squares) and on the angular distribution width (triangles) of flux arriving at the substrate. Also shown are experimentally determined deposition rates (circles). Angular distribution widths are full widths at half maximum (from Ref. 49).

proximity to the substrate will introduce a thickness variation, or ripple, with a period corresponding to the collimator grid spacing due to shadowing by the collimator vanes. For poor process design, this undesirable ripple can be substantial in amplitude. Such a pronounced thickness variation is shown experimentally and by simulation in Fig. 30 for an Al film sputter deposited using an aspect ratio 2 collimator (*49*). To reduce the ripple amplitude, the collimator could be placed further from the substrate or the pressure could be increased. However, both of these solutions will cause a broadening of the flux angular distribution incident on the substrate, that is, they increase the pressure–distance product between the collimator and the substrate, thus increasing flux scattering. These complex interdependencies may be evaluated effectively by simulation.

The effect of the shape of the target erosion profile is used as a final example of simulation of collimated sputtering. Most magnetron sources have a nonuniform erosion profile due to the placement of magnets behind the target. For high levels of collimation and relatively low pressures, flux reaching a particular point on the substrate is restricted to being mainly from the point on the target directly above it. Thus the shape of the target erosion profile will be reflected in the film thickness profile. The higher the collimator aspect ratio, the more significant this effect becomes. Figure 31 shows both the experimental and simulated nonuniformity across a wafer for sputtering from a 5 × 8-in. planar magnetron target with a well-defined annular erosion track (*50*). Al films were sputter deposited through an

FIG. 29. SIMBAD simulations for a 3 : 1 aspect ratio contact with (a) no collimator and (b) 3 : 1 aspect ratio collimator (Reprinted from *Can. Metall. Quart.*, **34**, T. Smy, S. K. Dew, and M. J. Brett, "Simulation and experimental analysis of refractory metal and aluminum deposition over high aspect ratio VLSI topography," pp. 195–202, 1995, with kind permission from Elsevier Science Ltd., The Boulevard, Langford Lane, Kidlington, 0X5 1GB, UK).

FIG. 29. Continued

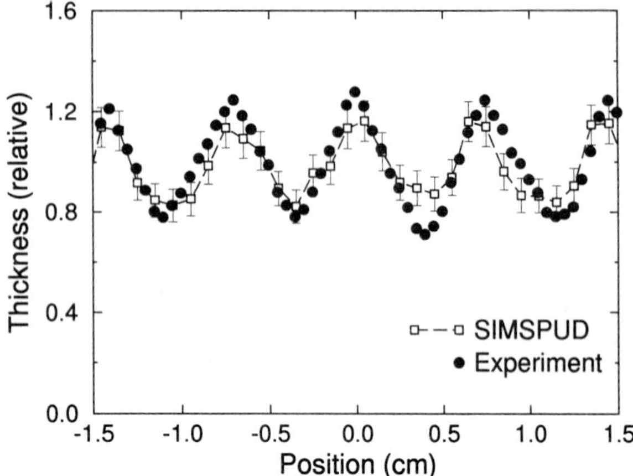

FIG. 30. Experimental and simulated film thickness profiles of aluminum films deposited using a collimator of aspect ratio 1 (from Ref. 49).

aspect ratio 2 square grid collimator. A high frequency thickness ripple, arising from the collimator itself, is superimposed over the gross nonuniformity created by the erosion track at the sputter target. A variety of process alterations could be made to improve film uniformity, including use of a target with a more uniform erosion profile, increased pressure, or decreased collimator aspect ratio. Once again there is a trade-off, because

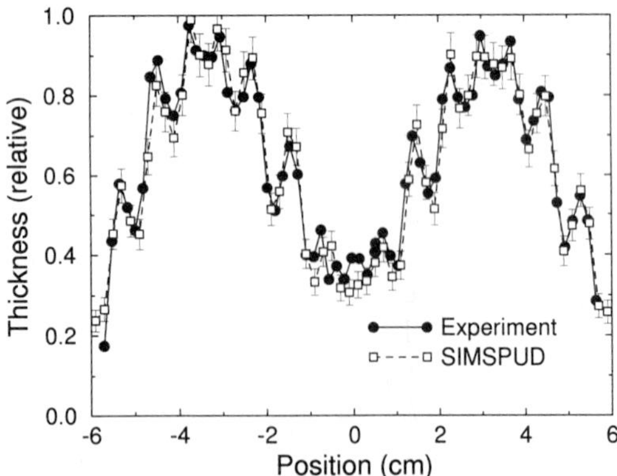

FIG. 31. Experimental and simulated film thickness nonuniformities of collimated sputtered films caused by the target erosion profile and the superimposed collimator ripple (from Ref. 50).

the latter two changes would also have the effect of increasing the angular distribution of flux incident at the substrate. The reader is directed to the references (*26,47–52*) for a full discussion of the relationships between collimator parameters, and also for simulation of the effects on film profiles of gradual filling of the collimator vanes by intercepted flux.

D. Simulation of Film Composition

Because many materials for ULSI metallization processes are alloys (i.e., Ti–W or Al–Cu) or compounds (i.e., TiN diffusion barriers), useful information for these processes includes how the composition of the deposited films may vary from that of the target, and how the composition may vary locally over topographical features. SIMBAD is capable of simulating the composition of alloy or reactively sputtering films, through the use of different disk types and respective sticking/resputtering coefficients for each element or constituent in the film. The relative deposition fluxes and incident angular distributions (determined by SIMSPUD) of each element must be provided in order to simulate deposition. Local film composition in the deposited film is determined simply by the ratio of numbers of disk types.

As an example, we examine the sputter deposition of Ti–W (10–90 wt. %), which is commonly used as a barrier or sticking layer in ULSI metalization. SIMSPUD was used to calculate incident flux angular distributions of Ti and W arriving at the substrate from a Ti–W target operating at 7 m Torr. Due to the much greater mass of W atoms, they are affected less by collisions with the Ar working gas than are the Ti atoms. Thus the angular distribution for incident W is much narrower than that for Ti, as illustrated in Fig. 32. For deposition over topography, the narrower distribution of W flux allows better penetration to the bottom of trenches or contact holes, resulting in a W-rich film at that point. Figure 33 shows the SIMBAD composition depiction of a Ti–W film deposited over a trench of aspect ratio 1, which accounts for the distribution differences. The lighter region at the bottom of the trench corresponds to W enrichment, whereas the trench sidewalls are Ti rich caused by the higher fraction of Ti atoms arriving obliquely at the substrate. The confirmation of these simulation results is described in Liu *et al.* (*25*). Clearly, if a 10–90% ratio of Ti–W is desired for the barrier at the bottom of the trench, the sputter target composition should be changed so that it contains less W. This composition issue may be complicated, however, by the likelihood that Ti may be preferentially resputtered by the much heavier W. In this regard, Fig. 34

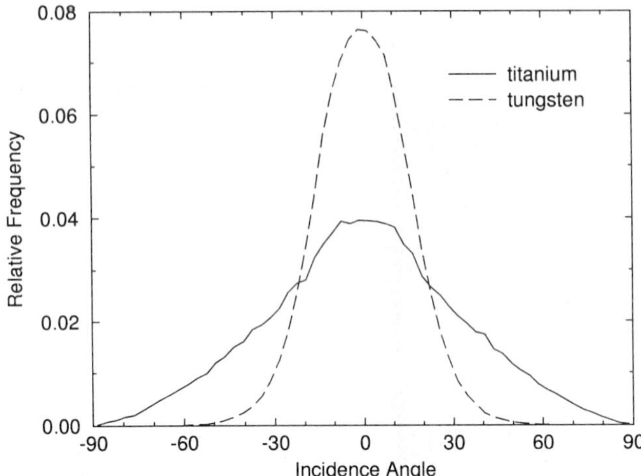

FIG. 32. Angular distributions of the sputtered atoms generated by SIMSPUD at 7 m Torr for Ti and W (after Ref. 25).

shows results of a Ti–W film sputtered onto an overhang structure. The film visible beneath the overhang is clear evidence that re-emission or resputtering of the deposited species can occur. Also, possible forward scattering or reflection of sputtered atoms incident at oblique angles on the sidewalls will further affect the film composition.

Reactive sputtering is a more complex process in which various atomic and molecular species are incident (with varying sticking coefficients) on the growing film. However, if these fluxes and parameters are quantified, then SIMBAD can be used to provide predictions of the film composition. A recent study using the SIMBAD model in this regard was made for films of TiN (*53*). TiN diffusion barrier films were deposited into contact holes of aspect ratio 2 by sputtering through a collimator for a metal Ti target operating in the non-nitrided mode with a N_2/Ar gas mixture. Composition analysis by auger electron spectroscopy and Rutherford backscattering (on scaled structures) demonstrated that films at the bottom of the contact were severely nitrogen depleted (i.e., $Ti_{1.78}N$), even through films on planar substrates were stoichiometric TiN.

To explain and model these results, the relative N/Ti flux incident on the substrate was determined from the experiment to be approximately 15. This value is an estimate due to the difficulties in adequately accounting for effects such as gettering, gas heating, sputter wind (*54,54*), and energetic neutral bombardment. SIMBAD was modified to incorporate a

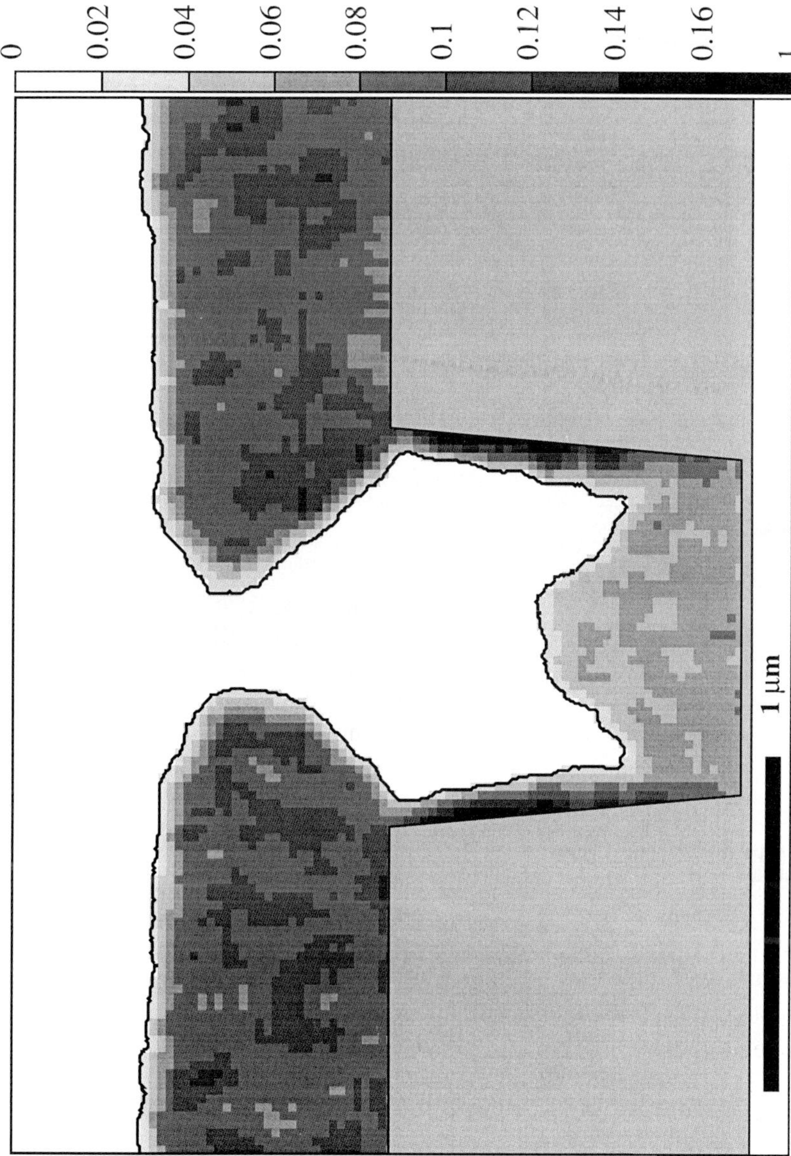

FIG. 33. The simulated variation of Ti concentration of Ti–W alloy films over trenches at 7 m Torr. The legend bar indicates relative Ti concentration in wt. % (from Ref. 25).

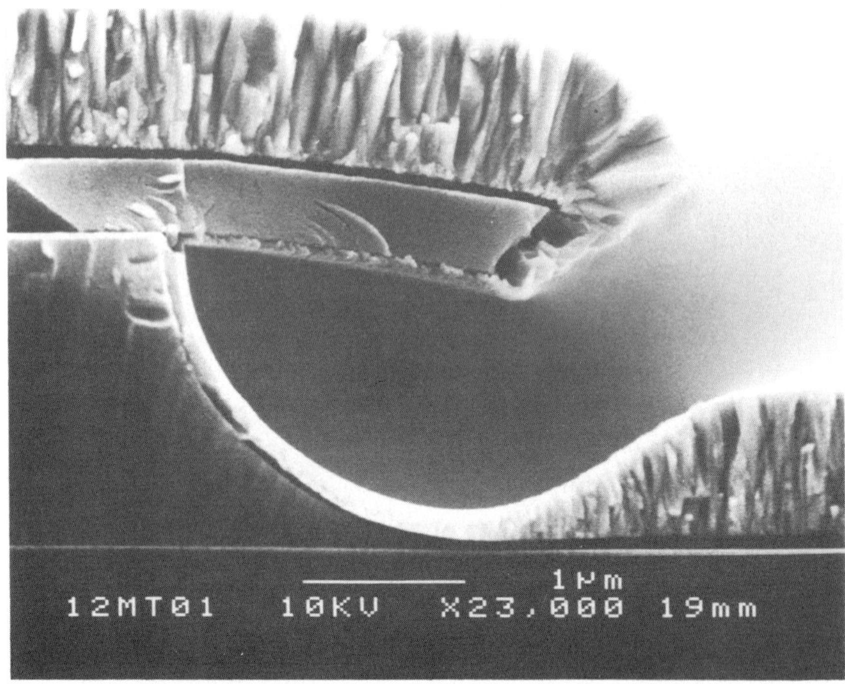

FIG. 34. SEM micrograph of a Ti–W alloy film sputtered over an overhang structure. A thin film is clearly visible on the underside of the overhang.

saturation-dependent sticking coefficient for nitrogen. In this model, the average sticking coefficient for nitrogen on pure Ti was one, but it decreases linearly with nitrogen fraction in the film. The sticking coefficient for Ti was always unity. The simulation in Fig. 35 shows the composition variation over topography, which agrees qualitatively with experiment in its prediction of a nitrogen-depleted film at the contact bottom. The depletion results from the fact that the nitrogen source is diffuse (gas), whereas the Ti flux coming through the collimator is highly directional. Thus sufficiently deep within topography, conditions are reached where there is adequate nitrogen to nitride the titanium fully. For a Ti target operating in the nitrided mode, both experiment and simulation showed nearly stoichiometric films everywhere due to the much higher ratio of N : Ti flux.

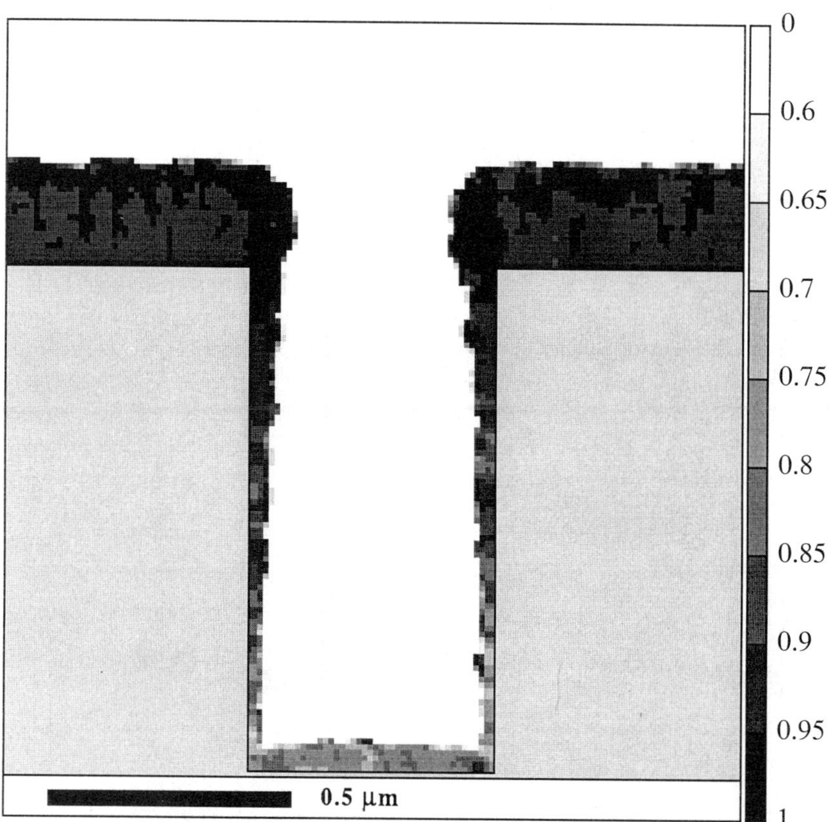

FIG. 35. SIMBAD simulation of the composition for TiN reactively sputtered from an unpoisoned Ti target into an aspect ratio of contact hole. The legend indicates the relative concentration of nitrogen (from Ref. 53)

E. High-Temperature Deposition and Bulk Diffusion

In this section, the sputter deposition of aluminum films onto elevated temperature substrates is examined. The increased diffusion length of Al adatoms will, in general, enable filling of concave submicron topographic features. However, geometric and energetic considerations such as surface curvature, wetting, and presence of bulk diffusion can create unexpected results for films deposited into extreme topography. SIMBAD has been extended so that it is a complete film model for high-temperature deposition employing algorithms for vacancy diffusion and allowing varying degrees of substrate wetting (*19,56*). Simulation of the phenomenon of grain boundary grooving (*57*) is controlled by a parameter which is the ratio of interfacial energy density of a grain boundary relative to the free surface energy. Cases are described for common metallization structures and processes where poor wetting of the substrate and 3-D curvature of the topographical feature leads to void formation, which can be overcome only by bulk (vacancy) diffusion.

A simple example of via filling leading to Al film planarization is illustrated first in Fig. 36 (*22*). Sputtering Al was deposited over 1-μm features at adatom diffusion lengths of 0.2, 0.6, and 0.93 μm, corresponding to nominal substrate temperatures of 245, 410, and 515°C, respectively. Ideally, surface wetting of the Al will be determined by the relative surface energies of the substrate and the film, but these simulations have assumed perfect wetting of the substrate by the aluminum. Since, in practice, a thin

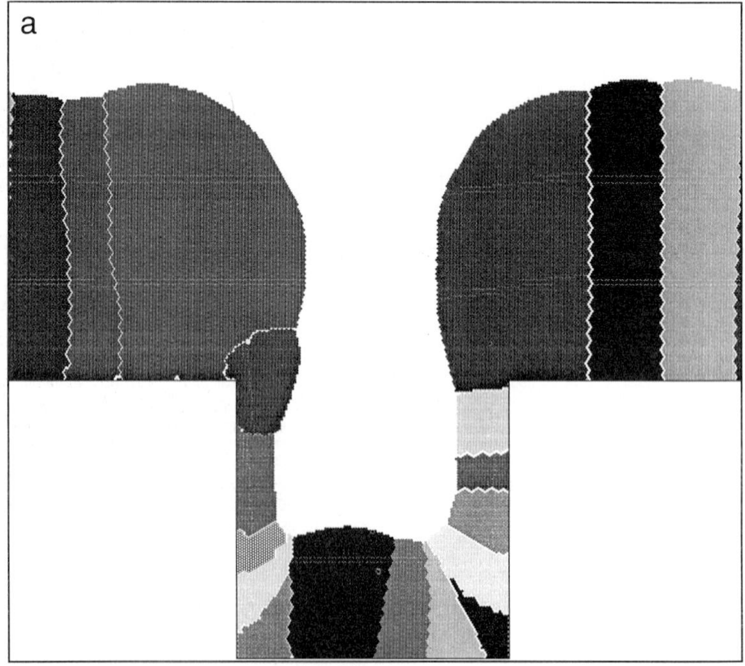

FIG. 36. Simulations of Al films sputter into into 1-μm features at substrate temperatures of (a) 245°C, (b) 410°C, and (c) 515°C.

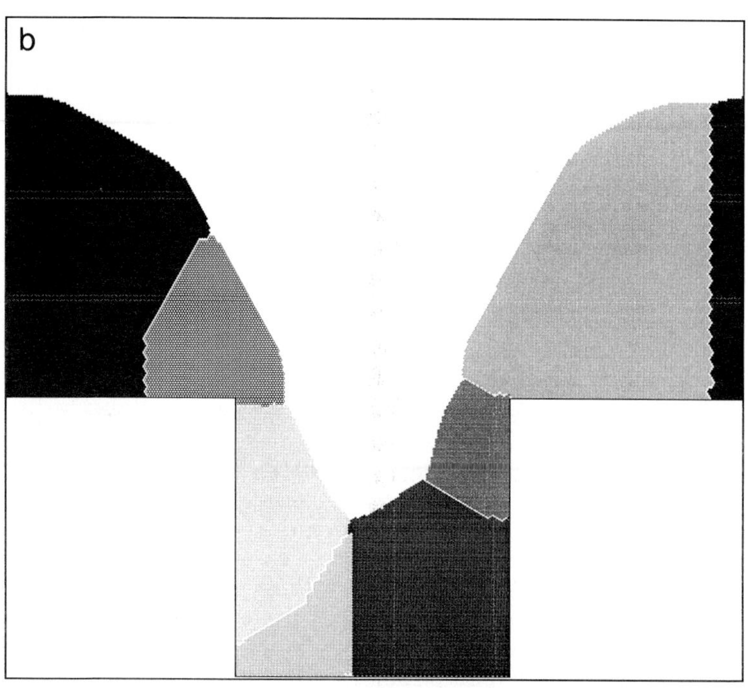

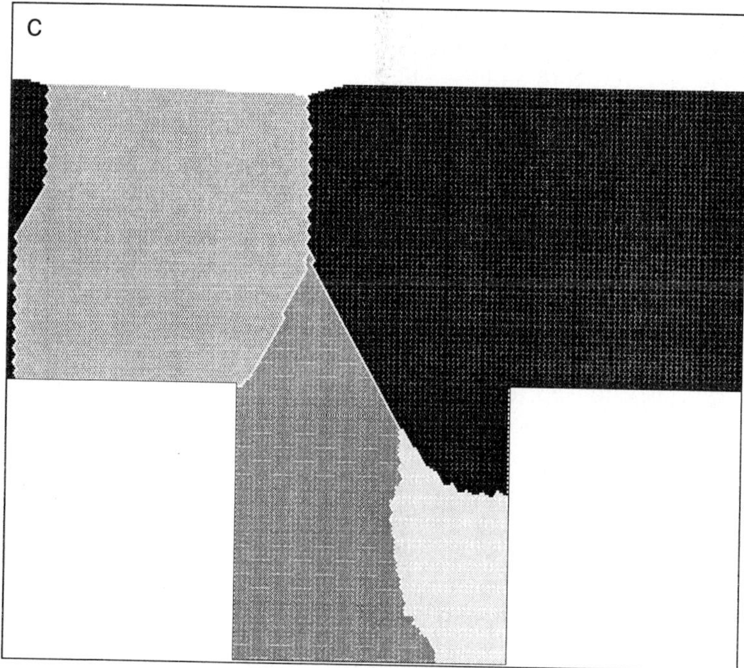

FIG. 36. Continued

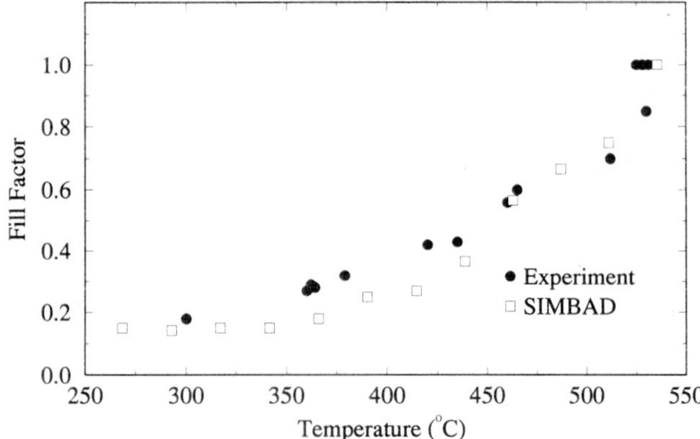

FIG. 37. The effect of substrate temperature on fill factor for 1 : 1 aspect ratio features 1 μm deep. Both SIMBAD and experimental results (*79*) are shown.

layer of Al is often initially sputtered at cold temperatures to enable wetting of the hot layer, this assumption is reasonable. Note that the lower temperature zone 2 structure and keyhole-shaped profile of Fig. 36a is in contrast to the larger grain and planarized film at higher temperatures. A quantitative verification of these results was made by comparing the temperature dependence of the fill factor in Fig. 37. Fill factor is defined as the ratio of the thickness of the film at the bottom of the feature to the sum of the feature depth and the film thickness. The SIMBAD predictions closely follow the experimental results (*22*) for Al sputtered into identical 1-μm features.

The curvature-dependent surface potential calculated by SIMBAD has allowed the simulation of diffusion for deposition temperatures that are high relative to the melting point of the film. However, under certain conditions, bulk diffusion of vacancies will dominate the evolution of the film profile (*56*). Figure 38 illustrates two common film and topography geometries. Figure 38a shows a topography structure similar to that of Fig. 36, in which the diffusion of matter is dominated by surface diffusion. Matter will be transported along the surface from areas of positive (convex) curvature to regions of negative (concave) curvature (such as the bottom of the contact). For Fig. 38b, with more severe topography, a void has formed in the film deposited in the contact hole. Matter transport will occur in a direction normal to the film surface, from the top of the film to the void. Surface diffusion cannot play a role in this transport because there is no surface path to the void. However, for high temperatures, vacancies are generated at the void surface and annealing (i.e., void removal) occurs by bulk diffusion vacancies.

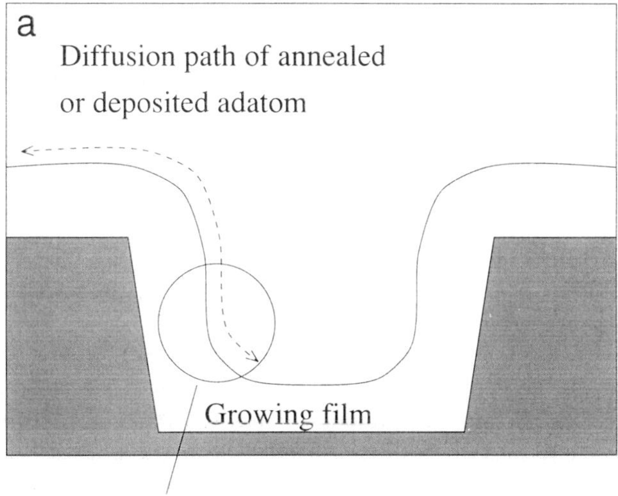

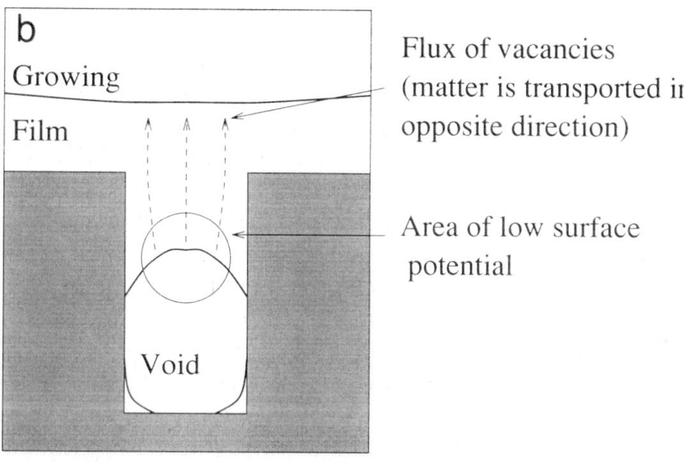

FIG. 38. Situations in which (a) surface diffusion would dominate the redistribution of surface atoms, and (b) the redistribution of matter from the top surface to the void would be dominated by diffusion of vacancies within the bulk (from Ref. 56).

The extension of SIMBAD to incorporate bulk diffusion required development of matter transport theory, described in detail in Smy *et al.* (*56*). The equation defining the flow of vacancies is:

$$\frac{\partial C}{\partial t} = -\nabla \cdot (F) = DN\nabla^2(C), \tag{8}$$

where F is the vacancy flux, C the vacancy concentration, D the self-diffusivity, and N the atomic density. For a diffusion of vacancies limited by surface generation, one may assume a constant vacancy concentration C_0 in the bulk. The vacancy flux normal to the surface F_n driven by surface curvature may then be defined at any point on the surface by:

$$F_n = -\frac{d\nu}{6}\left(-C_0 + N\exp - \frac{(E_0 + \gamma\kappa\, d\nu)}{kT}\right), \qquad (9)$$

where E_0 is the activation energy of vacancies from a flat surface, γ is the surface tension, κ is the surface curvature, d is the atomic jump distance, and ν is the atomic jump frequency. Thus Eq. (9) enables calculation of the bulk vacancy concentration and Eq. (10), given in a later section, defines the flux at any point on the surface. Disks are added or removed from the surface in a probabilistic manner determined by the flux at the surface. During vacancy diffusion (no deposition), matter is conserved as the net flux over the entire surface is zero.

These new algorithms for film growth were tested for a 3-D circular contact hole geometry (*56*). The growth sequence at high temperature for sputtered Al is shown in Fig. 39. Initially, matter transport is by surface diffusion and causes the film to accumulate near the mouth of the via. This unusual profile is created by the very high concave curvature of the contact in the horizontal plane perpendicular to the page. The contact corners are highly attractive to the diffusing atoms, particularly as the contact starts to close up and the curvature becomes more extreme. In Fig. 39b the contact has closed up and a void results at the bottom of the contact. This void may only fill by bulk diffusion, which is driven by the low surface potential of the void and the subsequence generation of a large number of vacancies which would flow to the film surface, which is essentially flat. The filled via, at a later time, is shown in Fig. 39c. Micrographs of sputtered aluminum films sputtered onto hot substrates corresponding to the conditions of Figs. 39b and 39c are shown in Fig. 40. These photographs of an 0.8-μm-wide contact hole clearly show void formation and the subsequent filling. The process of void filling is also strongly dependent on the wetting angle of the film with the substrate, which is in turn dependent on system cleanliness and film/substrate surface energies. Poor wetting (i.e., wetting angles much greater than 0°), can lead to enhanced void formation and increased difficulties in void filling. The interested reader is directed to Smy *et al.* (*56,58*) for a detailed discussion of these effects.

F. Interpolated Three-Dimensional SIMBAD Model

For films deposited near the edge of a large substrate or wafer, the finite size of the target or source leads to an asymmetric angular distribution of incident flux. This leads to an often undesirable asymmetric coverage for features located off axis near the wafer edge, and is exacerbated by 3-D shadowing by the feature and the growing film. For simulation of these effects, a full 3-D model of film growth over topography showing adequate microstructural detail would require enormous computer resources. As an alternative, within SIMBAD a model has been developed that incorporates the 3-D shadowing effects of topographic features but does not require a full 3-D data structure.

In the "interpolated" 3-D simulation model (*59*), a number of slices through the center of the topographic feature (i.e., contact or via) are taken. Simulation of film deposition on each of these slices is performed using the standard 2-D SIMBAD simulator, incorporating full 3-D incident flux distributions from SIMSPUD. To model the 3-D surface in a cylindrical coordinate system, the surface profiles of the 2-D simulations are interpolated between slices. For example, Fig. 41 shows the interpolated 3-D surface profile of a simulated film deposited into a cylindrical contact hole near the edge of a wafer (i.e., off axis). The interpolated 3-D surface is updated frequently and used in examination of subsequent disk trajectories to identify the 2-D slice on which incident disks impact. This approach accounts for asymmetric incident flux and shadowing by the topographic feature and the growing film.

An experimental study of film coverage asymmetry was performed for a system with a 12.5-cm-diameter target located 8 cm from the substrate (*59*). Figures 42a and 42b show cross-sections of W films sputtered into holes of aspect ratio 3. These contact holes were located at the center of the wafer, and at 8 cm from the wafer center (45° from the target center axis), respectively. The slices through the contact holes were made along radii from the wafer center. An expected nonuniformity in bottom and sidewall coverage is clearly shown. The further the feature is located from the central axis of the target, the larger the asymmetry in coverage. Figures 43a and 43b show the results of simulation of these structures by the interpolated 3-D model. The simulations not only show an accurate film surface profile, but they also properly depict the orientation of the film microstructure. The model also provides the film structure for slices through the contact in other directions relative to the wafer radii. The slice or orientation of cross-section illustrated in Fig. 42b represents the case of most asymmetry between the contact sidewalls.

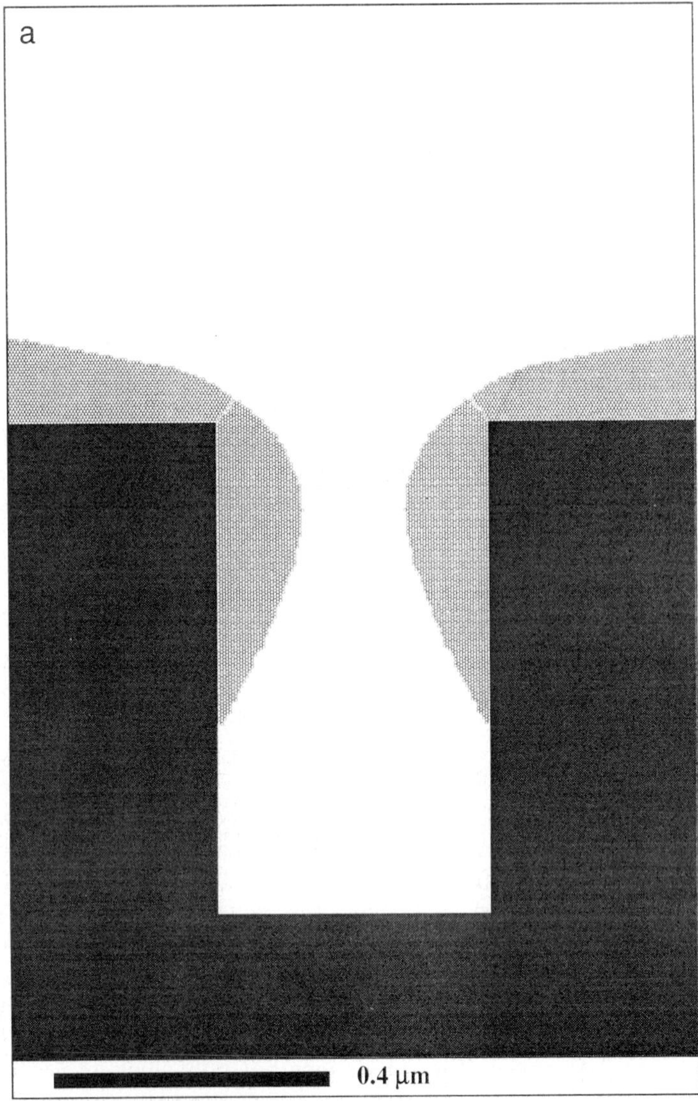

FIG. 39. Simulation of the film growth sequence for Al sputtered onto high-temperature substrates into 3-D circular contact holes (from Ref. 56).

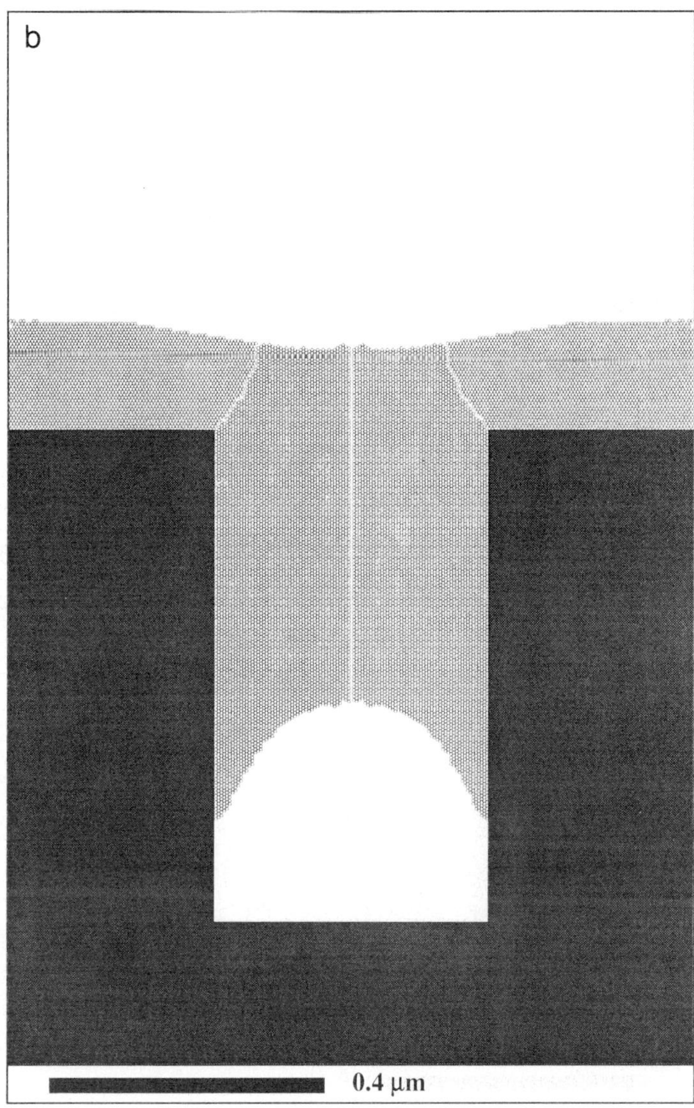

FIG. 39. Continued

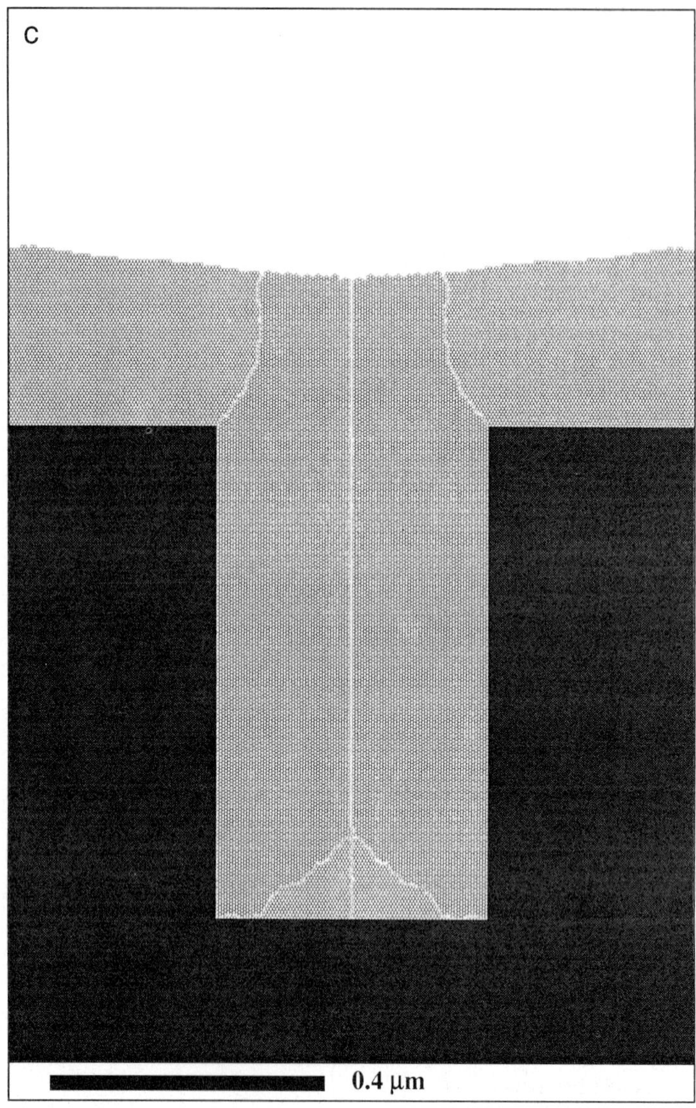

FIG. 39. Continued

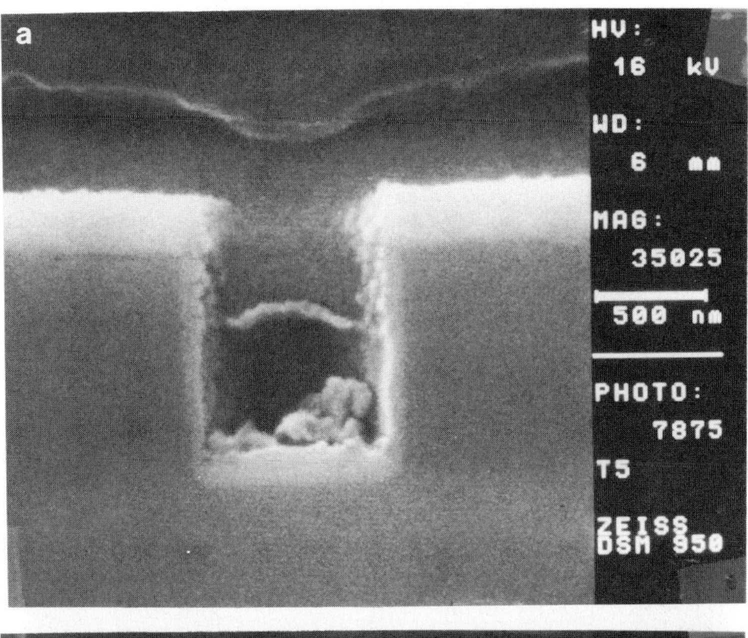

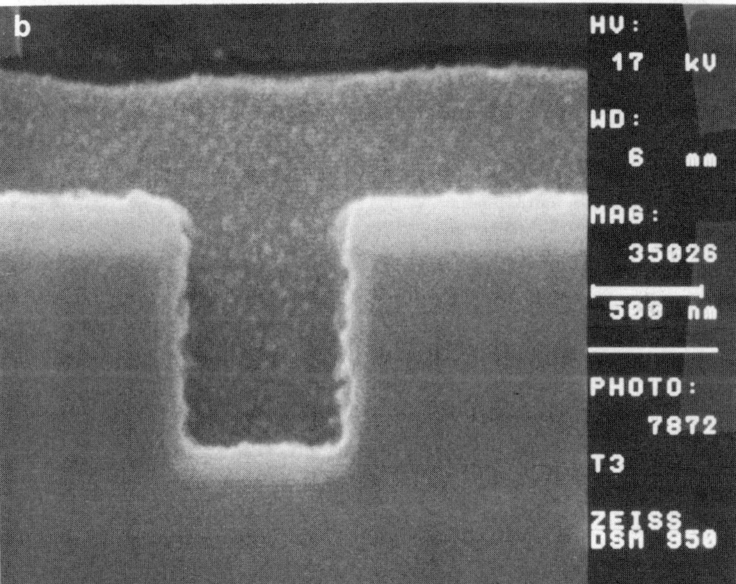

FIG. 40. SEM micrograph of (a) void formation in an Al film in a circular via hole and (b) complete filling of the via by Al. The corresponding simulations are Figs. 39(b) and 39(c), respectively (Reprinted from *Can. Metall. Quart.*, **34**, T. Smy, S. K. Dew, and M. J. Brett, "Simulation and experimental analysis of refractory metal and aluminum deposition over high aspect ratio VLSI topography," pp. 195–202, 1995, with kind permission from Elsevier Science Ltd., The Boulevard, Langford Lane, Kidlington, 0X5 1GB, UK).

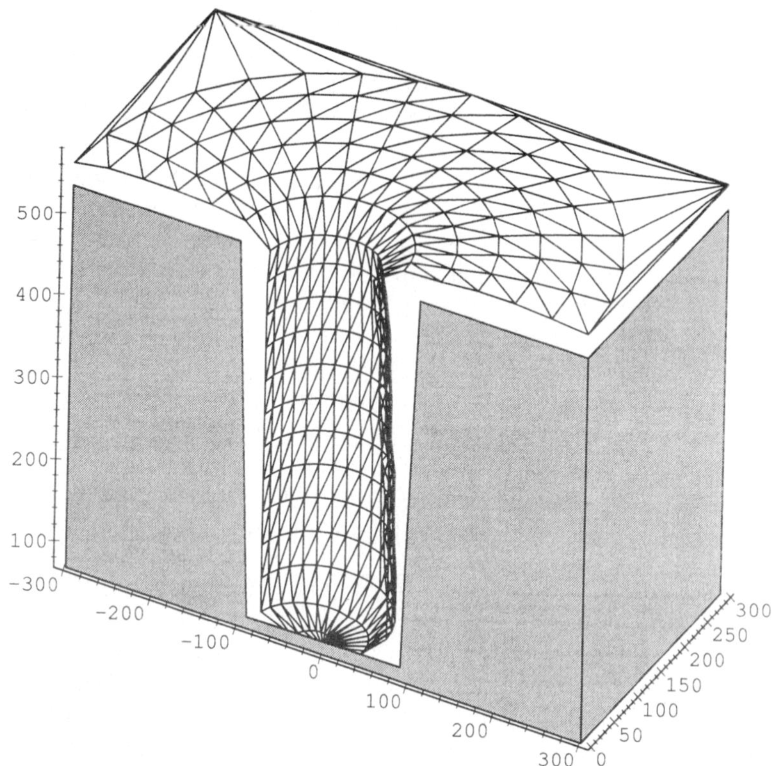

FIG. 41. The interpolated 3-D surface profile of a simulated film deposited into a cylindrical contact hole (from Ref. 59).

G. Etching Processes

The microstructure-induced porosity and the ability to manipulate disks in the SIMBAD representation of the film have enabled physically representative models of etching to be incorporated into SIMBAD. In earlier sections of this chapter, wet etch simulations (Sec. III.B) and the etching inherent in the bias sputter process (Sec. IV.B) were described. In this section, the application of SIMBAD to both plasma etching and sputter etching is discussed and compared with experiment. The reactive ion etching (RIE) process has also been simulated using SIMBAD, but readers are referred to Tait *et al.* (*60*) due to space limitations in this chapter.

For plasma etch simulations (*61*), the active gas species are represented by disks designated as etchant disks. Consistent with the interpretation of film disks, each etchant disk represents an average of a large number of etchant species. Gas phase collisions in the simulation are neglected since even at common etching pressures of 400 m Torr the mean free path of a gas molecule is significantly longer than the simulation dimensions. In reactive etching, free reactive gas atoms reach the substrate by diffusion, thus the flux of reactive species incident on the substrate may be represented by a cosine distribution. On impact at the film surface, etchant disks have a fixed probability of reacting and removing a film disk, a parameter referred to as the reaction coefficient. If an etchant disk fails to react, it is desorbed from the surface according to a thermal cosine distribution and proceeds along a linear trajectory until it once again contacts the film surface, or exits the simulation region. Second-order effects on the local reaction probability such as multiple echant species and nonlinear reaction paths are not currently included in this model.

The reaction coefficient is a critical parameter in this model and may be determined from a knowledge of the discharge. As an example, we consider etching of W in a CF_4/O_2 plasma (*61*). Etching is caused by free F atoms and the concentration of F has been found to be about 10^{15} atoms/cm^3 for a 200-W discharge of CF_4 at 45 Pa with 8% O_2 (*62*). This concentration corresponds to an impingement flux of $4 \times 10^{19}\ cm^{-2}s^{-1}$ of F onto the substrate. Etching experiments with this process determined a W atom removal rate of $2 \times 10^{16}\ cm^{-2}s^{-1}$, from which may be inferred a reaction coefficient of order 10^{-3}. Figure 44a shows an unetched cross-section of a W film that has been sputter deposited in a 1.5-μm-wide trench structure. After 30 s of plasma etching, the film is shown in Fig. 44b. Notice that the sidewalls and bottom trench corners have etched preferentially because of the lower density microstructure in that region. The simulations of the reactive etching are shown in Figs. 45a and 45b. It is only because SIMBAD has a microstructural-level depiction of the film that the faster etching is correctly modeled on the trench sidewalls and corners. The higher density planar film regions are less affected by the etch, similar to the results seen for wet etch in Sec. III.B.

In its simplest form, sputter etching is similar to bias sputtering but without concurrent deposition from the primary target. Ion disks are generated and may resputter the film according to the algorithms described in Sec. IV.B. To evaluate independently the accuracy of ion or sputter etch simulations, Ti and W films have been dc sputter deposited and then etched using an rf bias to the substrate, without breaking vacuum

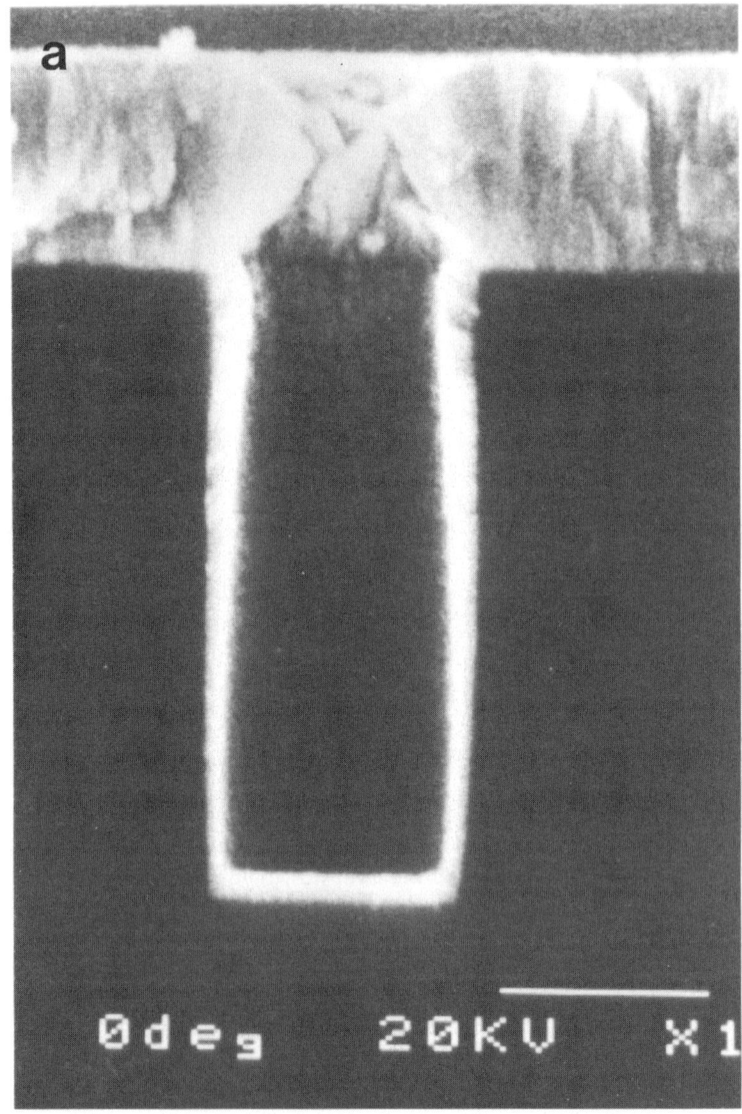

FIG. 42. SEM micrographs of W films sputtered into cylindrical contact holes (a) at the wafer center and (b) near the wafer edge (from Ref. 59).

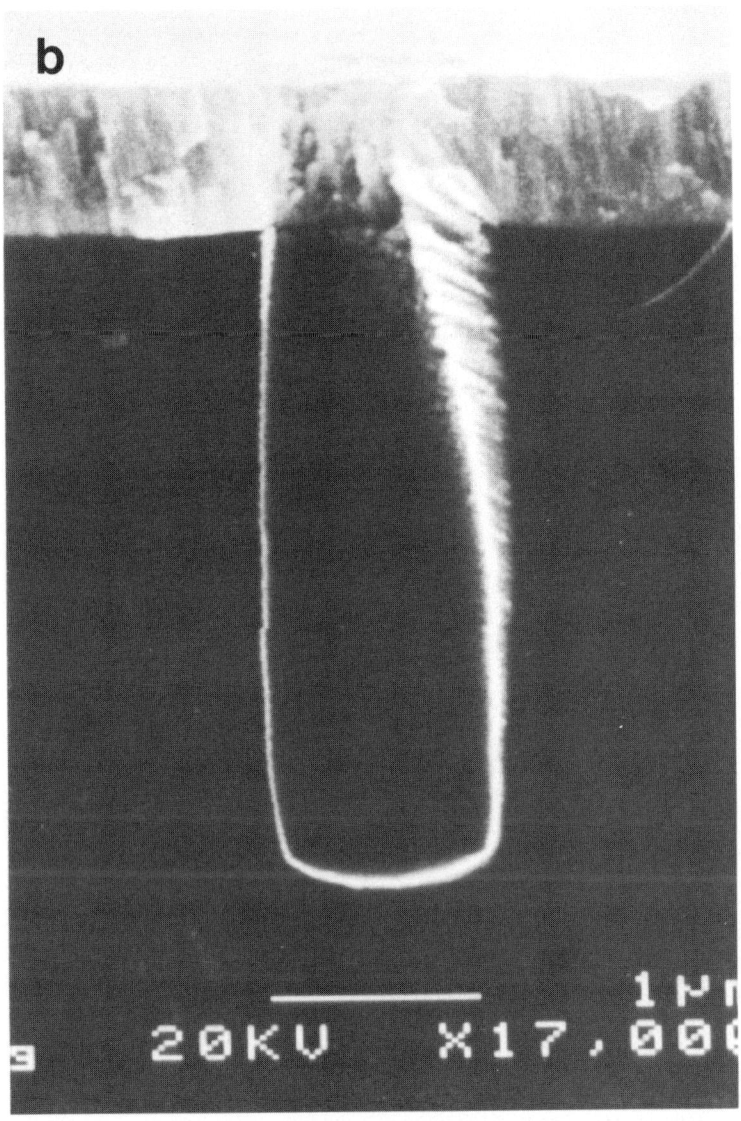

FIG. 42. Continued

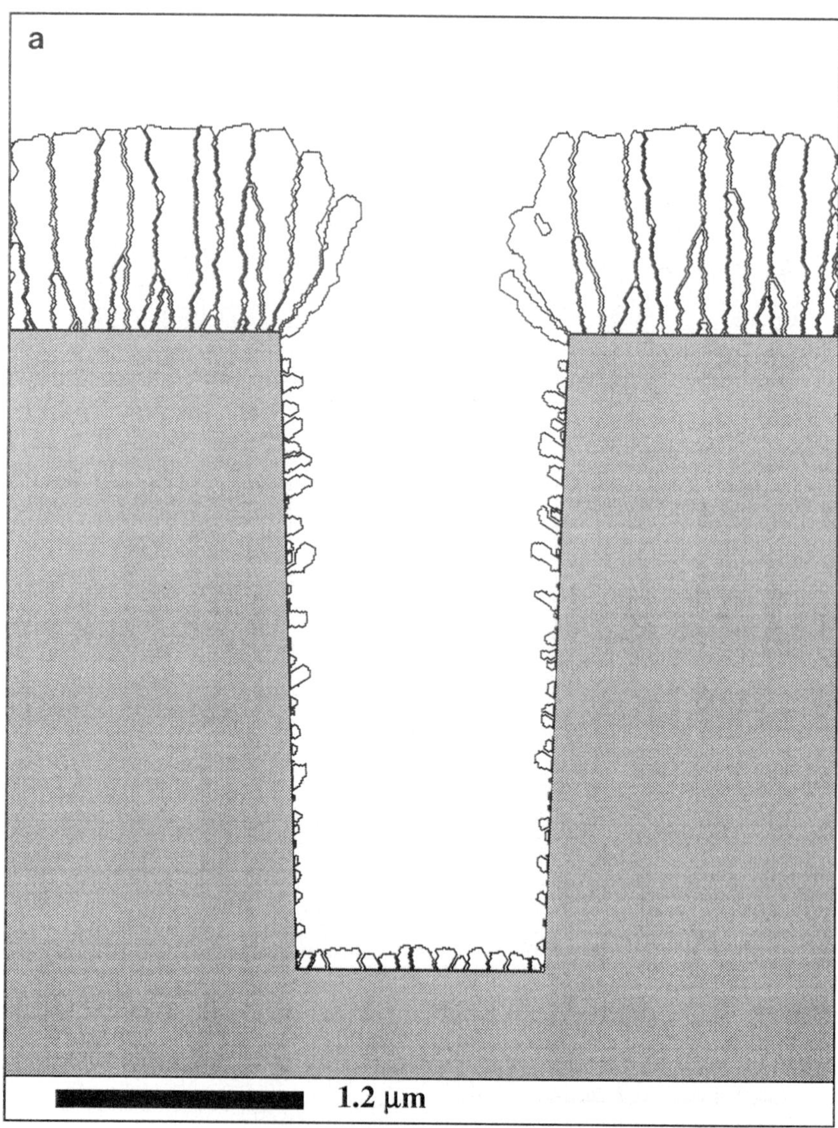

FIG. 43. Interpolated 3-D SIMBAD simulations of sputtered W films corresponding to Figs. 42(a) and 42(b), respectively (from Ref. 59).

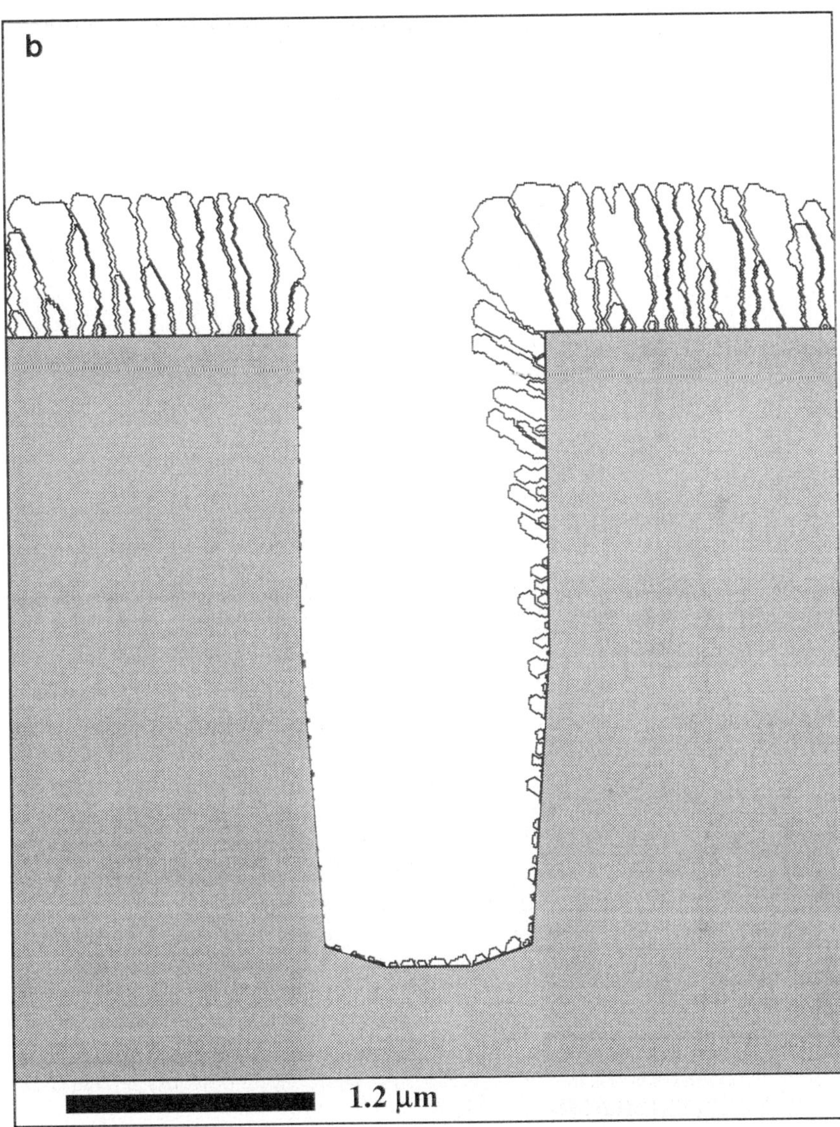

FIG. 43. Continued

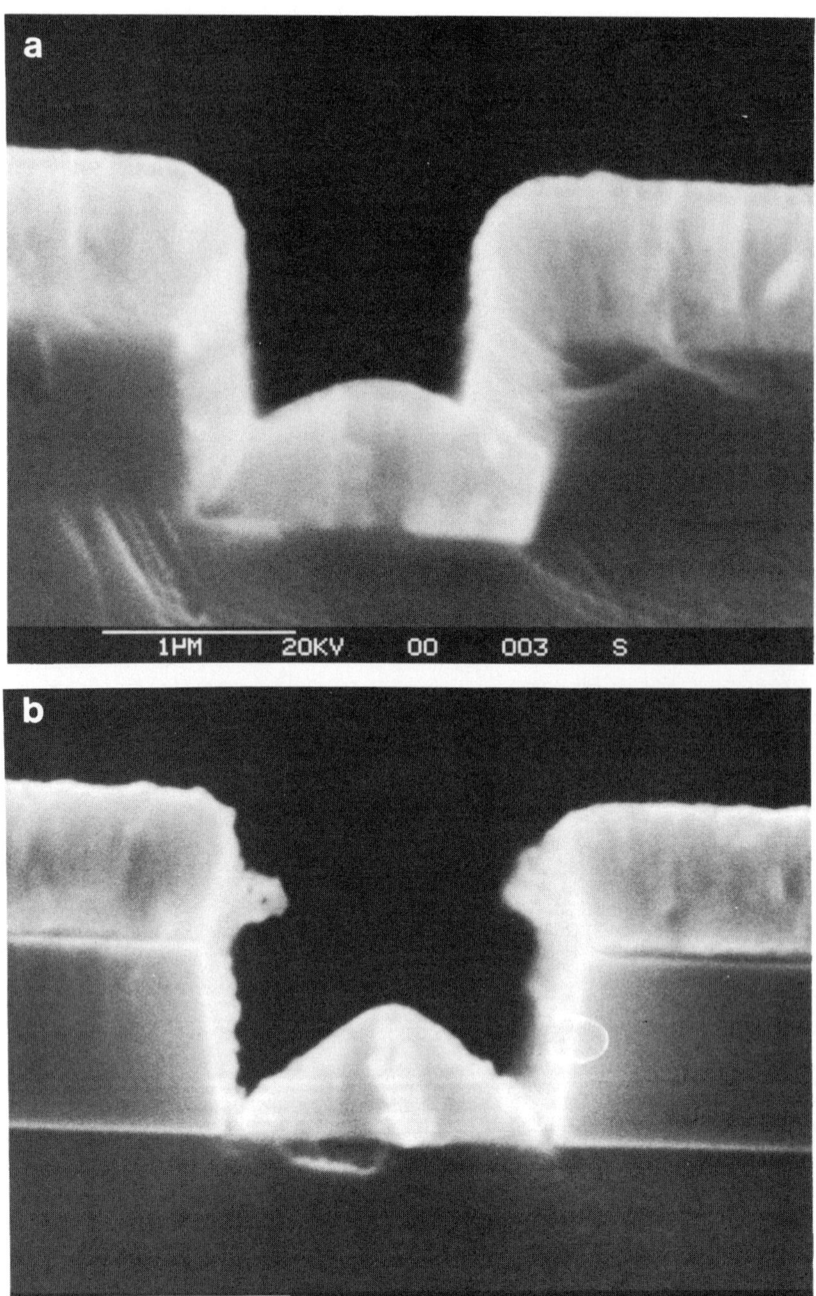

FIG. 44. SEM micrographs of sputtered tungsten films over a trench (a) as deposited and (b) after 30 s of plasma etching (from Ref. 61).

in the system (*63*). Figures 46a, 46b, and 46c show an as-deposited Ti film of thickness 1.3 μm, and Ti films after sputter etch removal of 0.25 and 0.5 μm, respectively. Since Ti is a refractory metal, and sputtering/etching was performed on unheated wafers, the films deposited over trenches show a detailed microstructure. The corresponding simulations of the sputter etch process are shown in Figs. 47a, 47b, and 47c. Both the final surface profile and the details of the microstructure are in excellent agreement with experiment. Multiple sputter-deposition/sputter-etch processes were also performed which yielded four-layer films in agreement with the simulation predictions, and were found to increase significantly the planarity of films.

V. New Developments in Film Growth Simulation

In multilevel metallization structures and for submicron device structures, details of the grain structure of deposited films are increasingly important in order to evaluate the film's performance, particularly because film dimensions are often comparable to that of an average grain. For instance, the effectiveness of diffusion barriers and the electromigration resistance of conductors are strongly dependent on microstructural details of the film. The grain structure is, in turn, a result of the interaction of interfacial energies and surface processes during nucleation and growth. In this section, we present GROFILMS (GRain Oriented FILm Microstructure Simulator), a new film growth model that explicitly incorporates interfacial energy calculations to create film microstructure predictions that include effects such as grain boundary grooving, crystal faceting, and surface wetting (*64*).

GROFILMS uses a string-like algorithm to describe the surface topography and internal microstructure of a thin film. Each grain is described by a sequence of line segments connecting nodes which contain local information about the surface. Boundary, interface, and surface nodes are distinguished. For film growth, particles of a given angular distribution (as determined by SIMSPUD) and representing a 2-D "volume" of material are incident on the substrate. Once energetics calculations are performed to determine where the particles will be incorporated, the surface evolves through advancement of nodes and possible creation of new boundaries or interfaces. Figure 48 shows the GROFILMS nodal structure, which is helpful for the following discussion, and shows the wetting and grain boundary grooving angles.

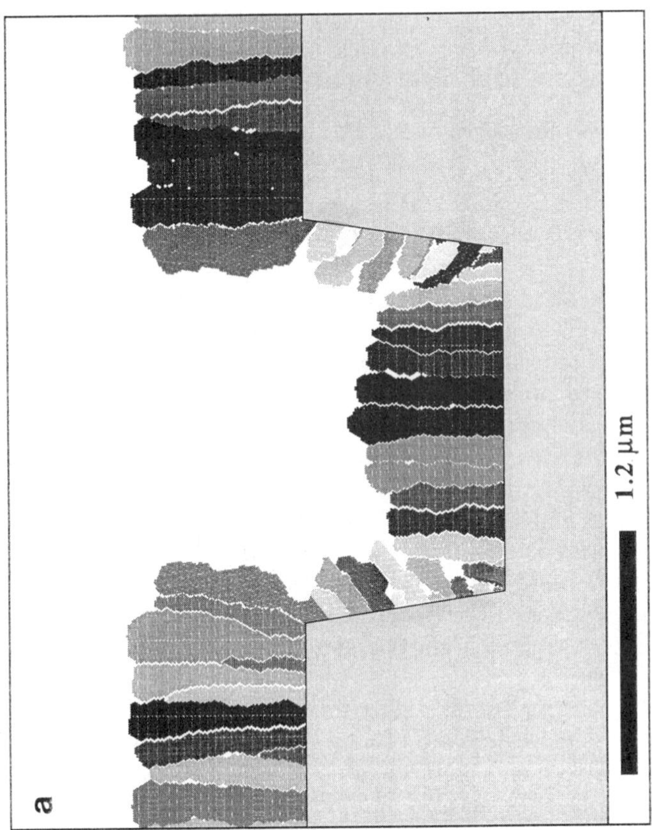

FIG. 45. Simulated film sputtered over a trench (a) as deposited and (b) after 30 s of plasma etching. These simulations correspond to the real films of Fig. 44 (after Ref. 61).

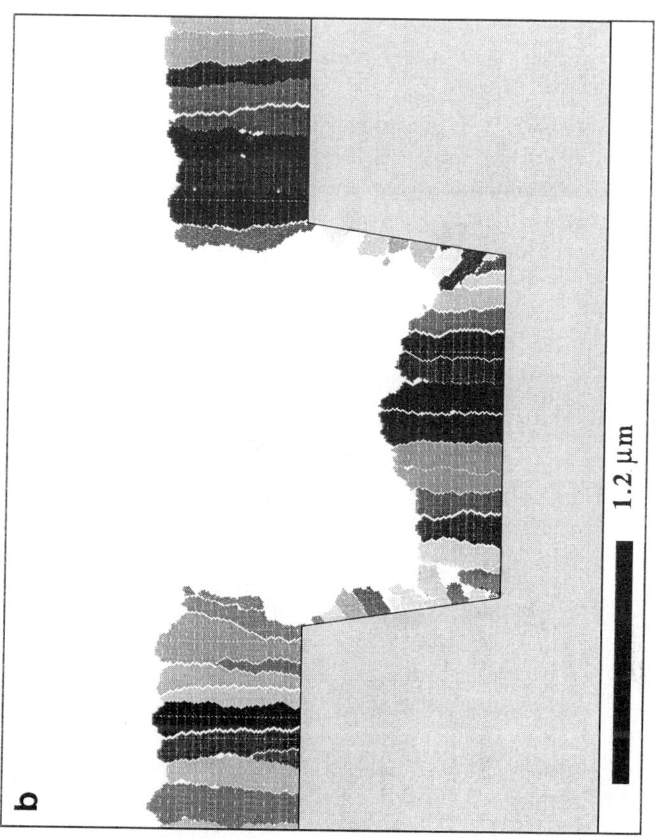

FIG. 45. *Continued*

The critical constraint during dcposition is that the incident volume of material will preferentially diffuse to a node which will result in the greatest minimization of the local free energy. At each node within a user-specified diffusion length, GROFILMS calculates the change in free energy if an infinitesimal amount of volume were added. For example, the free energy of formation of a droplet on a plane surface by heterogeneous nucleation is

$$\Delta G = V\Delta G_v + \gamma_{vf}A_{vf} + \gamma_{fs}A_{fs} + \gamma_{sv}A_{sv}, \tag{10}$$

where V is the volume, A is the area, and ΔG_v is the free energy change of formation per volume of the nucleus. The γ represents the interfacial surface tension between the vapor and film (vf), the film and substrate (fs), and the substrate and vapor (sv). At every node within one diffusion length of the impact site, $d(\Delta G)/dV$ is calculated to determine the most

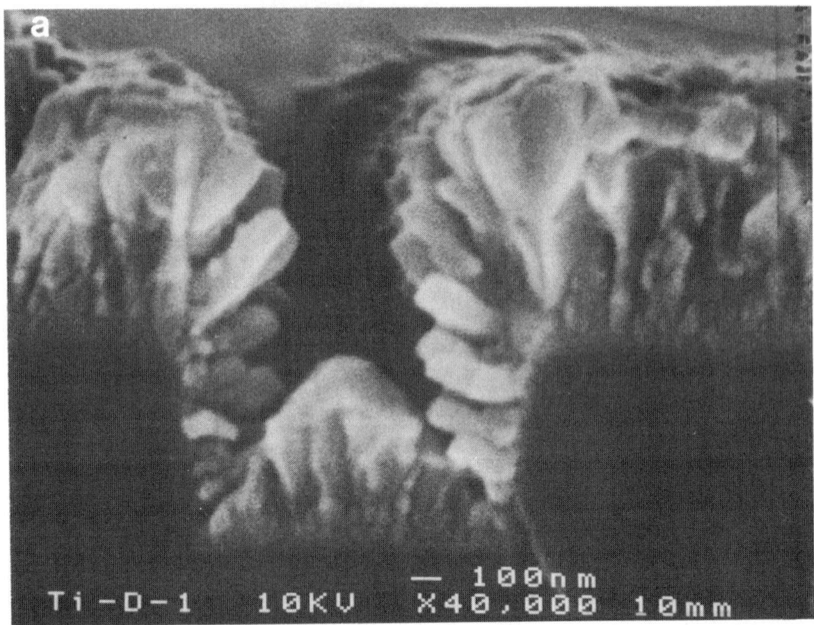

FIG. 46. SEM micrograph of Ti films sputter deposited over trenches: (a) 1.3-μm-thick film as deposited, (b) after sputter etching has removed 0.25 μm, and (c) after sputter etching has remove 0.5 μm (from Ref. 63).

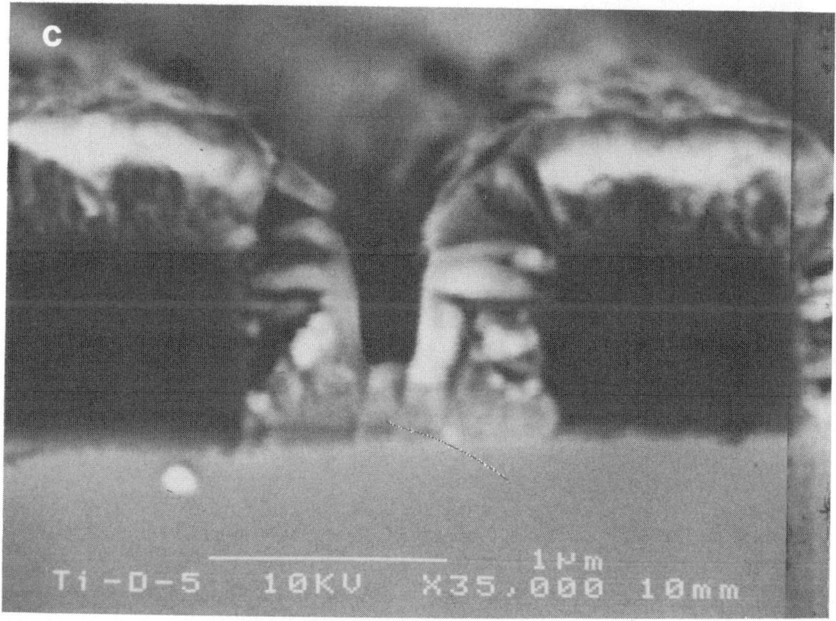

FIG. 46. Continued

favorable incorporation site. For a free surface, this derivative is simply proportional to the curvature of the surface, such that material will preferentially deposit in regions of high concavity.

At grain boundary nodes, the free energy change is dependent on the formation of local surfaces and is given by

$$\Delta G + V\Delta G_v + \gamma_{vf} A_{vf} + \gamma_{gb} A_{gb}, \tag{11}$$

where γ_{gb} is the grain boundary interfacial tension and A_{gb} is the area. The values for grain boundary tension will determine the equilibrium angle for the phenomenon of grain boundary grooving.

Faceting of crystals is determined by the variation of surface tension with crystal orientation, often defined by a Wulff plot (*65*). GROFILMS incorporates data from Wulff plots to generate the faceting phenomenon. For this case, the volume derivative of ΔG includes a change in surface tension with volume due to the dependent of γ and α, the angle that a

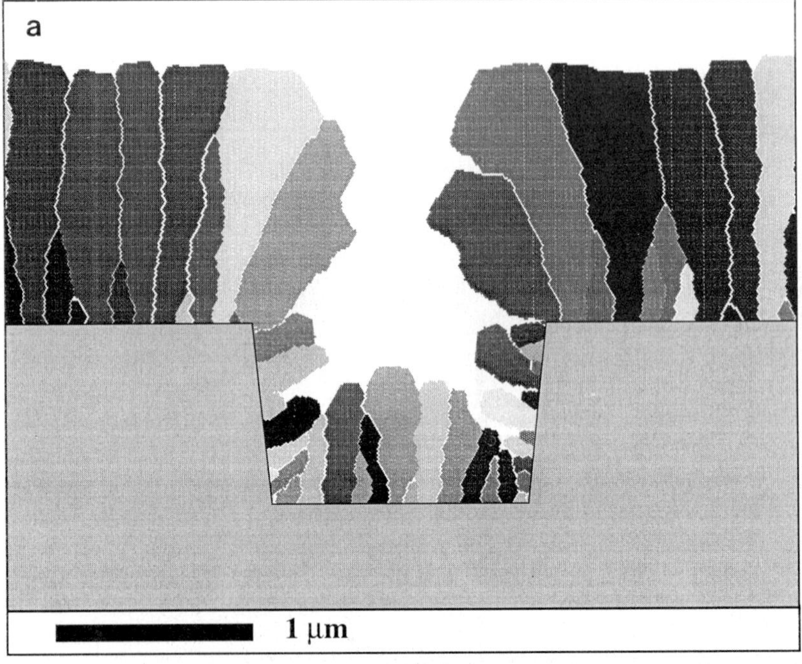

FIG. 47. SIMBAD simulations of Ti films corresponding to Fig. 46: (a) initial film, (b) after sputter etch of 0.25 μm, and (c) after sputter etch of 0.5 μm (from Ref. 63).

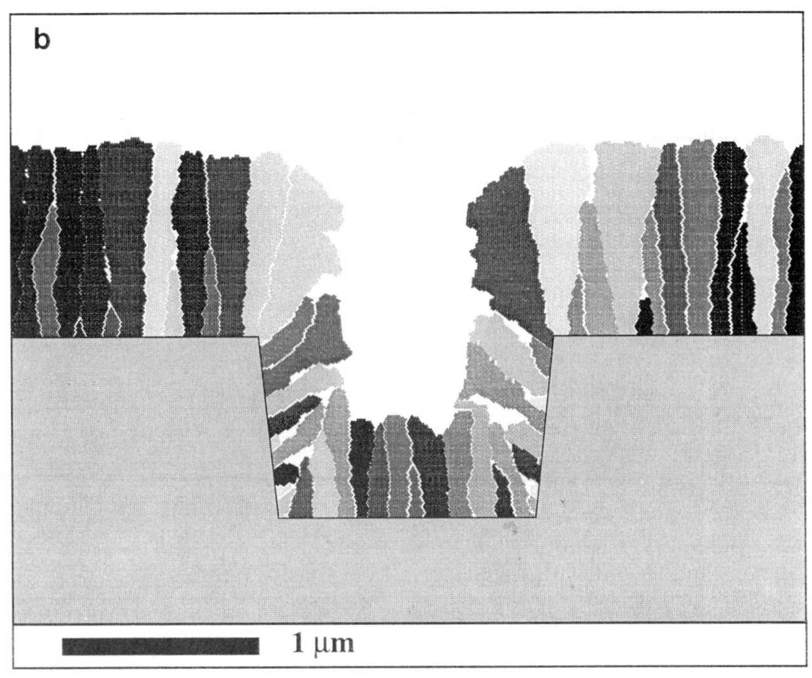

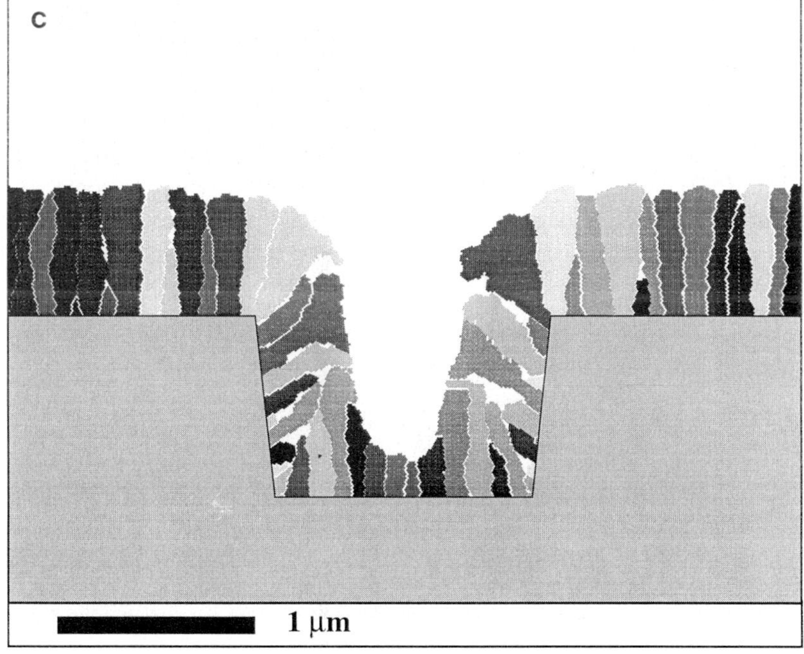

FIG. 47. Continued

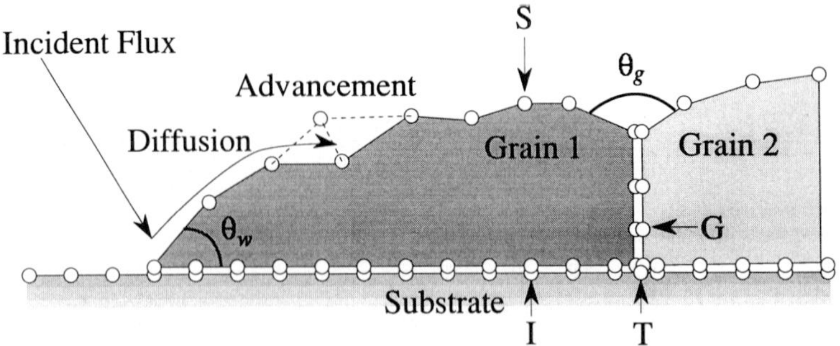

FIG. 48. The nodal structure of the GROFILMS film simulator, where θ_w and θ_g are the wetting angle and grain boundary grooving angles, respectively (from Ref. 66).

surface plane makes with respect to the grain orientation:

$$\frac{d}{dV}(\gamma(\alpha)A) = \left(\frac{d\gamma}{d\alpha}\right)\left(\frac{d\alpha}{dV}\right)A + \gamma(\alpha)\frac{dA}{dV}. \tag{12}$$

Since a crystal structure is three dimensional, a projection of the Wulff plot data onto a specific crystal plane must be used for a 2-D GROFILMS simulation.

Preliminary results from the new GROFILMS model are shown in Figs. 49, 50, and 51 (*64*). Grain boundary grooving between grains is shown in Fig. 49, whereas grain competition and (top surface) crystal faceting are shown in Fig. 50. Figure 51 shows an aluminum sputter deposition at high temperatures over topography, with poor wetting, which leads to the formation of undesiravle void regions along the surface that prevent filling. Although GROFILMS has the ability to predict evolution of film grain structure with a physically correct model, one of the current difficulties is

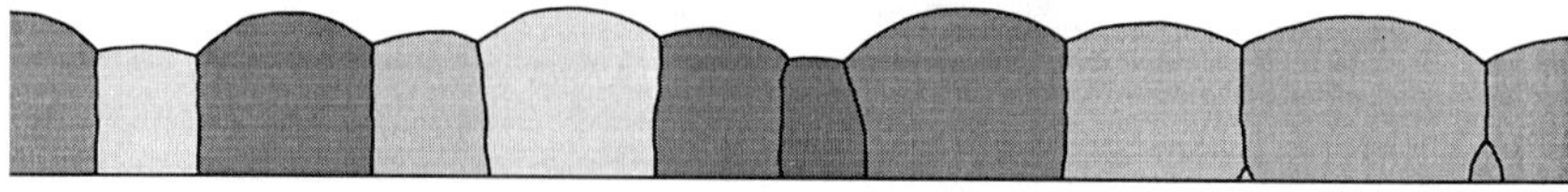

FIG. 49. Film simulation using GROFILMS showing grain boundary grooving, resulting from interfacial energy parameters $\gamma_{fv} = \gamma_{gb} = 1.4\,\mathrm{J/m^2}$ (Reprinted from *Thin Solid Films*, **266**, L. Friedrich, S. K. Dew, M. J. Brett, and T. J. Smy, "Thin film microstructure modelling through line segment simulation," pp. 83–88, 1995, with kind permission from Elsevier Science S. A., Lausanne, Switzerland).

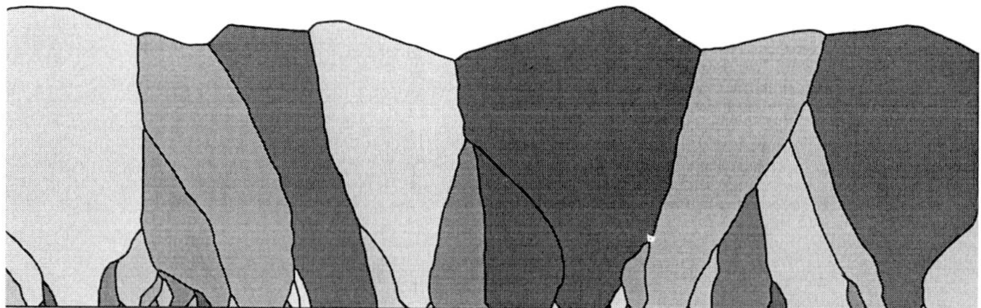

FIG. 50. GROFILMS simulation of a film deposited onto a high-temperature substrate, showing faceting on the surface of the film as a result of the growth of large grains (Reprinted from *Thin Solid Films*, **266**, L. Friedrich, S. K. Dew, M. J. Brett, and T. J. Smy, "Thin film microstructure modelling through line segment simulation," pp. 83–88, 1995, with kind permission from Elsevier Science S. A., Lausanne, Switzerland).

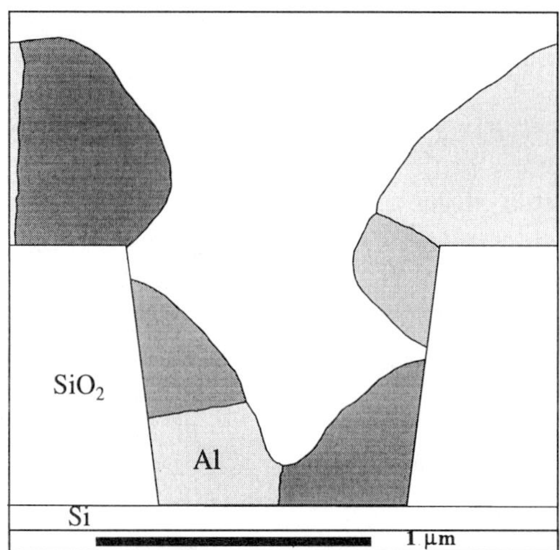

FIG. 51. GROFILMS simulation of high-temperature deposition of Al over topography. The poor wetting and large grain growth during deposition have lead to discontinuities in the film (Reprinted from *Thin Solid Films*, **266**, L. Friedrich, S. K. Dew, M. J. Brett, and T. J. Smy, "Thin film microstructure modelling through line segment simulation," pp. 83–88, 1995, with kind permission from Elsevier Science S. A., Lausanne, Switzerland).

that the surface tension parameters are not well known for some materials. Ongoing research is addressing this need, and further details verifying the algorithmic accuracy of GROFILMS, and comparison to experiment can be found in Friedrich *et al.* (*64,66*).

VI. Conclusions

This chapter has reviewed the model algorithm, verification, and application of the film growth simulator SIMBAD. The approach used in this model of chemical or physical vapor deposition is accretion of disks of film material at a substrate, which implicitly incorporates shadowing processes fundamental to microstructure determination and feature coverage, and provides a depiction of film microstructure and density. The model also enables incorporation of the growth processes of bulk and surface diffusion, resputtering, re-emission, nonunity sticking coefficient, substrate wetting, grain boundary grooving, grain faceting, and etching by ion bombardment, plasma, or wet etch procedures. The transport simulator SIMSPUD provides an energy, spatial, and angular description of sputtering flux in which itself will impact film coverage and microstructure through such deposition processes as, for example, collimated sputtering.

Space limitations in this work have not allowed a full description of applications of the SIMBAD film simulator. In the literature, interested readers may find discussion of use of SIMBAD to study film optical properties (*67*), film nodule or defect growth (*68–69*), electromigration (*70*), the structural anisotropy in film microstructure (*71*), collimator lifetime (*72*), spurious sputtered film features (*73*), effects of film density on interconnect heating (*74*), and ice-rime deposition or hailstone formation (*75*). Further inquiries regarding the SIMBAD simulator should be made to the authors (c.f. *76*).

Acknowledgments

The authors thank their collaborators and students whose work is included in this chapter. They include D. Liu, L. Friedrich, P. Li, K. Robbie, N. Tait, K. Sheergar, S. Winterton, T. Janacek, G. Braybrook, D. Hodul, W. Tsai, T. Sheng, M. Biberger, K. Chen, and S. Hsia. Financial support from Natural Sciences and Engineering Research of Canada, the Alberta Microelectronic Centre, and Varian Associates is gratefully acknowledged.

References

1. B. A. Movchan and A. V. Demchishin, *Fiz. Met. Metalloved.* **28**, 653 (1969).
2. J. A. Thornton, *Annu. Rev. Mater. Sci.* **7**, 239 (1977).
3. T. S. Cale and G. B. Raupp, J. Vac. Sci. Technol., B8, 1242 (1990).
4. W. G. Oldham, A. R. Neureuther, C. Sung, J. L. Reynolds, S. N. Nandgeonkar, *IEEE Trans. Electron Devices*, **ED-27**, 1455 (1980).
5. J. Ignacio, F. Ulacia, and J. P. McVittie, *J. Appl. Phys.* **65**, 1484 (1989).
6. T. Thurgate, *IEEE Transactions of the Computer-Aided Design of IC's and Systems*, **CAD-10**, 1101 (1991).
7. P. Meakin, P. Ramanlal, L. M. Sander, and R. C. Ball, *Phys. Rev. A* **34**, 5091 (1986).
8. D. Henderson, M. H. Brodsky, and P. Chaudhari, *Appl. Phys. Lett.* **25**, 641 (1974).
9. M. J. Vold, *J. Colloid Sci.* **18**, 684 (1963).
10. K. H. Müller, *J. Appl. Phys.* **58**, 2573 (1985).
11. H. Tsuge and S. Esho, *J. Appl. Phys.* **52**, 4391 (1981).
12. H. Oechsner, *Appl. Phys.* **8**, 185 (1975).
13. M. A. Vidal and R. Asomoza, *J. Appl. Phys.* **67**, 477 (1990).
14. M. J. Goeckner, J. A. Goree, and T. E. Sheridan, *IEEE Trans. Plasma Sci.* **19**, 301 (1991).
15. T. Motohiro, *J. Vac. Sci. Technol.* **A4**, 189 (1986).
16. R. E. Somekh, *J. Vac. Sci. Technol.* **A2**, 1285 (1984).
17. R. S. Robinson, *J. Vac. Sci. Technol.* **16**, 185 (1979).
18. W. D. Westwood, *J. Vac. Sci. Technol.* **15**, 1 (1978).
19. S. K. Dew, T. Smy, and M. J. Brett, *IEEE Trans. Electron Dev.* **39**, 1599 (1992)
20. R. N. Tait, T. Smy, and M. J. Brett, *Thin Solid Films* **187**, 375 (1990).
21. T. Smy, R. N. Tait, and M. J. Brett, *IEEE Trans. Computer-Aided Des.* **10**, 130 (1991).
22. S. K. Dew, Ph.D. Thesis, University of Alberta, Edmonton, Canada, 1992.
23. S. K. Dew, T. Smy, and M. J. Brett, *Jpn. J. Appl. Phys.* **33**, 1141 (1994).
24. R. C. Weast, Ed., "Handbook of Chemistry and Physics, 1st Student Edition," CRC Press, Boca Raton, Florida, 1988.
25. D. Liu, S. K. Dew, M. J. Brett, T. Smy, and W. Tsai, *J. Appl. Phys.* **75**, 8114 (1994).
26. D. Liu, S. K. Dew, M. J. Brett, T. Janacek, T. Smy, and W. Tsai, *J. Appl. Phys.* **74**, 1339 (1993).
27. G. Neumann and W. Hirschwald, *Z. Physik. Chem. Neue Folge*, **81**, 176 (1972).
28. D. Pramanik and A. N. Saxena, *Solid State Technology* **33(3)**, 73 (1990).
29. S. K. Dew, "Processes and Simulation for Advanced Integrated Circuit Metallization," Ph.D. Thesis, Department of Electrical Engineering, University of Alberta, Edmonton, Alberta, Canada, 1992.
30. J. M. Nieuwenhuizen and H. B. Haanstra, *Phillips Techn. Rev.* **27(3)**, 87 (1966).
31. R. N. Tait, T. Smy, and M. J. Brett, *Thin Solid Films* **226**, 196 (1993).
32. R. N. Tait, T. Smy, and M. J. Brett, *J. Vac. Sci. Technol.* **A8**, 1593 (1990).
33. R. N. Tait, S. Dew, T. Smy, and M. J. Brett, *J. Appl. Phys.* **70**, 4295 (1991).
34. T. Smy, S. K. Dew, and M. J. Brett, *Can. Metall. Quarterly* **34**, 195 (1995).
35. M. J. Cooke and G. Harris, *J. Vac. Sci. Technol. A* **7**, 3217 (1989).
36. J. C. Rey, L.-Y. Cheng, J. P. McVittie, and K. C. Saraswat, "Proceedings IEEE VLSI Multilevel Interconnection Conf.," IEEE Electron Devices Society, New York, p. 425, 1990.
37. S. K. Dew, T. Smy, and M. J. Brett, *J. Vac. Sci. Technol. B* **10**, 618 (1992).

38. S. K. Dew, T. Smy, and M. J. Brett, "Microstructure simulation of CVD refractory films," *in* Advanced Metallization for UHV Applications (V. V. S. Rana, R. V. Joshi, and I. Ohdomari, Eds.), Materials Research Society, Pittsburgh, p. 85, 1992.
39. J. L. Vossen and J. J. Cuomo, in *Thin Film Processes* (J. L. Vossen and W. Kern, Eds.), Academic Press, New York, 1978.
40. S. M. Rossnagel and J. Hopwood, *J. Vac. Sci. Technol. B* **12**, 449 (1994).
41. S. M. Rossnagel and J. Hopwood, *Appl. Phys. Lett.* **63**, 3285 (1993).
42. M. Hou and M. T. Robinson, *Appl. Phys.* **17**, 371 (1978).
43. W. Eckstein and J. P. Biersack, *Z. Physik* **B63**, 109 (1986).
44. "SAMPLE User Guide V.1.7," University of California, Berkeley, p. 99, 1989.
45. A. Wucher and W. Reuter, *J. Vac. Sci. Technol.* **A6**, 2316 (1988).
46. S. K. Dew, T. Smy, R. N. Tait, and M. J. Brett, *J. Vac. Sci. Technol.* **A9**, 519 (1991).
47. S. M. Rossnagel, D. Mikalsen, H. Kinoshita, and J. J. Cuomo, *J. Vac. Sci. Technol.* **A9**, 261 (1991).
48. T. Smy, S. K. Dew, and M. J. Brett, *Can. Metall. Quart.* **34**, 195 (1995).
49. S. K. Dew, D. Liu, M. J. Brett, and T. Smy, *J. Vac. Sci. Technol.* **B11**, 1281 (1993).
50. S. K. Dew, *J. Appl. Phys.*, **76**, 4857 (1994).
51. D. Liu, S. K. Dew, M. J. Brett, T. Janacek, T. Smy, and W. Tsai, *Thin Solid Films* **236**, 267 (1993).
52. R. N. Tait, S. K. Dew, W. Tsai, D. Hodul, M. J. Brett, and T. Smy, *J. Vac. Sci. Technol. B* (1996), in press.
53. W. Tsai, D. Hodul, T. Sheng, S. Dew, K. Robbie, M. J. Brett, and T. Smy, *Appl. Phys. Lett.* **67**, 220 (1995).
54. D. W. Hoffman, *J. Vac. Sci. Technol. A* **3**, 561 (1985).
55. S. M. Rossnagel, *J. Vac. Sci. Technol. A*, **6**, 19 (1988).
56. T. Smy, S. S. Winterton, S. K. Dew, and M. J. Brett, *J. Appl. Phys.* **78**, 3572 (1995).
57. W. W. Mullins, *J. Appl. Phys.* **28**, 333 (1957).
58. T. Smy, S. K. Dew, M. J. Brett, W. Tsai, M. Bilberger, K. C. Chen, and S. T. Hsia, Proc. of 1994 International VLSI Multilevel Interconnection Conf., Santa Clara, California, p. 371.
59. T. Smy, K. Sheergar, S. K. Dew, and M. J. Brett, "Proc. of 1995 International VLSI Multilevel Interconnection Conf.," Santa Clara, California, p. 670.
60. R. N. Tait, S. K. Dew, T. Smy, and M. J. Brett, *J. Vac. Sci. Technol. A* **12**, 1085 (1994).
61. R. N. Tait, S. K. Dew, T. Smy, and M. J. Brett, *J. Vac. Sci. Technol. A* **10**, 912 (1992).
62. C. J. Mogab, A. C. Adams, and D. L. Flamm, *J. Appl. Phys.* **49**, 3796 (1978).
63. P. Li, T. Smy, S. K. Dew, and M. J. Brett, *J. Elect. Mat.* **23**, 1215 (1994).
64. L. Friedrich, S. K. Dew, M. J. Brett, and T. J. Smy, *Thin Solid Films* **266**, 83 (1995).
65. J. M. Blakely, "Introduction to the Properties of Crystal Surfaces," Pergamon Press, Oxford, UK, p. 64, 1973.
66. L. J. Friedrich, K. Robbie, S. K. Dew, M. J. Brett, and T. J. Smy, in "Process Control, Diagnostics and Modeling in Semiconductor, Manufacturing" (M. Meyyappan, D. J. Economou, and S. W. Butler, Eds.), Electrochemical Soc., Pennington, N.J., Vol. 95-4, p. 541, 1995.
67. R. N. Tait, S. K. Dew, T. Smy, and M. J. Brett, "Modeling of Optical Thin Films, II," *Proc. SPIE 1324*, 112 (1990).
68. M. J. Brett, R. N. Tait, S. K. Dew, T. Smy, S. Kamasz, and A. H. Labun, *J. Mat. Sci.* **3**, 64 (1992).
69. R. N. Tait, T. Smy, S. K. Dew, and M. J. Brett, *J. Electr. Mat.* **24**, 935 (1995).
70. S. S. Winterton, T. Smy, and M. J. Brett, *J. Appl. Phys.* **73**, 2821 (1993).

71. R. N. Tait, T. Smy, and M. J. Brett, *J. Vac. Sci. Technol. A* **10**, 1518 (1992).
72. R. N. Tait, S. K. Dew, W. Tsai, D. Hodul, M. J. Brett, and T. Smy, *J. Vac. Sci. Technol. B* **B14**, 679 (1996).
73. T. Smy, M. Salahuddin, S. K. Dew, and M. J. Brett, *J. Appl. Phys.* **78**, 4157 (1995).
74. T. Smy, D. J. Reny, and M. J. Brett, *J. Vac. Sci. Technol. B* **10**, 2267 (1992).
75. E. P. Lozowski, M. J. Brett, R. N. Tait, and T. Smy, *Quart. J. Royal Meteorological Society* **117**, 427 (1991).
76. T. Janacek, Alberta Microelectronic Centre, Edmonton, Alberta, Canada.
77. D. K. Pandya, A. C. Rastogi, and K. L. Chopra, *J. Appl. Phys.* **46**, 2966 (1975).
78. N. G. Nadhodkin and A. I. Shaldervan, *Thin Solid Films* **10**, 109 (1972).
79. A. Aronson and I. Wagnar, "Advanced Aluminum Metallization. Part II. Planarization," *in* Advances in Magnetron Sputtering and Etching, Materials Research Corp., 1988.

Mathematical Methods for Thin Film Deposition Simulations

S. HAMAGUCHI

IBM T. J. Watson Research Center, Yorktown Heights, New York

I. Introduction

The surface evolution of solid materials is a problem commonly encountered in science and technology. For semiconductor device fabrication processes, resist development in lithography processes, planarization by chemical–mechanical polish, and etch and deposition of thin films in dry (e.g., plasma, ion beam) and wet (e.g., electrochemical plating) processes are a few examples. In a given process, the velocity of every point on the surface is determined by the underlying physical and/or chemical mechanisms of the surface evolution.

The task of predicting the topography of a new surface after a small time interval Δt using the information of surface velocities is *not* straight-

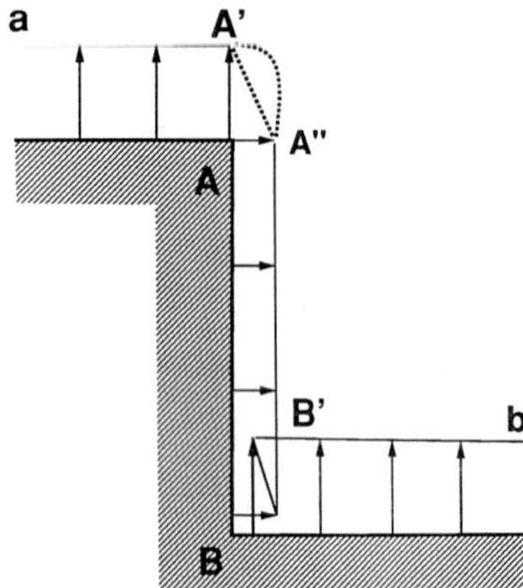

FIG. 1. Problems involved with the construction of a new surface under a deposition process. Arrows indicate known deposition rates on the initial surface. It is not clear from the deposition rate information how to connect A′ and A″ and how to handle the loop at B′ correctly.

forward. For example, let us ask the following questions. Figure 1 shows an attempt to construct a newly deposited film surface a–b after a small time interval. If the deposition material beam from the source is anisotropic, then the film thickness on the horizontal surfaces differs from that on the vertical surface. Since the surface normal is not well defined at sharp corner A, we may draw two arrows AA′ and AA″, the lengths of which are proportional to the deposition rates in the respective directions. The first question is how the two points A′ and A″ should be connected. Connecting A′ and A″ by a straight line segment seems to be a rather crude approximation, but how crude is it compared with the true solution? Does a small deviation from the true solution introduced in each time step lead to a grossly erroneous solution after finite time steps (i.e., numerical instability)?

Another problem of reconnecting moving points is the formation of loops, as is seen at point B′ in Fig. 1. Naïvely extending our knowledge of the motion of finite-area surfaces to a singular point such as point B′ in Fig. 1, one cannot avoid the formation of a loop no matter how small a time step is chosen. Since such loops are not physical, a possible remedy is to remove the loops (delooping). In the case of Fig. 1, we may choose the numerical solution to be a–A′–A″–B′–b. Is this a good approximation? What is the exact solution evolving from corner B anyway?

These questions are not mere "mathematical subtleties" that ordinary practitioners of modeling can safely ignore. In topography simulation, one often finds unexpected features emerging from corners in complex deposition processes or at material boundaries in multimaterial etching processes. A good understanding of the numerical algorithm often helps one distinguish numerical artifacts from physically correct solutions. Also good numerical algorithms can avoid such artifacts altogether. The fundamentals of surface evolution thus need to be understood not only by model builders but also by serious users of surface evolution simulators.

To understand the fundamentals, one first needs to consider the equation that governs the motion of evolving boundaries. As shown in the next section, an evolving surface satisfies a Hamilton–Jacobi type equation. What complicates the surface evolution problem, however, is that a solution to the Hamilton–Jacobi type equation is *not* necessarily the physically plausible solution that describes the evolving surface. In other words, the Hamilton–Jacobi type equation in general admits many spurious solutions (nonunique solutions) and we need *to choose the physically plausible solution*, using additional conditions (*1*).

The nonuniqueness of the solution is manifested in the uncertainty that we faced when we attempted to connect point A and point A′ or when we attempted to deloop at point B′ in Fig. 1. Several different ways of connecting A and A′—some of which seem to give reasonable shapes for a deposited film—can be admitted as solutions to the Hamilton–Jacobi type equation, but only one of them is the physically correct solution. Therefore, additional conditions are needed that select the correct solution from all possible solutions to the Hamilton–Jacobi type equation. Such conditions (which we later call entropy conditions) should be derived from physical mechanisms that are not taken into account in the derivation of the Hamilton–Jacobi type equation.

The goal of this article is to define the additional conditions (entropy conditions) and use them together with the Hamilton–Jacobi type equation to determine the surface evolution correctly under given conditions. We focus, in particular, on some fundamental aspects of surface evolution problems in the light of etching and deposition processes widely used in semiconductor manufacturing. For simplicity, we discuss only two-dimensional problems, where material surfaces are represented by boundary curves. Many basic ideas discussed in this article, however, can also be extended to three dimensions.

This article is organized as follows. In the next section, we discuss the mathematical formulation of moving boundary problems, deriving the Hamilton–Jacoci type equation and entropy conditions. Because most readers are expected to be interested in engineering applications of modeling and simulations, we avoid excessive mathematical rigor and use examples and pictures in an attempt to illustrate essential ideas behind the formulation. In Sec. III, the shock-tracking method for numerical simulation is presented. Up to Sec. III, we assume that surface velocities such as etching/deposition rates are given. The methods to obtain such rates in dry processes are discussed in Sec. IV. In Sec. V, some representative examples of the numerical simulation are presented. The simulations presented here are performed by SHADE (Shock-tracking Algorithm for Deposition and Etching), a simulation code based on the shock-tracking method *(1–5)*. The summary of this article is given in Sec. VI.

II. The Hamilton–Jacobi Type Equation

A. Boundary Motion

We now derive the equation that governs surface motion. For the coordinate system, we take the x-axis in the horizontal direction and the z-axis in the vertical direction. The y-coordinate is chosen accordingly to form the usual right-hand coordinate system. The system is then assumed to have the translational symmetry in the y direction. At time t, the surface (i.e., boundary curve) may be represented by an equation of the form $\psi(x, z, t) = 0$, where we assume that $\psi(x, z, t) > 0$ (< 0) represents the material (vacuum) side of the boundary. Etching and deposition processes may then be formulated as the time evolution of this surface. The velocity vector $\mathbf{C}$ of the boundary surface at point (x, z) may be written as

$$\mathbf{C} = C_n \hat{\mathbf{n}} + C_t \hat{\mathbf{t}}. \tag{1}$$

Here $\hat{\mathbf{n}}$ and $\hat{\mathbf{t}}$ denote the unit normal and tangent unit vectors, that is,

$$\hat{\mathbf{n}} = \frac{1}{\sqrt{\psi_x^2 + \psi_z^2}} \begin{pmatrix} \psi_x \\ \psi_z \end{pmatrix}, \qquad \hat{\mathbf{t}} = \frac{1}{\sqrt{\psi_x^2 + \psi_z^2}} \begin{pmatrix} -\psi_z \\ \psi_x \end{pmatrix},$$

with $\psi_x = \partial\psi/\partial x$ and $\psi_z = \partial\psi/\partial z$. Evidently $C_n > 0$ and $C_n < 0$ represent etching and deposition of the material in this formulation.

Since $\psi(x(t), z(t), t) \equiv 0$, differentiating this equation with respect to time t and substituting Eq. (1) [note that $(\partial x/\partial t, \partial z/\partial t) = \mathbf{C}$] into the resulting equation, we obtain

$$\psi_t + C\sqrt{\psi_x^2 + \psi_z^2} = 0. \tag{2}$$

Here we write $C = C_n$ for simplicity. Note that the tangential velocity component C_t does not appear in Eq. (2). The velocity along the curve does not alter its shape, and the motion of the curve is determined only by the normal component $C \equiv C_n$.

Equation (2) is the Hamilton–Jacobi equation if the function C depends only on time t, position (x, z), the unknown function ψ, and its first derivatives (ψ_x, ψ_z) *(6–9)*. In some processes, the function C can depend only on these variables. For general etch/deposition problems, however, C can be a more complicated function (see Sec. IV) and Eq. (2) is not necessarily a Hamilton–Jacobi equation in the strict sense. Thus we call Eq. (2) a Hamilton–Jacobi type equation in general.

In a special case where the "height" of the boundary curve z is uniquely determined by the horizontal position x (i.e., z is a "function" of x), Eq. (2) can be further simplified. In this case, the equation $\psi(x, z, t) = 0$ can be solved as $z = u(x, t)$, so we may write $\psi(x, z, t) = u(x, t) - z$ and Eq. (2) becomes

$$u_t + C\sqrt{1 + u_x^2} = 0. \tag{3}$$

This is the Hamilton–Jacobi equation if $C = C(t, x, u, u_x)$. For a more general C, we call the equation a Hamilton–Jacobi type equation.

Let us consider a beam that bombards the surface in the negative z direction and denote the slope of the boundary curve $p = u_x = -\psi_x/\psi_z$ (Fig. 2). In such a case, the function $f(p) = C\sqrt{1 + p^2}$ represents the amount of material (in volume) removed from the surface per unit beam flux and unit time. For a unit area perpendicular to the beam flux, $-u_t = f(u_x)$ (> 0) represents the volume of the material removed from the surface in a unit time. Therefore, $f(p)$ is proportional to the sputtering yield Y, that is, the flux (i.e., number of atoms) sputtered from the surface per single incident beam particle (ion or atom). We call $f(p)$ the flux function *(1,10)*.

B. A Spurious Solution

Now we demonstrate that Eqs. (2) and (3) indeed allows more than one solution with given initial conditions. Let us consider a simple deposition problem, using Eq. (3) with a constant rate $C = -1$ (isotropic deposition).

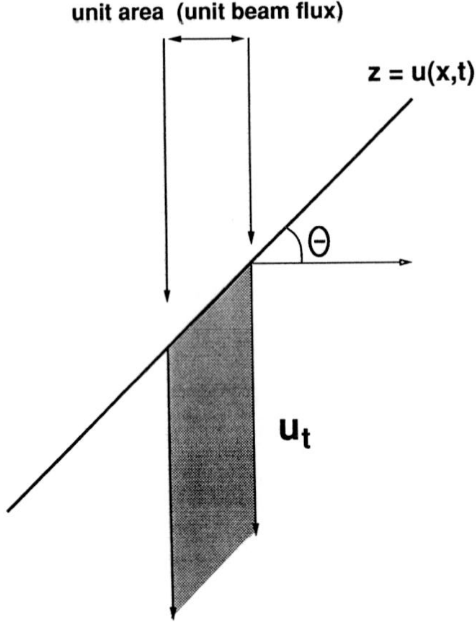

FIG. 2. The surface given by $z = u(x, t)$ and its time derivative u_t.

The initial profile consists of a horizontal surface and 45° surface as shown in Fig. 3, somewhat similar to corner A of Fig. 1. In terms of the function $z = u(x, 0)$, we have, at time $t = 0$,

$$u(x,0) = \begin{cases} 0 & x \leq 0 \\ -x & x \geq 0 \end{cases}. \tag{4}$$

It is easy to confirm that the following two solutions both satisfy Eq. (3)

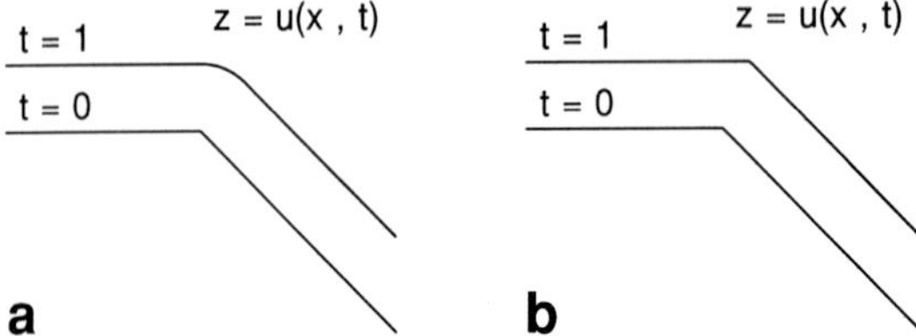

FIG. 3. Two different solutions that satisfy the Hamilton–Jacoci equation [Eq. (3)] and the initial condition of Eq. (4). (a) Solution I and (b) Solution II. At least one of these solutions must be unphysical. These solutions demonstrate that the Hamilton–Jacobi equation (or the method of characteristics that solves the equation) is not sufficient to determine the physically realizable solutions of moving boundary problems.

and the initial condition of Eq. (4):

$$\text{Solution I} \qquad u(x,t) = \begin{cases} t & x \le 0 \\ \sqrt{t^2 - x^2} & 0 \le x \le \dfrac{\sqrt{2}}{2}t, \\ -x + \sqrt{2}\,t & x \ge \dfrac{\sqrt{2}}{2}t \end{cases}$$

$$\text{Solution II} \qquad u(x,t) = \begin{cases} t & x \le (\sqrt{2} - 1)t \\ -x + \sqrt{2t} & x \ge (\sqrt{2} - 1)t \end{cases}.$$

These solutions at $t = 1$ are shown in Figs. 3a and 3b.

These two solutions demonstrate that neither Eq. (2) nor Eq. (3) suffices to determine the physically plausible solutions of moving boundary problems. The assumptions we have made so far (that the boundary curve is continuous and moves with the normal velocity C) are satisfied by both solutions I and II. The problem is that, at time $t = 0$, the slope of the initial curve (and therefore its normal direction) is not defined at the corner $(0, 0)$ in Fig. 3. So far no rule has been given as to how to evolve the surface in the neighborhood of such a corner. Thus one clearly needs additional conditions to determine the solution that represents what actually occurs under given etching/deposition conditions.

C. Characteristics and Slope Discontinuities

Although it is now clear that the initial value problem for the Hamilton–Jacobi type equation does not solve the surface evolution equation in itself, it is still meaningful to consider a method to construct solutions to Eq. (2) or (3). Since the Hamilton–Jacobi type equation is a first-order partial differential equation (PDE), its solutions can be constructed by the method of characteristics (*10,11*). If $C = C(t, x, z; p)$ with $p = u_x = \tan\theta$, then the characteristic equations of Eq. (2) or (3) may be given by

$$\frac{dx}{dt} = C_\theta \cos\theta + C\sin\theta, \tag{5}$$

$$\frac{dz}{dt} = C_\theta \sin\theta - C\cos\theta, \tag{6}$$

$$\frac{d\theta}{dt} = -C_x \cos\theta - C_u \sin\theta. \tag{7}$$

Here $C_\theta = \partial C / \partial \theta$. It is clear from these equations that if $C_\theta \neq 0$, the direction of a characteristic curve is not perpendicular to the boundary curve $z = u(x,t)$. Especially if the rate function C depends only on the slope (i.e., $C_x = C_u = 0$), then the slope $p = \tan \theta$ of the boundary curve remains constant (i.e., $d\theta/dt = 0$) along each characteristic curve, which thus becomes a straight line.

If the initial curve has a sharp corner (such as points A and B in Fig. 1), that is, a discontinuity of the slope $u_x(x,0)$, characteristic equations (5) through (7) do not lend themselves to solving Eq. (2) or (3) in the neighborhood of the discontinuity since $p = u_x$ is not defined on this point. In such a case, we go back to the original Hamilton–Jacobi equation and construct a possible solution that admits slope discontinuities. We first consider Eq. (3) for simplicity.

Let us assume that the sharp corner on the initial curve remains a sharp corner as the surface evolves (as in Fig. 3b, rather than Fig. 3a). Under this assumption, we denote the position of the sharp corner by $(X(t), Z(t))$, which of course satisfies $Z(t) = u(X(t), t)$. Using shorthand notation,

$$u_x(X(t) \pm 0, t) = \lim_{\epsilon \to 0} u_x(X(t) \pm \epsilon, t)$$

to represent the values of u_x (or other quantities) to the left ($x < X(t)$) and right ($x > X(t)$) of the discontinuity point, we have $u_x(X(t) - 0, t) \neq u_x(X(t) + 0, t)$ although the continuity of the function $u(x,t)$ ensures that $u(X(t) - 0, t) = u(X(t) + 0, t)$. From Eq. (3), clearly

$$u_t(X(t) - 0, t) + f(t, X(t), Z(t), p_l) = 0, \tag{8}$$

$$u_t(X(t) + 0, t) + f(t, X(t), Z(t), p_r) = 0, \tag{9}$$

where $p_l = u_x(X(t) - 0, t)$ and $p_r = u_x(X(t) + 0, t)$. By differentiating the equation $u(X(t) - 0, t) = u(X(t) + 0, t)$ with respect to time, t, we have

$$p_l \frac{dX(t)}{dt} + u_t(X(t) - 0, t) = p_r \frac{dX(t)}{dt} + u_t(X(t) + 0, t).$$

Substituting Eqs. (8) and (9) into the preceding equation yields

$$\frac{dX(t)}{dt} = \frac{f_l - f_r}{p_l - p_r}, \tag{10}$$

where $f_l = f(t, X(t), Z(t), p_l)$ and $f_r = f(t, X(t), Z(t), p_r)$. Equation (10) represents the x component of the velocity of the slope discontinuity (if the discontinuity continues to exist for $t > 0$).

Similarly for the z component, differentiating $Z(t) = u(X(t) - 0, t)$ $(= u(X(t), t))$ with respect to time t yields $dZ(t)/dt = p_l \, dX(t)/dt + u_t(X(t) - 0, t)$. Substituting Eqs. (8) and (10) into this expression, we obtain

$$\frac{dZ(t)}{dt} = \frac{p_r f_l - p_l f_r}{p_l - p_r}. \tag{11}$$

Equations (10) and (11) are called the *jump conditions* because they relate the size of the "jump" in the slope, that is, $p_l - p_r$ to the position of the discontinuity. In gas dynamics, similar discontinuity conditions for some physical quantities (such as the gas density) across a shock front are known as the Rankine–Hugoniot conditions (*10,12*).

The jump conditions for the more general Eq. (2) can be derived in a similar manner:

$$\frac{dX(t)}{dt} = \frac{\cos\theta_r C_l - \cos\theta_l C_r}{\sin(\theta_l - \theta_r)}, \tag{12}$$

$$\frac{dZ(t)}{dt} = \frac{\sin\theta_r C_l - \sin\theta_l C_r}{\sin(\theta_l - \theta_r)}, \tag{13}$$

where $p_l = \tan\theta_l$, $p_r = \tan\theta_r$, $C_l = C(t, x, u; p_l)$ and $C_r = C(t, x, u; p_r)$. It is easy to confirm that Eqs. (12) and (13) converge to characteristic Eqs. (5) and (6) in the limit of continuity (i.e., $\theta_l \to \theta_r$).

D. Viscosity Solutions

To determine which of the two solutions, I or II, in Sec. II.B. is physically realizable, some additional physical mechanisms must be considered. Such mechanisms are related to the microscopic (molecular-level) nature of the surface that smooths out too many irregularities of the surface topography and thus allows us to use a continuous curve to represent it in macroscopic scales (i.e., scales larger than molecular scales).

Under normal conditions, there is always some material transport along the surface of a solid (surface diffusion). If the temperature of the solid is sufficiently high (e.g., near its melting point), this transport becomes very large and can cause a significant change of the surface topography. The surface diffusion is known to be linearly dependent on the surface curvature κ (*13*) and thus smooths out the surface. The smoothing effect may be incorporated into Eqs. (2) and (3) by replacing C by $C - \nu\kappa$, where C

represents the surface velocity due to the etching and/or deposition and $-\nu\kappa$ due to the curvature driven diffusion (ν is typically a small positive constant). If the surface is given by $z = u(x,t)$, then

$$\kappa = \frac{u_{xx}}{(1 + u_x^2)^{3/2}},$$

and therefore Eq. (3) becomes

$$u_t^{(\nu)} + C\sqrt{1 + u_x^{(\nu)2}} = \frac{\nu u_{xx}^{(\nu)}}{1 + u_x^{(\nu)2}}. \tag{14}$$

Here we denote the solution of this equation $u^{(\nu)}(x,t)$ to emphasize its dependence on the constant ν.

Since we are interested in the limit of small diffusion, (i.e., $\nu \to 0$), we only retain the smoothing capability of the right-hand side of Eq. (14), that is, the second derivative $u_{xx}^{(\nu)}(x,t)$, and simplify the expression by ignoring the $u_x^{(\nu)}$ dependence in the denominator:

$$u_t^{(\nu)} + C\sqrt{1 + u_x^{(\nu)2}} = \nu u_{xx}^{(\nu)}. \tag{15}$$

Thus the right-hand side becomes the usual diffusion term and its diffusion coefficient is given by ν.

The solution to Eq. (15) is expected to be smooth for all x even if the initial conditions contain some slope discontinuities. The solution that we were originally interested in (i.e., topography evolution under etching/deposition conditions without surface diffusion) should also be the solution to Eq. (15) in the limit of vanishing diffusion (i.e., $\nu \to 0$). Setting $\nu = 0$ in Eq. (15) first and solving the resulting equation [i.e., Eq. (3)] lead to at least two incompatible solutions (solution I and II) per Sec. II.B. However, solving Eq. (15) first and taking the limit of its solution, that is, $u^{(0)} \equiv \lim_{\nu \to 0} u^{(\nu)}$, should yield the physically meaningful solution. [It is known that Eq. (15) has a unique solution.]

The limiting solution $u^{(0)} \equiv \lim_{\nu \to 0} u^{(\nu)}$ may contain some sharp corners, that is, its spatial derivatives may not exist at certain points, due to the nominal absence of the diffusion effect. The solution to Eq. (3) obtained from $\lim_{\nu \to 0} u^{(\nu)}$ is called the *viscosity solution*. The viscosity solution satisfies the original Hamilton–Jacobi equation (with $\nu = 0$) except at points where the solution is not differentiable. Such solutions with discontinuous gradients are called "weak" or "generalized" solutions, in contrast to the regular (i.e., differentiable) solutions.

The viscosity solution to the more general Eq. (2) may be defined in a similar manner. Introducing a small diffusion term in the right-hand side of Eq. (2), we obtain

$$\psi_t^{(\nu)} + C\sqrt{\psi_x^{(\nu)2} + \psi_z^{(\nu)2}} = \nu \Delta \psi^{(\nu)}, \tag{16}$$

where $\Delta = \partial^2/\partial x^2 + \partial^2/\partial z^2$ denotes the Laplacian. The viscosity solution is then given by $\psi^{(0)} \equiv \lim_{\nu \to 0} \psi^{(\nu)}$.

E. Entropy Conditions

To choose the physically meaningful solution from the set of solutions admitted by Eq. (3), one can use the relation $u^{(0)} \equiv \lim_{\nu \to 0} u^{(\nu)}$. However, it is not practical to derive the function $u^{(\nu)}$ for arbitrary ν from Eq. (3) just to take the limit. Fortunately, there is a method in two dimensions that allows us to construct the viscosity solution directly.

By differentiating both sides of Eq. (3) and writing $p = u_x$ (*14,15*), we obtain the equation for the slope $p = u_x$ as

$$p_t + \partial_x f(p) = 0, \tag{17}$$

where we assume that $f = C\sqrt{1 + p^2}$ depends only on p for simplicity. The function p is said to satisfy the conservation form (*10*). As in the case of Eq. (3), this equation may admit more than one solution. We call the solution physically plausible if the solution is given in the limit $p = \lim_{\nu \to 0} p^{(\nu)}$, where $p^{(\nu)}$ is the solution of

$$p_t^{(\nu)} + \partial_x f(p^{(\nu)}) = \nu p_{xx}^{(\nu)}.$$

It is known that such a solution (viscosity solution) is unique—there is only one physically plausible solution to Eq. (17).

From the theory of conservation laws, the following fact is known (*10,16*): Suppose the function $p(x,t)$ satisfies Eq. (17) everywhere except for some points on which p is discontinuous. Therefore, this function $p(x,t)$ is one of the admissible solutions of Eq. (17). Then function $p(x,t)$ is indeed the physically plausible solution if and only if p satisfies the following conditions.

If $p_r < p_l$, then the graph of $f(p) = C(p)\sqrt{1 + p^2}$ over the interval $[p_r, p_l]$ in the $(p, f(p))$ plane must lie *below* the chord connecting (p_r, f_r) and (p_l, f_l):

$$f(\alpha p_r + (1 - \alpha)p_l) \leq \alpha f_r + (1 - \alpha) f_l \tag{18}$$

for $0 \le \alpha \le 1$ (here $f_r = f(p_r)$ and $f_l = f(p_l)$ as before). If $p_r > p_l$, on the other hand, the graph of $f(p)$ over $[p_l, p_r]$ must lie *above* the chord connecting (p_l, f_l) and (p_r, f_r):

$$f(\alpha p_r + (1-\alpha)p_l) \ge \alpha f_r + (1-\alpha)f_l \tag{19}$$

for $0 \le \alpha \le 1$. The two conditions (18) and (19) may be written as a single system of inequalities for all p between p_l and p_r:

$$\frac{f_r - f}{p_r - p} \le \frac{f_l - f_r}{p_l - p_r} \le \frac{f_l - f}{p_l - p}. \tag{20}$$

In other words, if the function $p(x, t)$ satisfies both Eq. (17) and the initial conditions for all x and is continuous, then $p(x, t)$ is the physically plausible solution. If such a function $p(x, t)$ is discontinuous, then it is physically plausible if and only if p satisfies the entropy conditions of Eq. (20) for all the discontinuity points. A discontinuity that satisfies entropy conditions is called a "shock" in gas dynamics. Thus we also use the same term and call a physically plausible slope discontinuity a "shock" (*1,10,12*).

Using the scheme mentioned earlier we can construct the solution $u(x, t)$ that develops from a sharp corner [Oleinik construction (*16*)]. Suppose that, at time $t = 0$, there is a slope discontinuity of $y = u(x, 0)$, at which the slope $p = u_x$ changes from p_l to p_r ($p_r > p_l$) as x increases, as shown in Fig. 4. If the flux function $f(p) = C\sqrt{1 + p^2}$ is convex, as in Fig. 5(a), the graph of $f(p)$ lies below the chord connecting the points (p_l, f_l) and (p_r, f_r), and condition (19) is violated. Therefore, no sharp corner can develop from this slope discontinuity—instead, multiple characteristic lines emanate from the corner point A and the initial sharp corner evolves

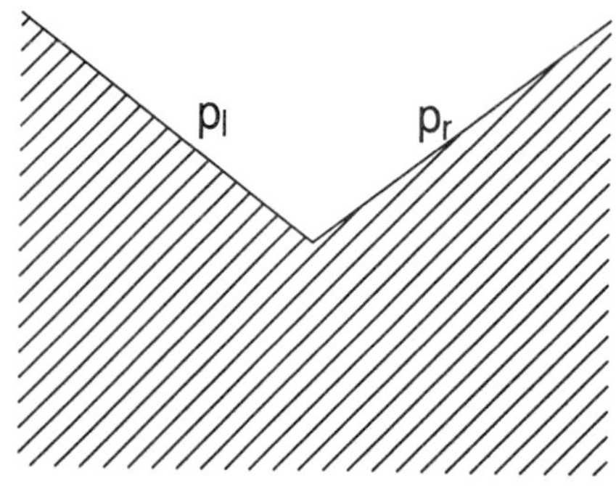

FIG. 4. A sharp corner with left and right slopes $p_l = \tan\theta_l$ and $p_r = \tan\theta_r$, respectively.

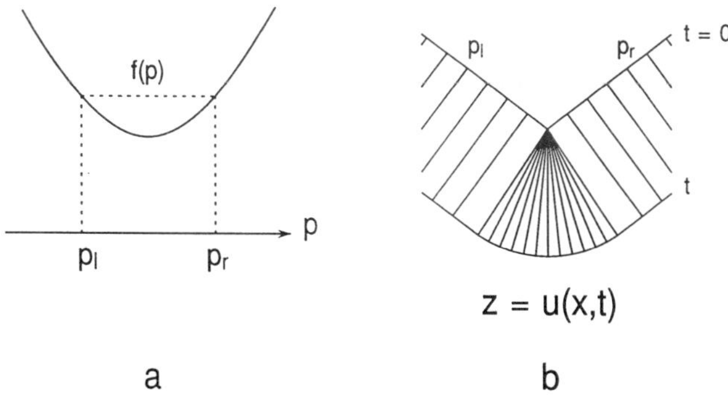

FIG. 5. The etching case of a convex flux function $f(p) = C\sqrt{1+p^2}$ (>0): (a) the flux function and (b) the evolution of the surface from $t = 0$ to t.

into a smooth arc. Such characteristics may be calculated from Eqs. (5) through (7) for all $p = \tan\theta$ in the range $p_l \leq p \leq p_r$, as shown in Fig. 5(b). Using the gas dynamics analogy again, we call such a smooth curve developing from a sharp corner a "rarefaction wave" (*1,10,12*).

On the other hand, if the flux function $f(p)$ is concave as in Fig. 6(a), the graph of $f(p)$ lies above the chord between the points (p_l, f_l) and (p_r, f_r), and condition (19) is satisfied. Therefore, the function $u(x, t)$ with this slope discontinuity is the legitimate solution and the slope discontinuity (sharp corner) propagates according to Eqs. (10) and (11). Figure 6(b) illustrates the characteristics and the propagation of the sharp corner.

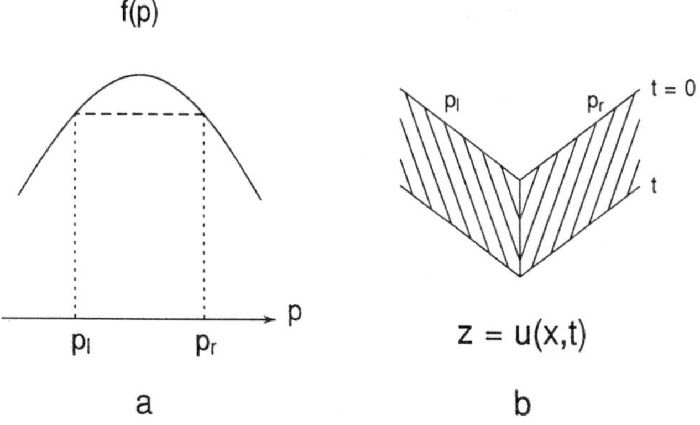

FIG. 6. The etching case of a concave flux function $f(p) = C\sqrt{1+p^2}$ (>0): (a) the flux function and (b) the evolution of the surface from $t = 0$ to t.

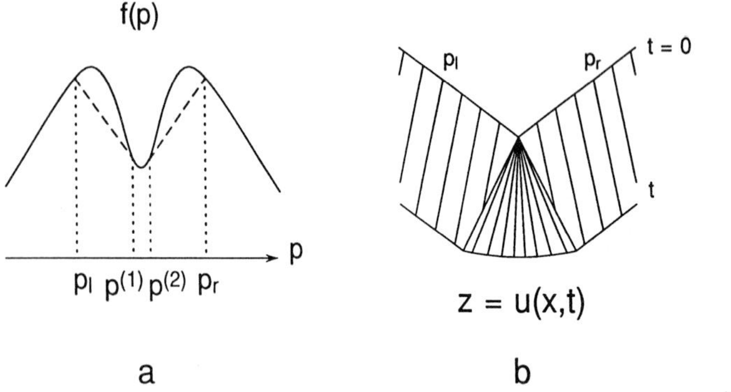

FIG. 7. The etching case of a nonconvex/nonconcave flux function $f(p) = C\sqrt{1+p^2}$ (>0): (a) the flux function and (b) the evolution of the surface from $t = 0$ to t.

A more complicated example is shown in Fig. 7(a), where the flux function $f(p)$ is neither convex nor concave. In this case, we construct a convex envelope (concave envelope for the case $p_r < p_l$), as illustrated in Fig. 7(a). It is clear from this figure that two sharp corners (shocks), associated with the intervals $[p_l, p^{(1)}]$ and $[p^{(2)}, p_r]$ develop from the original single corner. Denoting the trajectories of these two sharp corners by (X_1, Z_1) and (X_2, Z_2), we have from Eqs. (10) and (11):

$$\frac{dX_1(t)}{dt} = \frac{f^{(1)} - f_l}{p^{(1)} - p_l},$$

$$\frac{dZ_1(t)}{dt} = \frac{p^{(1)} f_l - p_l f^{(1)}}{p^{(1)} - p_l},$$

and

$$\frac{dX_2(t)}{dt} = \frac{f_r - f^{(2)}}{p_r - p^{(2)}},$$

$$\frac{dZ_2(t)}{dt} = \frac{p^{(2)} f_r - p_r f^{(2)}}{p_r - p^{(2)}},$$

where $f^{(1)} = f(p^{(1)})$ and $f^{(2)} = f(p^{(2)})$. Values of p between $p^{(1)}$ and $p^{(2)}$ give rise to a rarefaction wave propagating from the sharp corner of the initial curve. Figure 7(b) shows the resulting evolution of the surface in this case.

F. Geometric Interpretation of Entropy Conditions

A careful examination of the entropy conditions discussed earlier reveals that we may consider a slope discontinuity (i.e., sharp corner) of the function $u(x, t)$ as the limit of a smooth curve with the same convexity. Figure 8 shows the approximation of such a smooth curve by a piecewise linear function. Here N small line segments ($N = 4$ in the case of Fig. 8) constitutes the piecewise linear function that approximates the round corner, whose radius of curvature is given by ε. Each small line segment corresponds to the center angle $\delta\phi$. Here θ_n ($n = 0, 1, \ldots, N + 1$) denotes the slope angle of each line segment with $p_l = \tan\theta_0$ and $p_r = \tan\theta_{N+1}$ being the left and right slopes. The propagation of each slope discontinuity A, B, C, ... (see Fig. 9a) may be given by Eqs. (12) and (13). In the limit of $\delta\phi \to 0$ and $N \to \infty$, Eqs. (12) and (13) become the characteristic equations (5) and (6). Since we also consider the limit of infinitesimal ε (i.e., $\varepsilon \to 0$), the lengths of these arrows should be considered much larger than ε.

With nonzero ε, $\delta\phi$, and N, for example, if the velocity vector $\mathbf{v}_{01}$ from A and the velocity vector $\mathbf{v}_{23}$ from B in Fig. 9a intersect, the two nodes A and B converge to form a new node Q and the line segment AB will be eliminated after a small time interval. The new slope discontinuity then starts to propagate from Q, based on the left slope θ_0 and the right slope

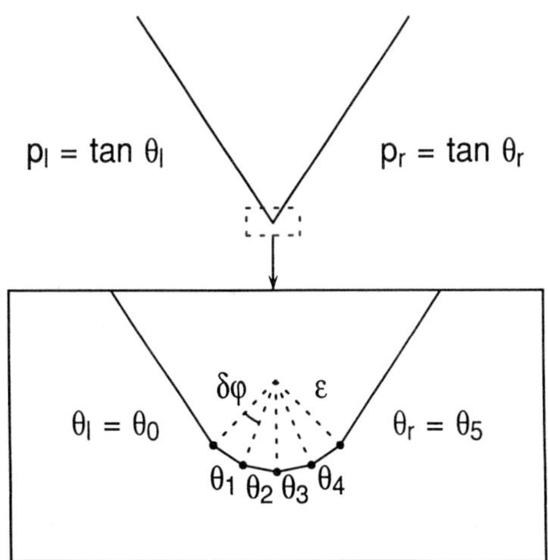

FIG. 8. An approximation of a sharp corner by a smooth curve and its discretization by angle $\delta\phi$. The radius of curvature of the discretized curve is ε.

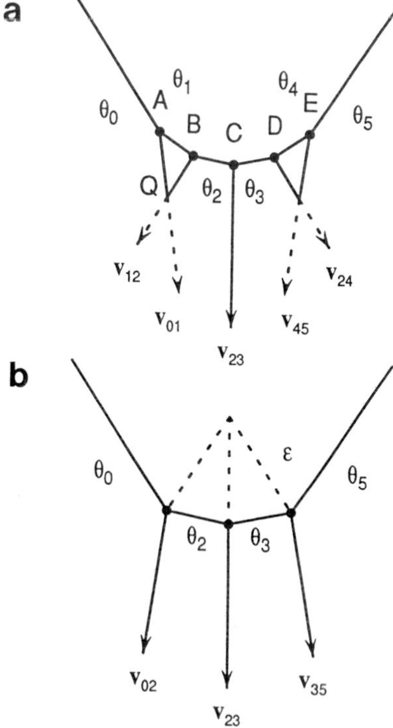

FIG. 9. The procedure of removing unneeded slope angles (e.g., θ_1 and θ_4). (a) Calculate the slope discontinuity propagation velocities (vectors) $\mathbf{v}_{nn+1}$ for all n. (b) Remove crossing vectors (e.g., $\mathbf{v}_{01}$ and $\mathbf{v}_{12}$) and the slope angle between them (e.g., θ_1). Construct the new vector $\mathbf{v}_{02}$ using θ_0 and θ_2 and determine whether this vector intersects another vector. If is does, repeat the same procedure.

θ_2 and its velocity vector is thus given by $\mathbf{v}_{02}$, as shown in Fig. 9b. [Here $\mathbf{v}_{nm} = (u_{nm}, v_{nm})$ denotes the velocity vector given by Eqs. (12) and (13) with the left and right slope angles given by θ_n and θ_m, that is,

$$u_{nm} = \frac{\cos\theta_m C_n - \cos\theta_n C_m}{\sin(\theta_n - \theta_m)},$$

$$v_{nm} = \frac{\sin\theta_m C_n - \sin\theta_n C_m}{\sin(\theta_n - \theta_m)},$$

where C_n and C_m are the normal velocities evaluated at the slopes θ_n and θ_m.] Since we consider the limit of small ε, the propagation of slope discontinuities thus obtained is essentially the same as that from the initial piecewise linear surface boundary given in Fig. 9b, where we have eliminated line segments AB and DE.

By eliminating all the crossing vectors (and corresponding line segments such as AB and DE in Fig. 9a) in this manner, we are left with the vectors that can emanate from these nodes even in the limit of small ε. By taking the limit of $\delta\phi \to 0$ and $N \to \infty$, we can now construct the correct topography that develops from the original sharp corner in Fig. 8. If no vectors cross, then each represents the true characteristics direction in this limit and the curve that emerges from the corner represents the smooth solution $u(x, t)$ (i.e., rarefaction wave). On the other hand, if some vectors cross, then slope discontinuities emerge from the corner, representing "shocks," that is, the physically meaningful solution $u(x, t)$ that contains some sharp corners.

Although we do not elaborate in this paper, one can easily confirm that the described method of constructing solutions is a generalization of the Oleinik construction, described in Sec II.F. In the method presented however, unlike the Oleinik construction, z does not need to be a (single-valued) function of x; that is, the method is applicable to a curve given by the general expression $\psi(x, z, t) = 0$.

III. The Shock-Tracking Algorithm

To calculate numerically the motion of a moving boundary, we implement the algorithm discussed in the previous section. We call it the shock-tracking algorithm (*1*) since it directly tracks the propagation of slope discontinuities.

First we represent the surface boundary by a piecewise linear function. Then each node on the piecewise linear function is treated as a real slope discontinuity if the slope angles of the adjacent sides, denoted as θ_l and θ_r, are not equal. Then, as in the previous section, we assume that all the slope angles between θ_l and θ_r are "hidden" at the node. For a given angle step $\delta\phi$, we may discretize these hidden angles, as illustrated in Fig. 8. For example, if $\theta_l < \theta_r$, we may introduce N hidden new angles such that $\theta_0 < \theta_1 < \theta_2 < \cdots < \theta_{N+1}$ with $\theta_0 = \theta_l$ and $\theta_{N+1} = \theta_l$. Using Eqs. (12) and (13) [or Eqs. (10) and (11) if applicable], we construct a possible slope discontinuity propagation vector $\mathbf{v}_{nn+1} = (dX/dt, dZ/dt)$ for every pair of angles θ_n and θ_{n+1} $(0 \leq n \leq N)$.

Following the method described in Sec. II.F, we now assume that there is a small but finite radius of curvature ε associated with this node (Fig. 8).

The length of vector $\mathbf{v}_{nm}$ represents the speed of slope discontinuity propagation. Therefore, in a magnified picture with finite ε, such as Fig. 9, the length of each vector $\mathbf{v}_{nm}$ may be considered infinite compared with ε in the limit of $\varepsilon \to 0$. The ordered propagation vectors $\mathbf{v}_{nm}$ $(0 \leq n, m \leq N - 1)$ with their lengths being infinite are then tested whether they intersect each other. If none intersects, all these vectors may be considered to be the characteristics emanating from the node. If some intersect, then those intersecting vectors are eliminated and new vectors are constructed in the same manner as is illustrated in Fig. 9. Newly constructed vectors in this way are again tested to determine whether they intersect each other or other vectors from the same original node. Repeating this construction process, we are eventually left with the propagation vectors that do not intersect each other (see Fig. 9b). This shock-tracking algorithm thus ensures that each node of the piecewise linear function (which represents the surface boundary) emanates a finite number of new nodes in a manner consistent with the original Hamilton–Jacobi equation [Eq. (2)] with the entropy conditions.

The shock-tracking algorithm presented here is appropriate for 2-D solid-surface moving boundary problems. As a numerical algorithm, it differs from the so-called front-tracking method that is widely used in gas dynamics simulations (*17,18*) although both methods correctly capture shocks (i.e., discontinuous physical quantities) in their respective systems.

IV. Deposition and Etch Rates

A. Collisionless Transport

Up to the previous section, we assumed that surface velocities (i.e., rate functions) were given and then considered the motion of the surface. In this section, we now derive such surface velocities for dry processes. To determine the local surface velocity **C**, we first calculate how many atoms are deposited and re-emitted at each point on the surface (*2,4*). For simplicity, we assume that the surface diffusion of materials is negligibly small. Although the boundary is two dimensional, the transport in the gas phase is three dimensional since atomic motion in the gas phase is not confined to the xz-plane.

For dry processes, mean free paths of all atoms and ions in the gas phase are much larger than the typical geometric scale (such as trench widths). Therefore, it is appropriate to consider collisionless transport of those species in the gas phase. Suppose the velocity distribution function of species α at position x is given by $f_\alpha(\mathbf{v}, \mathbf{x})$. Then in the absence of external forces, the transport equation in the gas phase is given by the steady-state Boltzmann (or Vlasov) equation

$$\mathbf{v} \cdot \nabla f_\alpha(\mathbf{v}, \mathbf{x}) = 0, \tag{21}$$

which simply states that the distribution function f_α is constant along a straight path in the direction of the velocity vector $\mathbf{v}$.

Equation (21) is now solved together with the boundary conditions. Consider a position $\mathbf{X}$ on the surface, where the unit normal vector of the surface is given by $\hat{\mathbf{N}}$. Suppose a single atom of species β with velocity $\mathbf{v}'$ hits the surface at position $\mathbf{X}$. Then the number of atoms of species α ejected from position $\mathbf{X}$ with the velocity $\mathbf{v}$ by this single collision may be denoted by $\mathscr{Y}_{\hat{\mathbf{N}}}^{\alpha\beta}(\mathbf{v}, \mathbf{v}')$. The superscript $\alpha\beta$ indicates the emitted (or sputtered) species and the adsorbed (or bombarding) species, and the subscript $\hat{\mathbf{N}}$ explicitly indicates that this quantity also depends on the surface normal. For simplicity, the dependence on $\mathbf{X}$ is not explicitly indicated here. Note that $\mathscr{Y}_{\hat{\mathbf{N}}}^{\alpha\beta}(\mathbf{v}, \mathbf{v}')$ may take nonzero values only if $\mathbf{v} \cdot \hat{\mathbf{N}} > 0$ and $\mathbf{v}' \cdot \hat{\mathbf{N}} < 0$. Suppose the velocity distribution function at position $\mathbf{X}$ on the surface is given by $f^\beta(\mathbf{v}, \mathbf{X})$, then the total number of atoms of species α ejected with velocity $\mathbf{v}$ is given by integrating $\mathscr{Y}_{\hat{\mathbf{N}}}^{\alpha\beta}(\mathbf{v}, \mathbf{v}')$ multiplied by the flux of species β [i.e., $f^\beta(\mathbf{v}', \mathbf{X})|\mathbf{v}' \cdot \hat{\mathbf{N}}|$] over all $\mathbf{v}'$ satisfying $\mathbf{v}' \cdot \hat{\mathbf{N}} < 0$. The same number can also be expressed as $f^\alpha(\mathbf{v}, \mathbf{X})(\mathbf{v} \cdot \hat{\mathbf{N}})$ (the number of atoms of species α emitted from a unit area per unit time), so we obtain the following relation:

$$f^\alpha(\mathbf{v}, \mathbf{X}) = \frac{1}{(\mathbf{v} \cdot \hat{\mathbf{N}})} \sum_\beta \int_{\mathbf{v}' \cdot \hat{\mathbf{N}} < 0} \mathscr{Y}_{\hat{\mathbf{N}}}^{\alpha\beta}(\mathbf{v}, \mathbf{v}') f^\beta(\mathbf{v}', \mathbf{X}) \, |\mathbf{v}' \cdot \hat{\mathbf{N}}| \, d\mathbf{v}' \tag{22}$$

for $\mathbf{v} \cdot \hat{\mathbf{N}} > 0$. This equation gives the boundary conditions on the solid surface. The other boundary conditions that need to be specified are the velocity distribution functions in the gas phase representing fluxes coming directly from the process tool.

Suppose we obtain the distribution function $f^\alpha(\mathbf{v}, \mathbf{X})$ on the surface by solving Eq. (21) with such boundary conditions. Then the total flux of species α that the surface receives at t is given by

$$\begin{aligned} \mathscr{F}^\alpha = & \int_{\mathbf{v}' \cdot \hat{\mathbf{N}} < 0} f^\alpha(\mathbf{v}', \mathbf{X}) \, |\mathbf{v} \cdot \hat{\mathbf{N}}| \, d\mathbf{v}' \\ & - \sum_\beta \int_{\mathbf{v}' \cdot \hat{\mathbf{N}} < 0} \mathscr{Y}_{\hat{\mathbf{N}}}^{\alpha\beta}(\mathbf{v}, \mathbf{v}') f^\beta(\mathbf{v}', \mathbf{X}) \, |\mathbf{v} \cdot \hat{\mathbf{N}}| \, d\mathbf{v}'. \end{aligned} \tag{23}$$

The second term of Eq. (23) represents the sputtering or re-emission due to the arrival of species β (which may include the same species α) at the position **X**. Denoting the number density of the solid phase of the material α by ρ_α, we may express the deposition rate D (recall that the normal velocity of the surface is defined to be positive for etching) as

$$D = -\frac{1}{\rho_\alpha}\mathscr{F}^\alpha.$$

The transfer probability $\mathscr{Y}_{\hat{\mathbf{N}}}^{\alpha\beta}(\mathbf{v}, \mathbf{v}')$ also has the following physical meanings. If the incoming atoms (or ions) have high energy and thus sputter the material α from the surface, $\mathscr{Y}_{\hat{\mathbf{N}}}^{\alpha\beta}(\mathbf{v}, \mathbf{v}')$ is the differential sputtering yield. To obtain the usual sputtering yield, one simply integrates this over **v**, i.e.,

$$Y = \int_{\mathbf{v}\cdot\hat{\mathbf{N}}>0} \mathscr{Y}_{\hat{\mathbf{N}}}^{\alpha\beta}(\mathbf{v}, \mathbf{v}')\, d\mathbf{v}, \tag{24}$$

which generally depends only on the incident energy $\mathscr{E}' = m_\beta |\mathbf{v}'|^2/2$ with m_β being the atomic mass of species β and the incident angle θ, where $\cos\theta = (\mathbf{v}' \cdot \hat{\mathbf{N}})/|\mathbf{v}'|$.

If the incoming species and sputtered species are the same, the sputtering yield $Y(\mathscr{E}, \theta)$ calculated earlier is called the *self-sputtering* yield, where the reflected atoms of the incident beam cannot be distinguished from the species sputtered from the surface. On the other hand, if the incident species and ejected species are the same, say, β, but different from the surface material α, then $\mathscr{Y}_{\hat{\mathbf{N}}}^{\beta\beta}(\mathbf{v}, \mathbf{v}')$ represents the differential reflection (or scattering) probability of species β on the surface α.

If the incoming species α has sufficiently low energy, then $\mathscr{Y}_{\hat{\mathbf{N}}}^{\alpha\alpha}(\mathbf{v}, \mathbf{v}')$ is the differential re-emission probability. The integration of this quantity over **v** generally depends on neither $\mathscr{E}'$ nor θ, yielding a constant:

$$S = 1 - \int_{\mathbf{v}\cdot\hat{\mathbf{N}}>0} \mathscr{Y}_{\hat{\mathbf{N}}}^{\alpha\alpha}(\mathbf{v}, \mathbf{v}')\, d\mathbf{v} \tag{25}$$

is known as the sticking coefficient for such low energy incident species.

B. The Two-Energy Model

It is practically impossible to solve Eq. (21) with the boundary conditions given earlier under general conditions. Thus we introduce some simplifying assumptions here. First we assume that there are only two energy levels: high energy ions that are produced in the process tool and bombard the surface and low energy atoms that are either supplied by the tool source or re-emitted (i.e., sputtered or desorbed) from the surface.

As an example, let us consider metal deposition by the ionized magnetron sputtering system (*4,19,20*). As is explained in more detail in Sec. V, metal species arrive at the substrate surface as either ions or neutral atoms in this system. We assume that the distribution functions of these ion and neutral fluxes are known. The deposited metal film may also be resputtered by the bombardment by metal ions and/or operating gas (such as Ar) ions. For simplicity, we assume that all the incident ions are directional, going in the direction of the negative z direction. Also we ignore the operating gas and its ions in the following discussion for simplicity (single-species model).

Although this example seems very specific to a particular application, it contains many features common to various etching and deposition systems, which may also be modeled using the discussion presented in this section with minor modification. For example, oxide deposition by a CVD system may be modeled essentially by replacing metal by oxide although sputtering yields may be small. On the other hand, deep-trench etching with surface passivation may be modeled by increasing the sputtering yield such that the Si substrate can be etched and treating the neutral atoms as a passivation material (*2,3,21*).

Let us denote now the distribution function of the deposition material (metal in the example above) by $f(\mathbf{v}, \mathbf{x})$, omitting the superscript indicating the species for simplicity. In the two-energy model, the distribution function of the deposited material may be given as the sum of two components: one being the high-energy ion component and the other the low-energy neutral component as

$$f(\mathbf{v}, \mathbf{x}) = f^{\text{ion}}(\mathbf{v}, \mathbf{x}) + f^{\text{neut}}(\mathbf{v}, \mathbf{x}).$$

The incoming and outgoing fluxes are defined as

$$\mathscr{F}_{\text{in}}(\mathbf{X}) = \int_{\mathbf{v}\cdot\hat{\mathbf{N}}<0} f(\mathbf{v}, \mathbf{X})\, |\mathbf{v}\cdot\hat{\mathbf{N}}|\, d\mathbf{v},$$

$$\mathscr{F}_{\text{out}}(\mathbf{X}) = \int_{\mathbf{v}\cdot\hat{\mathbf{N}}>0} f(\mathbf{v}, \mathbf{X})(\mathbf{v}\cdot\hat{\mathbf{N}})\, d\mathbf{v}.$$

In this model, the total incoming flux consists of low- and high-energy components, whereas the total outgoing flux is assumed to be of low energy:

$$\mathscr{F}_{\text{in}} = \mathscr{F}_{\text{in}}^{\text{ion}} + \mathscr{F}_{\text{in}}^{\text{neut}},$$

$$\mathscr{F}_{\text{out}} = \mathscr{F}_{\text{out}}^{\text{neut}}.$$

Since we know the ion distribution function $f^{\text{ion}}(\mathbf{v}, \mathbf{x})$, the flux $\mathscr{F}_{\text{in}}^{\text{ion}}$ may be calculated from the distribution function directly:

$$\mathscr{F}_{\text{in}}^{\text{ion}}(\mathbf{X}) = \int_{\mathbf{v}\cdot\hat{\mathbf{N}}<0} f^{\text{ion}}(\mathbf{v}, \mathbf{X})\,|\mathbf{v}\cdot\hat{\mathbf{N}}|\,d\mathbf{v}. \tag{26}$$

The neutral (i.e., low-energy) incoming flux $\mathscr{F}_{\text{in}}^{\text{neut}}$, on the other hand, depends on the outgoing (re-emitted) fluxes $\mathscr{F}_{\text{out}}$ from other sites on the surface:

$$\mathscr{F}_{\text{in}}^{\text{nert}} = \mathscr{F}_{\text{in}}^{\text{ss}} + \mathscr{F}_{\text{in}}^{\text{so}}, \tag{27}$$

where $\mathscr{F}_{\text{in}}^{\text{ss}}$ denotes the surface-to-surface flux, that is, flux originated from other sites on the surface as re-emission, and $\mathscr{F}_{\text{in}}^{\text{so}}$ is the neutral flux directly coming from the plasma source. As in the case of the high-energy ion flux $\mathscr{F}_{\text{in}}^{\text{ion}}$, since we know the neutral distribution function $f^{\text{so}}(\mathbf{v}, \mathbf{x})$ in the source region, the magnitude of the flux $\mathscr{F}_{\text{in}}^{\text{so}}$ at every point on the surface may be calculated from the distribution function, in a manner similar to Eq. (27):

$$\mathscr{F}_{\text{in}}^{\text{so}}(\mathbf{X}) = \int_{\mathbf{v}\cdot\hat{\mathbf{N}}<0} f^{\text{so}}(\mathbf{v}, \mathbf{X})\,|\mathbf{v}\cdot\hat{\mathbf{N}}|\,d\mathbf{v}. \tag{28}$$

The outgoing flux may also be split to two components as

$$\mathscr{F}_{\text{out}} = \mathscr{F}_{\text{out}}^{\text{des}} + \mathscr{F}_{\text{out}}^{\text{sput}}, \tag{29}$$

where $\mathscr{F}_{\text{out}}^{\text{des}}$ denotes the desorption flux and $\mathscr{F}_{\text{out}}^{\text{sput}}$ denotes the flux sputtered by the high-energy ion flux $\mathscr{F}_{\text{in}}^{\text{ion}}$:

$$\mathscr{F}_{\text{out}}^{\text{des}}(\mathbf{X}) = \int_{\mathbf{v}\cdot\hat{\mathbf{n}}>0} d\mathbf{v} \int_{\mathbf{v}'\cdot\hat{\mathbf{N}}<0} d\mathbf{v}'\mathscr{Y}_{\hat{\mathbf{N}}}(\mathbf{v}, \mathbf{v}')f^{\text{neut}}(\mathbf{v}', \mathbf{X})\,|\mathbf{v}'\cdot\hat{\mathbf{N}}|, \tag{30}$$

$$\mathscr{F}_{\text{out}}^{\text{sput}}(\mathbf{X}) = \int_{\mathbf{v}\cdot\hat{\mathbf{n}}>0} d\mathbf{v} \int_{\mathbf{v}'\cdot\hat{\mathbf{N}}<0} d\mathbf{v}'\mathscr{Y}_{\hat{\mathbf{N}}}(\mathbf{v}, \mathbf{v}')f^{\text{ion}}(\mathbf{v}', \mathbf{X})\,|\mathbf{v}'\cdot\hat{\mathbf{N}}|. \tag{31}$$

Since we already know the sputtering yield and ion distribution function $f^{\text{ion}}(\mathbf{v}, \mathbf{x})$, $\mathscr{F}_{\text{out}}^{\text{sput}}$ of Eq. (31) can also be easily calculated.

For the desorption flux $\mathscr{F}_{\text{out}}^{\text{des}}$, we further assume that the distribution function $f_{\text{out}}^{\text{des}}(\mathbf{v}, \hat{\mathbf{X}})$ for the re-emitted metal atoms is given by the "generalized" cosine law (2) regardless of the direction of incident fluxes:

$$f_{\text{out}}^{\text{des}}(\mathbf{v}, \hat{\mathbf{X}}) = f_\nu(v, \hat{\mathbf{X}})\cos^\nu\Theta \qquad \text{for} \quad \mathbf{v}\cdot\hat{\mathbf{N}} \geq 0, \tag{32}$$

where Θ is the angle between vectors $\mathbf{v}$ and $\hat{\mathbf{N}}$, and ν (> -1) is a parameter. Note that the distribution function $f(\mathbf{v}, \mathbf{x})$ in general is assumed to be independent of y due to the translational symmetry of the system in the y direction.

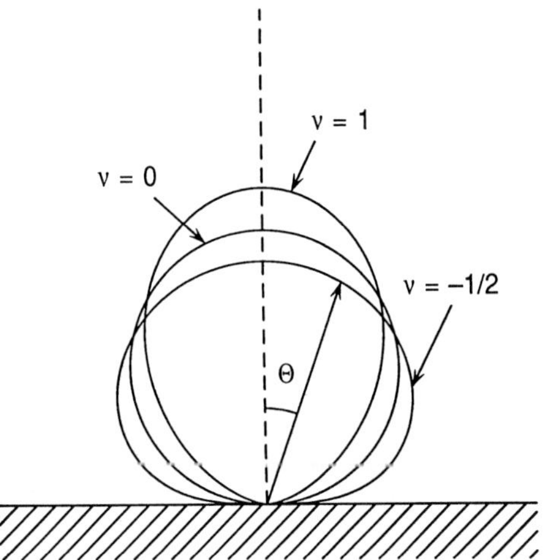

FIG. 10. The angular distribution of re-emission: $\nu = 0$ is the cosine distribution, $\nu > 0$ is an over-cosine distribution, and $-1 < \nu < 0$ is an under-cosine distribution.

Under condition (32), the transfer probability (i.e., differential desorption yield) $\mathscr{Y}_{\hat{\mathbf{N}}}$ is independent of $\mathbf{v}'$ (*22*). Performing the integration of Eq. (30) with Eq. (25) thus yields the relation

$$\mathscr{F}_{\text{out}}^{\text{des}}(\mathbf{X}) = [1 - S(\mathbf{X})]\mathscr{F}_{\text{in}}^{\text{neut}}(\mathbf{X}). \tag{33}$$

Note that the sticking probability (coefficient) S generally depends on position $\mathbf{X}$.

The total outgoing flux of re-emitted atoms from the unit area at $\hat{\mathbf{X}}$ may be given by

$$\mathscr{F}_{\text{out}}^{\text{des}}(\hat{\mathbf{X}}) = \int_0^{\infty} dv \int_{(\mathbf{v}\cdot\hat{\mathbf{n}})>0} d\Omega\, v^3 f_\nu(v, \hat{\mathbf{X}}) \cos^{\nu+1}\Theta, \tag{34}$$

where $d\Omega$ denotes the solid angle element. From the Θ dependence of the integrand of Eq. (34), the re-emission distribution is called *cosine* when $\nu = 0$. Likewise, an over-cosine distribution may be obtained by choosing $\nu > 0$ and an under-cosine distribution by $-1 < \nu < 0$ (Fig. 10). Performing the integration in Θ, we obtain

$$\int_0^{\infty} v^3 f_\nu(v, \hat{\mathbf{X}})\, dv = \frac{(\nu + 2)}{2\pi} \mathscr{F}_{\text{out}}^{\text{des}}(\hat{\mathbf{X}}). \tag{35}$$

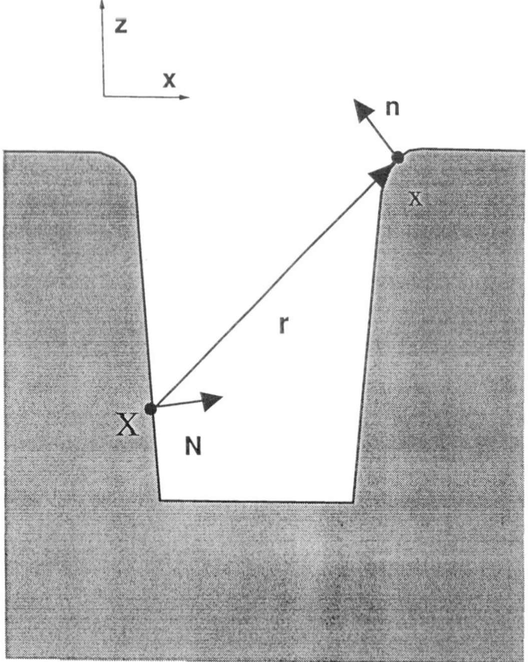

FIG. 11. The coordinate system.

Let the coordinates of point **X** on the surface be $(X, 0, Z)$, where we have chosen the origin of the coordinate system in such a way that **X** is on the xz-plane. We also choose another point $\mathbf{x} = (x, y, z)$ on the surface, where the unit normal vector of the surface is given by **n** (see Fig. 11). If **x** lies on a line of sight from **X**, it follows from Eq. (21) that, for any distribution function f, we have

$$f(\mathbf{v}, \mathbf{X}) = f(\mathbf{v}, \mathbf{x}) \tag{36}$$

for all $v = |\mathbf{v}|$.

Here we must note that **x** is no longer independent of **v** since the vector $\mathbf{r} = \mathbf{x} - \mathbf{X}$ is antiparallel to the velocity vector $\mathbf{v}'$ (i.e., $\mathbf{v} = -v'\hat{\mathbf{r}}$ with $\hat{\mathbf{r}} = \mathbf{r}/|\mathbf{r}|$). Using the usual polar coordinate system around the z-axis, we may relate **x** to **v** as

$$\cos\theta = \frac{\tilde{z}}{\mathbf{r}}, \qquad \tan\phi = \frac{y}{\tilde{x}},$$

where

$$\tilde{x} = x - X,$$
$$\tilde{z} = z - Z,$$
$$r = \sqrt{\tilde{x}^2 + y^2 + \tilde{z}^2}\,.$$

The volume element is then given (2) by

$$d\mathbf{v} = v^2 \sin\theta \, dv \, d\theta \, d\phi = \frac{v^2}{r^2} |\hat{\mathbf{r}} \cdot \hat{\mathbf{n}}| \, ds \, dy \, dv, \tag{37}$$

where $ds = \sqrt{dx^2 + dz^2}$ is the line element on the boundary curve.

The surface-to-surface flux introduced in Eq. (27) is given by

$$\mathcal{F}_{\text{in}}^{\text{ss}}(\mathbf{X}) = \int_{\mathbf{v}\cdot\hat{\mathbf{N}}<0} d\mathbf{v} f^{\text{ss}}(\mathbf{v}, \mathbf{X}) \, |\mathbf{v} \cdot \hat{\mathbf{N}}|, \tag{38}$$

where $f^{\text{ss}}(\mathbf{v}, \mathbf{X})$ denotes the distribution function of atoms that are re-emitted from the surface. Using Eq. (37), we may rewrite Eq. (38) as

$$\mathcal{F}_{\text{in}}^{\text{ss}}(\mathbf{X}) = \int_{\mathcal{B}} ds \int_{-\infty}^{\infty} dy \int_{0}^{\infty} dv g(\mathbf{X}, \mathbf{x}) v^3 f_{\text{out}}(\mathbf{v}, \mathbf{x}) \, \frac{(\hat{\mathbf{r}} \cdot \hat{\mathbf{N}}) \, |\hat{\mathbf{r}} \cdot \hat{\mathbf{n}}|}{r^2}, \tag{39}$$

where $\mathcal{B}$ denotes the surface boundary at the xz-plane, and $f_{\text{out}}(\mathbf{v}, \mathbf{x})$ denotes the distribution function for the atoms re-emitted (i.e., sputtered or desorbed) at position $\mathbf{x}$. The integration $\int_{\mathcal{B}} ds \int_{-\infty}^{\infty} dy$ represents the integral over the entire surface. The function $g(\mathbf{X}, \mathbf{x})$ is the visibility factor defined by

$$g(\mathbf{X}, \mathbf{x}) = \begin{cases} 1 & \text{if } \mathbf{x} \text{ is visible from } \mathbf{X}. \\ 0 & \text{otherwise}. \end{cases} \tag{40}$$

For 2-D structures, it is evident that we may write

$$g(\mathbf{X}, \mathbf{x}) = g(\mathbf{X}, \boldsymbol{\xi})$$

with $\boldsymbol{\xi} = (x, 0, z)$, that is, the projection of $\mathbf{x}$ to the xz surface.

As in Eq. (29), we may write $f_{\text{out}} = f_{\text{out}}^{\text{des}} + f_{\text{out}}^{\text{sput}}$, where $f_{\text{out}}^{\text{des}}$ is the desorption distribution function [given by Eq. (32)] and $f_{\text{out}}^{\text{sput}}$ is the sputtering distribution function. Then we obtain from Eq. (39)

$$\mathscr{F}_{\text{in}}^{\text{ss}}(\mathbf{X}) = \int_{\mathscr{B}} ds \int_{-\infty}^{\infty} dy \int_{0}^{\infty} dv\, g(\mathbf{X},\mathbf{x}) v^3 f_{\text{out}}^{\text{des}}(\mathbf{v},\mathbf{x}) \frac{(\hat{\mathbf{r}}\cdot\hat{\mathbf{N}})\,|\hat{\mathbf{r}}\cdot\hat{\mathbf{n}}|}{r^2} + \mathscr{G}_{\text{in}}^{\text{sput}}(\mathbf{X}), \tag{41}$$

where

$$\mathscr{G}_{\text{in}}^{\text{sput}}(\mathbf{X}) = \int_{\mathscr{B}} ds \int_{-\infty}^{\infty} dy \int_{0}^{\infty} dv\, g(\mathbf{X},\mathbf{x}) v^3 f_{\text{out}}^{\text{sput}}(\mathbf{v},\mathbf{x}) \frac{(\hat{\mathbf{r}}\cdot\hat{\mathbf{N}})\,|\hat{\mathbf{r}}\cdot\hat{\mathbf{n}}|}{r^2} \tag{42}$$

is the incoming neutral flux due to ion sputtering. The sputter distribution function $f_{\text{out}}^{\text{sput}}(\mathbf{v},\mathbf{x})$ is related to the incoming ion distribution function $f^{\text{ion}}(\mathbf{v},\mathbf{x})$ through Eq. (22):

$$f_{\text{out}}^{\text{sput}}(\mathbf{v},\mathbf{x}) = \frac{1}{(\mathbf{v}\cdot\hat{\mathbf{N}})} \int_{\mathbf{v}\cdot\hat{\mathbf{N}}<0} \mathscr{Y}_{\hat{\mathbf{N}}}(\mathbf{v},\mathbf{v}') f^{\text{ion}}(\mathbf{v}',\mathbf{x})\,|\mathbf{v}'\cdot\hat{\mathbf{N}}|\,d\mathbf{v}'. \tag{43}$$

From Eqs. (32) and (35), the first term of Eq. (41) may be rewritten as

$$\frac{(\nu+2)}{2\pi} \int_{\mathscr{B}} ds \int_{-\infty}^{\infty} dy\, g(\mathbf{X},\boldsymbol{\xi}) \mathscr{F}_{\text{out}}^{\text{des}}(\boldsymbol{\xi}) \frac{(\hat{\mathbf{r}}\cdot\hat{\mathbf{N}})\,|\hat{\mathbf{r}}\cdot\hat{\mathbf{n}}|^{\nu+1}}{r^2}, \tag{44}$$

where $\cos\Theta = |\hat{\mathbf{r}}\cdot\hat{\mathbf{n}}|$ is used. Carrying out the integration in y then yields

$$\frac{A_{\nu+4}(\nu+2)}{\pi} \int_{\mathscr{B}} ds\, g(\mathbf{X},\boldsymbol{\xi}) \mathscr{F}_{\text{out}}^{\text{des}}(\boldsymbol{\xi}) \frac{(\hat{\mathbf{R}}\cdot\hat{\mathbf{N}})\,|\hat{\mathbf{R}}\cdot\hat{\mathbf{n}}|^{\nu+1}}{R}. \tag{45}$$

Here $\mathbf{R} = \boldsymbol{\xi} - \mathbf{X}$, $R = |\mathbf{R}|$, $\hat{\mathbf{R}} = \mathbf{R}/R$, and

$$A_{\nu+4} = \int_0^{\infty} \frac{dt}{(t^2+1)^{(2+\nu/2)}} = \Gamma\left(\frac{\nu+3}{2}\right)\Gamma\left(\frac{1}{2}\right)\Big/ 2\Gamma\left(\frac{\nu}{2}+2\right).$$

For the cosine distribution ($\nu = 0$), we have $A_4 = \pi/4$.

Substituting Eq. (33) into Eq. (45) and using Eq. (27), we obtain

$$\mathscr{F}_{\text{in}}^{\text{neut}}(\mathbf{X}) = \mathscr{F}_{\text{in}}^{\text{so}}(\mathbf{X}) + \mathscr{G}_{\text{in}}^{\text{sput}}(\mathbf{X}) + \frac{A_{\nu+4}(\nu+2)}{\pi} \int_{\mathscr{B}} ds[1 - S(\boldsymbol{\xi})] \mathscr{F}_{\text{in}}^{\text{neut}}(\boldsymbol{\xi}) K_\nu(\mathbf{X},\boldsymbol{\xi}), \tag{46}$$

where $K_\nu(\mathbf{X},\boldsymbol{\xi})$ is the kernel defined by

$$K_\nu(\mathbf{X},\boldsymbol{\xi}) = g(\mathbf{X},\boldsymbol{\xi}) \frac{(\hat{\mathbf{R}}\cdot\hat{\mathbf{N}})\,|\hat{\mathbf{R}}\cdot\hat{\mathbf{n}}|^{\nu+1}}{R}. \tag{47}$$

From Eqs. (26), (28), and (42), the fluxes $\mathscr{F}_{\text{in}}^{\text{ion}}$, $\mathscr{F}_{\text{in}}^{\text{so}}$, and $\mathscr{G}_{\text{in}}^{\text{sput}}$ are all known. Thus integral equation (46) may be used to determine the low-energy incoming flux $\mathscr{F}_{\text{in}}^{\text{neut}}$.

C. Ion and Neutral Fluxes

With knowledge of the ion distribution function $f^{\text{ion}}(\mathbf{v}, \mathbf{x})$, one can determine the ion flux $\mathscr{F}_{\text{in}}^{\text{ion}}(\mathbf{X})$ on the water surface, using Eq. (28). If the ion flux is unidirectional and monoenergetic, that is,

$$f^{\text{ion}}(\mathbf{v}) = n_0 \delta(\mathbf{v} - \mathbf{v}_0), \tag{48}$$

where n_0 denotes the ion density in the source region, we obtain

$$\mathscr{F}_{\text{in}}^{\text{ion}}(\mathbf{X}) = J^{\text{ion}} \cos\theta, \tag{49}$$

as long as position $\mathbf{X}$ is directly exposed to the ion beam. Here $J^{\text{ion}} = n_0 |\mathbf{v}_0|$ is the surface flux on a horizontal surface and θ denotes the angle formed by vector $\mathbf{v}_0$ and surface normal $\hat{\mathbf{N}}$ [i.e., $(\mathbf{v}_0 \cdot \hat{\mathbf{N}}) = -|\mathbf{v}_0| \cos\theta$].

We now evaluate the first term of Eq. (46), namely incoming neutral flux from the source. Using Eqs. (28) and (36), we obtain

$$\mathscr{F}_{\text{in}}^{\text{so}}(\mathbf{X}) = \int_{\mathbf{N}\cdot\mathbf{v}<0} (\mathbf{N} \cdot \mathbf{v}) g(\mathbf{X}, \mathbf{x}) f^{\text{so}}(\mathbf{v}, \mathbf{x})\, d\mathbf{v}, \tag{50}$$

where $f^{\text{so}}(\mathbf{v}, \mathbf{x})$ is the neutral distribution function at position $\mathbf{x}$ in the source region. Suppose the distribution function f^{so} is isotropic—independent of the direction of $\mathbf{v}$—and independent of position $\mathbf{x}$: $f^{\text{so}}(\mathbf{v}, \mathbf{x}) = f^{\text{so}}(v)$. Using the usual polar angles θ and ϕ around the z direction again, we can write Eq. (50) as

$$\mathscr{F}_{\text{in}}^{\text{so}}(\mathbf{X}) = 2 \int_0^{\theta_{\text{max}}} d\theta \int_{\phi_r(\theta)}^{\phi_l(\theta)} d\phi \int_0^{\infty} dv \times (N_x \sin^2\theta \cos\phi + N_z \sin\theta \cos\theta) v^3 f^{\text{so}}(v), \tag{51}$$

where $\mathbf{N} = (N_x, 0, N_z)$. In Eq. (51), θ_{max} is the maximum angle that θ can take, which is usually less than $\pi/2$ due to the flux collimation (see the next section). The limits of the azimuthal angle $\phi_r(\theta)$ and $\phi_l(\theta)$, which are functions of θ, arise from the two-dimensional nature of the structure, as shown in Fig. 12. The integration over ϕ can be easily carried out.

In particular, if the sample is horizontal and flat, then we have $N = (0, 0, 1)$, $\phi_r(\theta) = 0$, and $\phi_l(\theta) = \pi$. Therefore, this flat-surface neutral flux

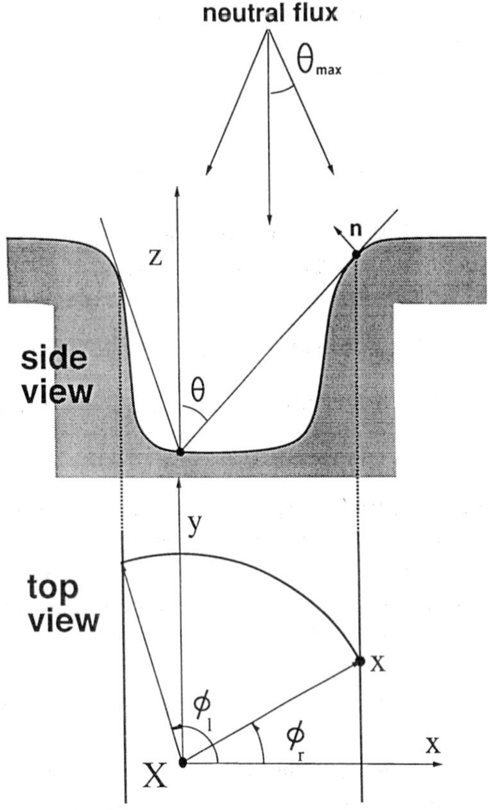

FIG. 12. The polar coordinate system and the neutral flux with cutoff angle θ_{max}.

$\mathscr{J}^{flat}(\theta_{max}) = \mathscr{F}_{in}^{so}(\mathbf{X})$ is independent of $\mathbf{X}$ and given by

$$\mathscr{J}^{flat}(\theta_{max}) = \frac{\pi}{2}(1 - \cos 2\theta_{max}) \int_0^{v} dv v^3 f^{so}(v).$$

Using this expression, Eq. (51) for general $\mathbf{X}$ may be recast as

$$\mathscr{F}_{in}^{so}(\mathbf{X}) = \frac{4\mathscr{J}^{flat}(\theta_{max})}{\pi(1 - \cos 2\theta_{max})} \times \int_0^{\theta_{max}} d\theta \int_{\phi_r(\theta)}^{\phi_l(\theta)} d\phi (N_x \sin^2 \theta \cos \phi + N_z \sin \theta \cos \theta). \quad (52)$$

In the case of vertical beam ions, where the ion distribution function is given by Eq. (48), the outgoing sputter distribution function given in Eq. (43) can be expressed as

$$f_{\text{out}}^{\text{sput}}(\mathbf{v},\mathbf{x}) = \frac{J^{\text{ion}}}{(\mathbf{v}\cdot\hat{\mathbf{N}})}\,\mathscr{Y}_{\hat{\mathbf{N}}}(\mathbf{v},\mathbf{v}_0)\cos\theta. \tag{53}$$

Then one can evaluate the sputter flux $\mathscr{G}_{\text{in}}^{\text{sput}}$ by substituting Eq. (53) into Eq. (42). Without specifying the function $\mathscr{Y}_{\hat{\mathbf{N}}}(\mathbf{v},\mathbf{v}_0)$, however, one cannot further simplify the resulting equation.

For the sake of simplicity, we assume that the sputter distribution function $f_{\text{out}}^{\text{sput}}$ is also given by the generalized cosine function regardless of the ion beam's angle of incidence:

$$f_{\text{out}}^{\text{sput}}(\mathbf{v},\mathbf{x}) = f_{\nu}^{\text{sput}}(v,\boldsymbol{\xi})\cos^{\nu}\Theta \qquad \text{for} \quad \mathbf{v}\cdot\hat{\mathbf{n}} \geq 0, \tag{54}$$

where $\boldsymbol{\xi} = (x,0,z)$, Θ is the angle between vectors $\mathbf{v}$ and $\hat{\mathbf{n}}$ and ν (> -1) is a parameter, as before.

Since

$$\mathscr{F}_{\text{out}}^{\text{sput}}(\mathbf{X}) = \int_{\mathbf{v}\cdot\hat{\mathbf{N}}>0} f_{\text{out}}^{\text{sput}}(\mathbf{v},\mathbf{X})(\mathbf{v}\cdot\hat{\mathbf{N}})\,d\mathbf{v}, \tag{55}$$

as in the case of Eq. (35), we obtain from Eq. (54)

$$\int_0^{\infty} v^3 f_{\nu}^{\text{sput}}(v,\boldsymbol{\xi})\,dv = \frac{(\nu+2)}{2\pi}\mathscr{F}_{\text{out}}^{\text{sput}}(\boldsymbol{\xi}). \tag{56}$$

Therefore, substituting Eq. (54) into Eq. (42) and performing the integration yields

$$\mathscr{G}_{\text{in}}^{\text{sput}} = \frac{A_{\nu+4}(\nu+2)}{\pi}\int_{\mathscr{B}} ds\, g(\mathbf{X},\boldsymbol{\xi})\mathscr{F}_{\text{out}}^{\text{sput}}(\boldsymbol{\xi})\frac{(\hat{\mathbf{R}}\cdot\hat{\mathbf{N}})\,|\hat{\mathbf{R}}\cdot\hat{\mathbf{n}}|^{\nu+1}}{R}, \tag{57}$$

as in Eq. (45).

In the case of beam ions given by Eq. (48), we obtain from Eqs. (24) and (31)

$$\mathscr{F}_{\text{out}}^{\text{sput}}(\mathbf{X}) = Y(\mathbf{v}_0)\mathscr{F}_{\text{in}}^{\text{ion}}(\mathbf{X}).$$

In terms of the angle θ formed by the surface normal and the direction of incident ions and the ion kinetic energy $\mathscr{E}$, we may write $Y(\mathbf{v}_0) = Y(\theta,\mathscr{E})$. As we discussed in Sec. II.A, the etch rate C (i.e., normal surface velocity) is proportional to $Y/\cos\theta$ with the slope angle θ. Figure 13 gives a typical

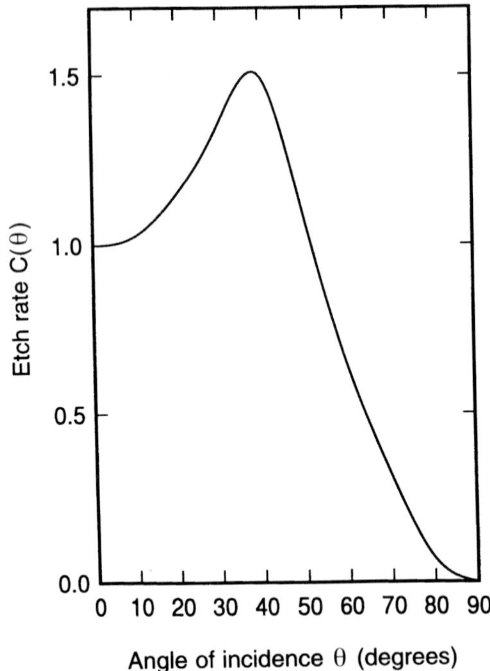

FIG. 13. The typical etch-rate dependence on the angle of incidence (slope angle) θ.

etch-rate dependence on the slope angle θ. It is seen that the etch rate has its maximum at an off-normal angle of beam incidence (*23–25*).

V. Sample Simulations

For representative examples, we consider the evolution of a microfeature in the ionized magnetron sputter metal deposition process. The ionized magnetron sputter metal deposition system is a combination of the conventional magnetron sputter deposition system and an inductively coupled plasma source. In this system, metal atoms are generated by the magnetron sputtering from the target cathode. Some of these metal atoms are then ionized in an Ar plasma generated by the radio-frequency induction (RFI) coil. By varying the plasma conditions, one can control the level of metal ionization. Such metal ions are then extracted to the substrate surface by the bias voltage. Details of the experiments can be found in Refs. *19* and *20*.

Both ion and neutral fluxes thus contribute to the deposition process. As in Sec. IV, the ion flux may be considered directional due to the bias voltage, whereas the metal neutral flux may be considered isotropic with a cutoff collimation angle θ_{max}. The cutoff angle may be due to a collimator actually placed in the tool or to the finite size of the target cathode. With a sufficiently high bias voltage, metal and Ar ions can also resputter the deposited metal from the sample surface and thus both etching and deposition take place simultaneously. Resputtered metal atoms are redeposited on the surface with a sticking probability S.

Figure 14 is a simulation result showing the effects of increased sputtering at the sample surface (*4*). Each boundary curve represents the metal film profile at equal time intervals and the shaded area represents the initial trenches. The simulation was done by using the numerical code SHADE (Shock-tracking Algorithm for Deposition and Etching), which is based on the moving-surface algorithm method discussed in Sec. III. In Fig. 14, the ratio of the ion flux to neutral flux is assumed to be 1 and the total sputtering yield Y at slope angle $\theta = 0$ is 0 for Fig. 14a, 0.6 for Fig. 14b, and 0.8 for Fig. 14c. Note that the total sputtering yield Y includes the sputtering by Ar ions. That is, for a single metal ion impinging on the surface, Y metal atoms are sputtered, whether the sputtering is due to the metal ion or other Ar ions. By interpreting the sputtering yield Y in this

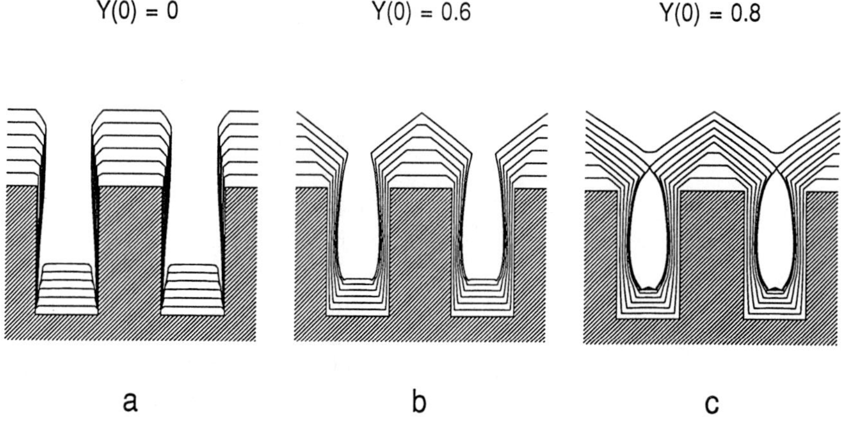

FIG. 14. Simulation results of metal deposition into trenches. The sputtering yields at zero slope angle are (a) 0, (b) 0.6, and (c) 0.8 (from Ref. *4*).

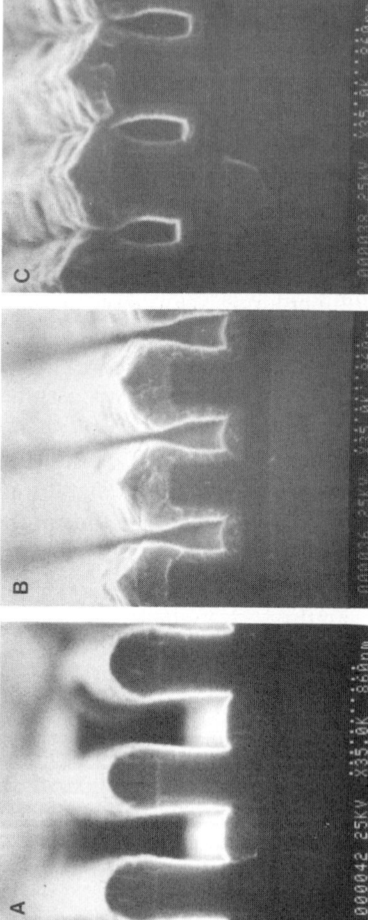

FIG. 15. SEM pictures of deposited metal (Al) films under similar conditions. The bias voltages are (a) 20 V or less, (b) about 80 V, and (c) 120 V or more (from Ref. *4*).

way, one can use the single-species model that we discussed in Sec IV. Here, however, we assume that the sputtering yields $Y_{metal}(\theta)$ and $Y_{Ar}(\theta)$ due to metal and Ar ions, respectively, have the same functional form for the angle dependence given in Fig. 13. As is clearly seen in Fig. 14, as the film thickness becomes comparable to the trench width, the atoms sputtered from one side of the trench are collected on the opposite side. This leads to a lateral buildup of resputtered materials, which can eventually result in closing of the trench (Fig. 14c).

The same effects are observed experimentally, as shown in the scanning electron mircoscopy (SEM) pictures of Fig. 15 (*4*). The film shown in Fig. 15a was deposited with ion energies of 20 eV or less, which can cause a directional deposition without causing major resputtering at the surface. Figure 15b is the case for ion energy of about 80 eV and Fig. 15c for 120 eV or more, both of which show the effect of increasing sputtering yields. It is clear that the simulation results show the correct tendency of the deposition profiles with increasing ion energies.

VI. Summary

We have discussed the theory of moving solid boundaries and numerical algorithms to simulate them. In Sec. II, the problem description (i.e., Hamilton–Jacobi type equation *and* entropy conditions) is presented. We have emphasized the fact that the Hamilton–Jacobi type equation (and therefore the ray-tracing method or method of characteristics) in itself cannot determine the physically plausible solid surface evolution. It is important to recognize that the additional physical mechanism (i.e., infinitesimal surface diffusion) is needed to derive the entropy conditions, which together with the Hamilton–Jacobi type equation determine the physical solution. A geometric method to construct the physical solution is presented in Sec. II.F, based on the entropy conditions of Eq. (20) or Oleinik construction. The shock-tracking algorithm (*1*) presented in Sec. III explicitly uses this geometric construction method to determine the correct moving boundary solution.

Earlier work on etching/deposition topography simulation focused primarily on the method of characteristics or ray tracing (*26–31*), but paid little attention to the entropy conditions. Without the entropy conditions, however, some other procedures must be invoked to avoid seemingly unphysical solutions (*32,33*). In many cases, such procedures are introduced based on the modeler's intuition of what the correct physical solutions should be, rather than additional physical processes such as surface diffusion.

In the latter half of this article, we have treated more specific applications, that is dry processes for etching and deposition, and presented some representative examples. The spatial scale treated in these problems is mesoscopic, that is, sufficiently smaller than macroscopic scales such as mean free paths of gaseous atoms, but sufficiently larger than molecular scales. For semiconductor dry processes, the representative scale for microfeatures is typically 1 μm or less, in which one can still approximate the solid by a continuum, rather than aggregates of molecules. In this mesoscopic continuum model, ion and neutral fluxes impinging on the surface are also treated as continuums. Under some realistic assumptions, such fluxes are easily evaluated, as shown in Sec. IV. For simplicity, we have only discussed the two-energy/single-species model, but extending this model to a multienergy/multispecies model is straightforward.

Since the goal of this article is to present the basic theory of moving boundary problems in dry processes, no attempt was made to discuss details of various processes used in industry. For instance, readers interested in more details of the examples shown in Sec. V are referred to Refs. *4, 5, 19*, and *20*.

Acknowledgments

The author thanks S. M. Rossnagel for helpful discussions on ionized magnetron sputter deposition processes.

References

1. S. Hamaguchi, M. Dalvie, R. T. Faruoki, and S. Sethuraman, *J. Appl. Phys.* **74**, 5172 (1993).
2. S. Hamaguchi and M. Dalvie, *J. Vac. Sci. Technol. A* **12**, 2745 (1994).
3. S. Hamaguchi and M. Dalvie, *J. Electrochem. Soc.* **141**, 1964 (1994).
4. S. Hamaguchi and S. M. Rossnagel, *J. Vac. Sci. Technol. B* **13**, 183 (1995).
5. S. Hamaguchi and S. M. Rossnagel, *J. Vac. Sci. Technol. B* in press (1996).
6. E. D. Conway and E. Hopf, *J. Math. Mech.* **13**, 939 (1964).
7. M. G. Crandall and P. L. Lions, *Trans. Am. Math. Soc.* **277**, 1 (1983).
8. M. G. Crandall, L. C. Evans, and P. L. Lions, *Trans. Am. Math. Soc.* **282**, 487 (1984).
9. M. G. Crandall and P. L. Lions, *Math. Comp.* **43**, 1 (1984).
10. P. D. Lax, "Hyperbolic Systems of Conservation Laws and the Mathematical Theory of Shock Waves," SIAM Regional Conference Series in Applied Mathematics, SIAM, Philadelphia, Pennsylvania, 1973.

11. F. John, "Partial Differential Equations." Springer-Verlag, New York, 1982.
12. J. Smoller, "Shock Waves and Reaction–Diffusion Equations." Springer-Verlag, New York, 1983.
13. Jean Philibert, "Atom Movements Diffusion and Mass Transport in Solids," Les Éditions de Physique, Les Ulis Cedex A, France, 1991.
14. D. S. Ross, *J. Electrochem. Soc.* **135**, 1235 (1988).
15. D. S. Ross, *J. Electrochem. Soc.* **135**, 1260 (1988).
16. O. A. Oleinik, *Ups. Mat. Nauk.* **12**, 3 (1957); English transl. in *Amer. Math. Soc. Transl. Ser. 2* **26**, 95 (1963).
17. J. Glimm, *Comm. Pure Appl. Math.* **18**, 697 (1965).
18. A. J. Chorin, *J. Comp. Phys.* **25**, 253 (1977).
19. S. M. Rossnagel and J. Hopwood, *Appl. Phys. Lett.* **63**, 3285 (1993).
20. S. M. Rossnagel and J. Hopwood, *J. Vac. Sci. Technol. B* **12**, 449 (1994).
21. G. S. Oehrlein, J. F. Rembetski, and E. H. Payne, *J. Vac. Sci. Technol. B* **8**, 1199 (1990).
22. In terms of the transfer probability, Eq. (32) is equivalent to

$$\mathscr{Y}_{\hat{\mathbf{N}}}(\mathbf{v},\mathbf{v}') = \frac{f_\nu(\mathbf{v},\hat{\mathbf{X}})}{F_{\mathrm{in}}^{\mathrm{neut}}(\hat{\mathbf{X}})} \cos^{\nu+1}\Theta$$

Note that $\mathscr{Y}_{\hat{\mathbf{N}}}$ is independent of $\mathbf{v}'$ here.
23. H. H. Andersen and H. L. Bay, *in* "Sputtering by Particle Bombardment I" (R. Behrisch, ed.), p. 145. Springer-Verlag, New York, 1981.
24. R. E. Lee, *J. Vac. Sci. Technol.* **16**, 164 (1979).
25. P. C. Zalm, *in* "Handbook of Ion Beam Processing Technology" (J. J. Cuomo, S. M. Rossnagel, and H. R. Kaufman, eds.), p. 78. Noyes, Park Ridge, New Jersey, 1989.
26. M. J. Nobes, J. S. Colligon, and G. Carter, *J. Mater. Sci.* **4**, 730 (1969).
27. G. Carter, J. S. Colligon, and M. J. Nobes, *J. Mater. Sci.* **6**, 115 (1971).
28. D. J. Barber, F. C. Frank, M. Moss, J. W. Steeds, and I. S. T. Tsong, *J. Mater. Sci.* **8**, 1030 (1973).
29. J. P. Ducommun, M. Cantagrel, and M. Marchal, *J. Mater. Sci.* **9**, 725 (1974).
30. R. Smith, G. Carter, and M. J. Nobes, *Proc. R. Soc. London Ser. A* **407**, 405 (1986).
31. C. W. Jurgensen and E. S. G. Shaqfeh, *J. Vac. Sci. Technol. B* **7**, 1488 (1989).
32. I. V. Katardjiev, *J. Vac. Sci. Technol. B* **7**, 3227 (1989).
33. I. V. Katardjiev, G. Carter, and M. J. Nobes, *J. Phys. D* **22**, 1813 (1989).

A Process Model for Sputter Deposition of Thin Films Using Molecular Dynamics

C.-C. FANG AND V. PRASAD

Process Modeling Laboratory, State University of New York, Stony Brook, New York

AND

R. V. JOSHI, F. JONES, AND J. J. HSIEH*

IBM T. J. Watson Research Center, Yorktown Heights, New York

* Present address: Novellus Systems, 2970 North First Street, San Jose, CA 95134.

I. Introduction

A. Introductory Remarks

Material coatings have been used for a long time to change the surface characteristics of industrial products. For example, the use of thin films in the construction of resistors goes back to the early days of this century (*1*). Diamond coatings have been very successful in increasing the lifetime of tools used for machining as well as improving the surface quality of the machined products. Nitride films have been used in solar cells as anti-reflection coatings and magnetic films have found applications in the area of random-access storage. The use of thin films has grown more rapidly in recent years because of their widespread applications in the manufacture of electronic devices. Deposition of insulators and doped materials in the fabrication of integrated circuits is considered to be a key technology in the manufacture of thin film resistors and VLSI/ULSI circuits. Indeed, the rapid growth of microelectronic devices has brought the coating process to new levels of progress. Instead of building a single component device, deposition processes are now being developed and used to fabricate a large number of devices at submicron levels directly above the substrate. The high-density packing of devices such as resistors and capacitors, and their line connections on the substrate, has stimulated extensive research and development in the area of thin film formation processes.

Thin films, particularly those deposited on a solid substrate from the vapor phase, are quite different from the bulk material as far as their structure and properties are concerned. Research interest in the electrical, optical, and mechanical properties of these films has therefore grown rapidly during the past few years. Special attention has been given to the relationship between the microstructure and material properties due to their strong dependence on the deposition method (*2*) and process parameters (*3*). For example, almost all films deposited by using the thermal evaporation methods are found to be rich in defects. If the structure of the deposited film is porous, its adhesion to the substrate and surrounding material can be weak (*4*). In some cases the stresses in these films can be sufficiently high to cause mechanical failures. Many times, the film either buckles or develops cracks during the growth process, and its refractive index can be low. Even if the deposited film is structurally sound, it may exhibit a concave or convex deformation depending on the internal stresses within the film and the substrate. The deposited film with residual stresses generally results in a higher production cost because it may require postprocessing, for example, annealing to relieve the intrinsic stresses.

B. Sputter Deposition of Thin Films

Sputter deposition processes such as magnetron sputtering and ion beam-assisted sputterings are well known for their ability to produce high-quality conformal coatings on substrate surface features as well as less gas contamination of the film. These physical vapor deposition methods generally involve bombardment by energetic particles such as ion and high-energy neutrals reflected from the target. Previous research in this area demonstrates that the film properties such as intrinsic stress, resistance, and refractive index can be controlled by changing the process conditions. For example, the film morphology and properties can be modified by varying the apparatus geometry (*5*), the substrate temperature (*6,7*), the inert gas species (*8*), the substrate orientation (*9*), the voltage bias (*10*), and the background gas pressure (*11*). Significant improvements in material and electrical properties can therefore be achieved by controlling the process parameters.

One of the major applications of the sputtering technique in semiconductor manufacturing is to deposit barrier or metal layers into a contact or via. For low-aspect-ratio contacts and vias, the step coverage of sputtered films is found to be reasonably good. As the critical dimensions of the electronic devices fall below 0.5 μm and the aspect ratios of these contact holes become large, the manufacturing technologies get more and more complex. Joshi and Brodsky (*12*) have demonstrated the use of a collimated sputter deposition technique that allows sufficient step coverage on a 0.25-μm contact. The low-pressure/long-throw sputtering technique has also shown promise in terms of producing reasonably good coverage at the bottom (*13*).

Sputter deposition is often used in the microelectronics industry because of its ability to produce dense films at temperatures that are lower than that required for CVD and evaporation processes. As a result, substantial research efforts have been devoted to develop an understanding of the sputtering process at the target (*14*). However, research on the growth of sputtered films under ion bombardment and with impurity incorporation is generally limited to experimental work. The mechanism of intrinsic stress formation and how the process conditions affect the step coverage by sputter deposition are not well understood. This is primarily due to the difficulty in correlating the process parameters and measured quantities with the microstructure of the film.

C. Discrepancy between Theory and Experiments

A satisfactory theory for the origin of intrinsic stresses in sputter-deposited thin films has not been found as yet. The experimental results with wide variations in data have not helped either. Primary reasons for this variation in experimental observations may be the nonuniqueness of the experimental methods and geometric configurations, and also the special character of the deposited thin films, for example, coatings with extremely small thicknesses. Even though experimental ambiguity exists, several models have been proposed to explain the origin of stresses in thin films and their creation mechanisms (*15–18*).

Murbach and Wilman (*15*) were the first to attempt to explain the internal stresses using a thermal model. According to them, the tensile stresses in a film condensed from vapor in a high vacuum are a result of the cooling of the deposit after its formation at the recrystallization temperature of the metal to the temperature at which annealing occurs. Since a difference exists between the thermal expansion coefficients of the solid and liquid, the film, which must pass through a liquid state during condensation, may gain either the compressive or tensile stress because of the phase change phenomena. The intensity of the tensile stress induced must therefore depend on the difference between the final deposition temperature and the recrystallization temperature. They also suggested that the mechanism that creates the compressive stress results from the codeposition process leading to an occlusion or intercrystalline adsorption of foreign materials in the metal deposits. Two serious objections have been raised to this theory. First, it is always found that the stress in individual crystallites is a function of the island size and, secondly, the temperature rise for the film in comparison to the substrate is very small and any heat gained from the energy particles is dissipated very quickly. Hoffman and coworkers (*16*) have shown that much of the stress is annealed out if the film is heated to a temperature above that which the substrate was held at during the deposition process. The argument is that a high density of imperfection, such as vacancies in the film, can cause either tensile or compressive stress.

Klokholm and Berry (*17*) tried to explain the intrinsic stresses in evaporated metal films by the annealing of a disorder model. According to them, the intrinsic stresses in metallic films may be caused by the mismatch at the film/substrate interface as well as the annealing and shrinkage of the disordered material underneath the surface of the growing film. However, under these conditions the intrinsic stresses are always tensile. The explanation of the exceptional case, that is, the creation of compressive stress, is lacking in this theory.

Hoffman and Thornton (*18*) later attempted to describe the intrinsic stresses in sputter-deposited thin films with the help of the theory of atomic peening mechanism. They studied the stresses of the chromium films grown in a magnetron-type dc sputtering chamber at a low pressure and bias sputtering, and postulated that the compressive stresses are caused by the atomic peening of the deposited film under the impact of high-energy atoms (energetic neutral from the target) or ions. The energetic ion or atom being deposited above the substrate brings the previously deposited atoms closer to each other by the cascade collision. Point defects such as vacancies are eliminated and the film becomes denser. The stress in the film is then compressive due to a compact microstructure. An examination of the sputtered chromium films, however, showed no evident change in microstructure or chemical contamination to account for the transition from tensile to compressive stresses. This comprehensive work with Al, V, Zr, and Nb (*19*) has also indicated that a small amount of background gas (argon or krypton) can be trapped in the film during the deposition process, a phenomenon not observed earlier.

Even though there is no evidence in the Hoffman and Thornton work to show that the compressive stress may be caused by the gas impurity, the entrapped gas is widely cited as the major cause for compressive stresses in sputtered metal films (*10,20,21*). For example, the stresses of the rf diode sputter-deposited TiC film are believed to be related to the trapped argon concentration in the film (*22*). Annealing of the film for 1 h at temperatures up to 500°C does not decrease the compressive stress, even though the Poisson compressive stress due to surface preening during the deposition process should decrease since the film is free to expand orthogonally in the plane of the substrate. The immobile gas impurity at this temperature (1/4 of the melting temperature) may be the main reason for the film to hold compressive stresses, although this theory has never been substantiated.

Thin film structures were first classified by Movchan and Demchishin (*23*) for thick metals and oxide deposits. In terms of T/T_m, where T is the substrate temperature and T_m is the melting temperature of the coating, three characteristic structural zones were identified. Zone 1 ($T/T_m < 0.3$) is associated with the shadow effect due to the roughness of the substrate and the growing film. In this range, the adatom has very little surface mobility and intends to stick to the place it arrives. This results in a preferred growth direction very similar to the coating flux. The film then consists of a columnar structure which is poorly bonded. Zone 2 ($0.3 < T/T_m < 0.5$) is characterized by an evolutionary growth due to the adatom intercrystalline boundaries, which consist of dense columnar grains with dislocations and defects. In the high-temperature zone 3 ($T/T_m > 0.5$), the film consists of equiaxed grains and a bright surface resulting from the bulk diffusion processes such as recrystallization. Later, Thornton (*5*) identified a transition zone T between zones 1 and 2, where the film consists of a dense array of poorly defined fibrous grains that may have been suppressed by the ion bombardment. The use of an additional variable, the working gas pressure, explains the influence of energetic particle bombardment (*18*). The film structures in various zones have been found to depend on the interplay of the atom shadowing, adatom diffusion, and surface and volume recrystallization. The zone concept has been quite useful in classifying the film structures, and the film properties have been defined by the structure features of the corresponding zone.

It is evident that extensive experiments have been performed to study the film growth behavior and intrinsic stresses in the film, and numerous theories have been proposed to explain the stress phenomena. However, the answers are still not satisfactory and the theories are always in question.

D. Simulation of Thin Film Deposition

In the study of sputter deposition processes, it may sometimes be difficult or even impossible to carry out experiments at all temperatures and pressures. For example, the study of the subtle details of fast ion in plasma seems to be unrealistic. The formation of the intrinsic stresses in the film at a submicron level is difficult to probe experimentally, but can be studied readily by computer simulations. The use of a computer model can also improve the efficiency of developing a new process or just modifying the existing ones. The resulting decrease in cycle time can cut the cost of development significantly. A computer model can greatly improve our understanding of the basic principle of the film growth mechanism so that an efficient and optimized process can be developed faster and at a lower cost.

In many applications including sputter deposition, the thickness of the film is below several thousand angstroms. Also, the mean free path of the deposited atoms in a low-pressure chamber can be comparable to the surface feature of the substrate. Macroscopic models based on the diffusion of species are therefore not suitable for a description of the relationship between the operating conditions and the properties of the film. Note that it is not the small surface feature but also the size of the grains and point defects which determine the macroscopic properties of the film. Simulations of the film growth and its structure using particle methods therefore seem to be more appropriate for sputter deposition processes. It is also possible to simulate the formation of the film under fast ion bombardment when a microscopic model is used. Here, we present a molecular dynamics (MD) (*24*) model suitable for the study of sputter deposition of thin films.

E. Molecular Dynamics Modeling

Several computer simulations of thin film nucleation and growth have been reported in recent years. Henderson *et al.* (*25*) and Kim *et al.* (*26*) reported the deposition of three-dimensional hard spheres onto a surface with subsequent relaxation to the nearest cradle. Brett *et al.* (*27*) have allowed the incident particle to relax to a site of lowest potential energy to study step coverage and microstructure. These models allow the individual evaporated particles to diffuse to the nearest cradle site. Simulations including various adatom relaxation schemes have been reviewed by Dirks and Leamy (*28*) and Leamy *et al.* (*29*). These models predict the formation of columnar structure, based on the self-shadowing and column orientation with respect to the substrate surface according to the tangent rule.

The only model to simulate the stresses in sputter-deposited thin films by Müller (*30*) needs to be mentioned here. He studied the relationship among the intrinsic stresses, the incident energy, and the ion bombardment for a high melting temperature material. His calculations predicted the microstructure and tensile stress in the film for incident adatom energies up to 2 eV. The columnar structure and voids within the films caused by the shadow effect and cone competition growth helped in explaining the mechanism of the intrinsic tensile stresses at high working pressures. However, the formation of interstitials, which are believed to be the source of compressive stresses in the film, and their influence on the film structure and its properties were not predicted by this model.

In a molecular dynamics model, a potential energy function resulting from the interaction of the particles is defined first, from which the conservative forces for atoms can be evaluated. Next, the initial positions and momentum states are assigned to the starting set of atoms. The many-body system is then allowed to move for a small time and new positions and momenta are determined, providing the particle trajectories as a function of time. The sputter deposition process is an example of film growth in an open system, in which the particles can enter or leave. To model this process, three kinds of particles need to be included in the model: (1) atoms constituting the growing film and the substrate, (2) background gas atoms that represent a source of impurity to the growing film, and (3) energetic ions. The energetic ions are used to simulate the ion-assisted growth.

The molecular dynamics model presented here consists of the following:

- a potential energy function between interacting particles,
- an algorithm to calculate the conservative force which is derivable from the potential energy function exerted on a given atom by its neighbors,
- a scheme to integrate the equations of motion using conservative and frictional forces,
- an algorithm that allows the exchange of energy between the deposited atoms and a constant temperature heat bath,
- an algorithm that permits the expansion and contraction of the simulation volume at a constant pressure, and
- an algorithm to calculate intrinsic stresses of the film as a function of its microstructure, which can be highly irregular or very uniform depending on the background pressure, adatom energy, ion bombardment energy, substrate temperature, and impurity levels, etc.

By using this MD model, it is possible to study the relationship between the film properties and the microstructure more realistically. Even though the degree of quantitative agreement between the predicted values and experimental data may be limited by the size of the model (the number of particles considered in a MD simulation is generally small), the qualitative information obtained from this model can be highly valuable.

This MD numerical model can be used to examine the growth of thin films in two or three dimensions. Since a large number of particles must be considered in a simulation cell to model a thin film deposition process realistically, this poses the challenging task of developing an efficient and cost-effective algorithm. Following are the primary considerations that must be made in developing an efficient numerical algorithm to simulate the sputter deposition processes:

- a higher order discretization scheme with nonuniform time interval to perform efficient integration,
- a fast search process for the neighbor atoms, which updates the table locally,
- a robust temperature control algorithm to allow fast interaction between the deposited atoms and the substrate acting as a heat bath,
- possibility of a deformation between the film and the substrate due to the external conditions such as the applied stress and/or temperature,
- a methodology to characterize the local stress phenomena in terms of the local microstructure of the film,
- an appropriate scheme to calculate the average stress in the film, and
- provision for the ion bombardment and the gas impurity in the model.

Each of these items is given special attention, resulting in significant improvement in many aspects of the molecular dynamics modeling of thin film formation phenomena. Although the model developed here is capable of predicting 3-D deposition phenomena, primarily 2-D computations are performed to study extensively the effect of process parameters on film structure and intrinsic stresses. We focus on the relationships among the microstructure, the adatom energy, the ion bombardment, the gas entrapment, and the stresses in thin films. Of particular interest are the phenomena of gas/ion reflection, penetration, and entrapment during the deposition process and their influence on film properties. Since the computer time increases exponentially with the number of atoms considered for the deposition, the size of the simulation cell strongly depends on the performance of the computer.

The fundamentals of the molecular dynamics model necessary to predict the film microstructure and intrinsic stresses are described in Sec. II and the simulation results for sputter deposition on plane substrates are presented in Sec. IV.

F. Modeling of Sputtering from the Target

The film microstructure and its properties strongly depend on the energy and incident angle of the atoms arriving at the substrate. It is, therefore, important to study the phenomena associated with the transport of atoms from target to the substrate through the background gas, and predict the conditions of these atoms arriving at the substrate surface. A particle method based on Monte Carlo calculations has been developed to study the effect of working pressure. The model can also be used to examine the effect of a collimating filter on the transport of the sputtered atoms. The details of this model are outlined in Sec. III and the results for selected cases/processes are presented in Sec. V.

II. Fundamentals of Molecular Dynamics

Here we describe the molecular dynamics model and its special features, the numerical algorithm, and the simulation scheme for sputter deposition of thin films. Computational results to investigate the suitability of various sorting and constant temperature control algorithms are also presented in this section. A stress parameter that can demonstrate the relationship between the local film stress and the microstructural phenomena, such as defects, voids, and gas impurity, is defined and a methodology to calculate the local and global stresses in the film is outlined.

A. Potential Function

It is assumed that the atoms interact with each other through a function $\Phi = (r_1 \cdots r_N)$ that describes the potential energy of a system of N atoms as a function of their positions. This interatomic potential is assumed to have the following form:

$$\Phi = \sum_{i} V_1(r_i) + \sum_{i<j} V_2(r_i, r_j) + \sum_{i<j<k} V_3(r_i, r_j, r_k) + \dots, \tag{1}$$

where r_i is the position of the ith particle and V_m is an "m-body potential." Presently, only the two-body potential functions have been considered to describe the interactions between the rare-gas atoms, simple metals, and ions. We use the Lennard–Jones potential function (*31*) to describe the strong repulsive and weak attractive forces between two particles:

$$U(r) = 4\varepsilon\left[\left(\frac{\sigma}{r}\right)^{12} - \left(\frac{\sigma}{r}\right)^{6}\right], \tag{2}$$

where ε, σ, and r are the potential well depth, lattice parameter, and distance between two atoms, respectively. Since the long-range attractive forces of the Lennard–Jones particles are weak, the interactive range of the potential function is often truncated in order to save computational time (*32*). By integrating the potential energies of N atoms within the truncated range expressed in terms of the nearest neighbor distance R_0 in a perfect crystal at zero temperature, an equilibrium value for σ as a function of R_0 can be obtained when the total potential energy function is minimal. The potential between the atoms of two different materials (a, b), for example, between the film and the substrate or the gas, can be expressed (*33*) by adjusting σ and ε:

$$\sigma_{ab} = 0.5(\sigma_{aa} + \sigma_{bb}), \tag{3}$$

$$\varepsilon_{ab} = \sqrt{\varepsilon_{aa}\varepsilon_{bb}}\ . \tag{4}$$

To consider the effect of ion bombardment, we use the Moliére potential (*34*), which describes the interaction between the ions and the atoms:

$$U(r) = \frac{Z_1 Z_2 e^2}{r}\phi(r), \tag{5}$$

$$\phi(r) = 0.35e^{-0.3r/a} + 0.55e^{-1.2r/a} + 0.10e^{-6.0r/a}, \tag{6}$$

$$a = 0.8853a_0/(Z_1^{1/2} + Z_2^{1/2})^{2/3}, \tag{7}$$

where a_0 is the Bohr radius, and Z_1 and Z_2 are the atomic numbers of the ion and the atom, respectively. The force, F_i, of particle i is then calculated from

$$F_i = \sum_{j=1,\, j\neq i,\, |r_{ij}|<R_c}^{N} -\frac{\partial U(r)\vec{r}_{ij}}{\partial r_{rj}\,|r_{ij}|}, \tag{8}$$

where r_{ij} is the distance between particles i and j, and R_c is the truncation range. Even though these potentials can be applied to any arbitrary configuration of atoms and ions, they do not describe accurately the systems other than the simplest closed-shell type of configurations such as the inert gases (*14,35*). It is believed that a two-body potential function incorporating the experimental results, for example, the embedded atom method (EAM) (*36–38*), can be more accurate. However, the experimental measurements of stresses in the sputter-deposited thin films demonstrate that the general phenomena of intrinsic stresses do not depend strongly on the type of the material and, hence, an idealized, simple two-body potential function can be successfully employed to develop a model for thin film deposition and to study its mechanical properties. Requirements of extensive computational resources when a more sophisticated potential is used also provide justification for a model based on simple two-body potentials.

B. Sorting Algorithms

In a MD model, each atom is allowed to interact with the surrounding particles through their potentials. It is therefore necessary to know the distance between each pair of particles so that the body forces acting on each one of them can be calculated. These force calculations generally account for a large portion of the computational time. To improve the speed of calculations, it is assumed that whenever the long-distance force among the atoms is weak, a short cutoff range can be used to limit the number of atoms that can interact with any atom. Computational efforts required to evaluate the forces on each atom are then reduced significantly. However, a large portion of CPU time is then spent on identifying and locating the particles that are within this cutoff range, at each time step. The commonly employed linked list method (*32*) and the method of light (*39*) are more effective in sorting than computing the distances between various particles. However, even if a linear sorting algorithm such as the linked list method is used to reduce the CPU time, significant computational effort is still needed to identify the neighboring atoms. To resolve this inefficiency in sorting at each time step, Verlet (*40*) introduced a table to store each particle's neighbor atoms and updated the table only at selected constant intervals. The calculation of forces between atoms is then performed by inspecting the neighbor table instead of looping around all atoms. Recently, the performance of the Verlet table in comparison with the linked list method has been well demonstrated for a closed system (*41*), especially for the case when the table is updated using either a conservative criterion (update the table at selected intervals) or a systematic method (*42*) (update the table when the maximum displacement exceeds the skin depth).

In an energetic system such as that of the thin film deposition, the problem of inefficient sorting is much more severe. For example, when an energetic atom is deposited above a substrate, the information on neighbors must be updated frequently because of the fast movement of the incident particles and the atoms that have collided. Indeed, whenever a neighbor table is used, the reconstruction of the table using a brute force algorithm requires substantial computations, particularly when the number

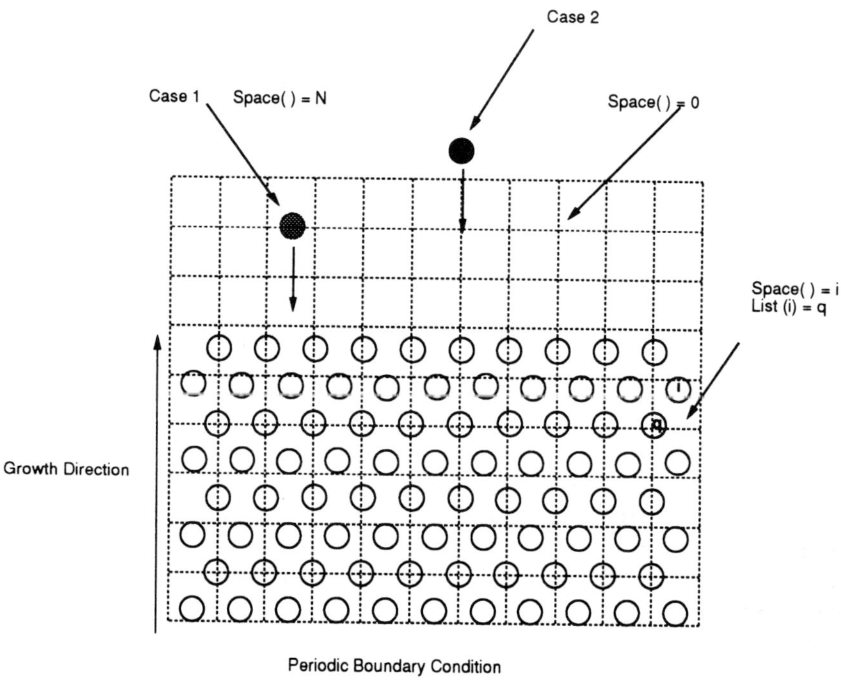

FIG. 1. Schematic of the molecular dynamics simulation cell. Case 1, an atom being deposited on a cradle site; case 2, an atom being deposited on top of another atom, an atomic site.

of atoms is large. It is therefore necessary to develop a method which updates the table efficiently.

A "position index method" has recently been developed that can update the table locally (*43*). In this scheme, only those atoms that have changed their neighbor atoms need to be relocated. Here, the simulation space is divided into a set of small cells (Fig. 1). These cells cover the possible positions of the atoms within the simulation space so that an open system can be studied by using a smaller amount of computer memory. For the calculations reported here, we have considered the length of the cell to be equal to the nearest neighbor distance. It is, however, possible that several particles may coexist within one cell. Within these cells, either the particles' numbers are stored or a specific number such as zero is given to indicate an empty cell. Also, a list is created to store the additional particle numbers whenever the cell contains more than one particle.

After the space cells are created, sorting is first carried out globally to create a complete neighbor table. Sorting of the neighbor atoms associated with a specific atom is performed by scanning the surrounding space cells of a center cell where this atom settles. The sorting process is made twice as fast by scanning the lower left side cells only. The force evaluation is then carried out using the neighbor table. After the new positions of atoms are obtained by integrating the equation of motion, the space cells need to be updated (at each time step) to determine the new neighbor atoms. The cost to maintain the space array is negligible compared to the reconstruction of neighbor tables during the simulation. Since the space array is updated at each time step, the search for neighbor particles for an atom that have changed their cell indices can be carried out by scanning the surrounding cells. The neighbor list is therefore updated locally and the reconstruction of a complete neighbor list is no longer necessary. The method is believed to be useful whenever more than one atom is deposited simultaneously and/or the incident energy of the incoming particle is large.

The use of a neighbor table with the linked list and position index methods has been evaluated and compared with the Verlet's table method. In these simulations, the sorting range is selected first and the atoms interacting with a specific atom within this range are identified and stored in a table by using these sorting algorithms. The forces between atoms within the force truncated range are then calculated by considering the neighbor atoms in this table. The neighbor table is updated whenever an atom has crossed the skin depth (i.e., the difference between the sorting range and the force truncated distance). Figure 2 presents the performance of various algorithms with and without the neighbor table. It is interesting to observe that the Verlet's table method is superior to the linked list method without the neighbor table whenever the number of atoms is small. However, the linked list method is more efficient than Verlet's table if the number of atoms is large because the implementation of the linked list method is more computation intensive than the looping around of all particles. When the number of atoms is small, the brute sorting of Verlet's table does not require many computations. However, an increase in the number of atoms increases the computation time by $O(N \times N)$ in brute sorting. A linear algorithm such as a linked list method is therefore superior to Verlet's table method. The performance of the position index method is similar to that of the linked list method except that it is more efficient in sorting because it uses smaller scanning areas.

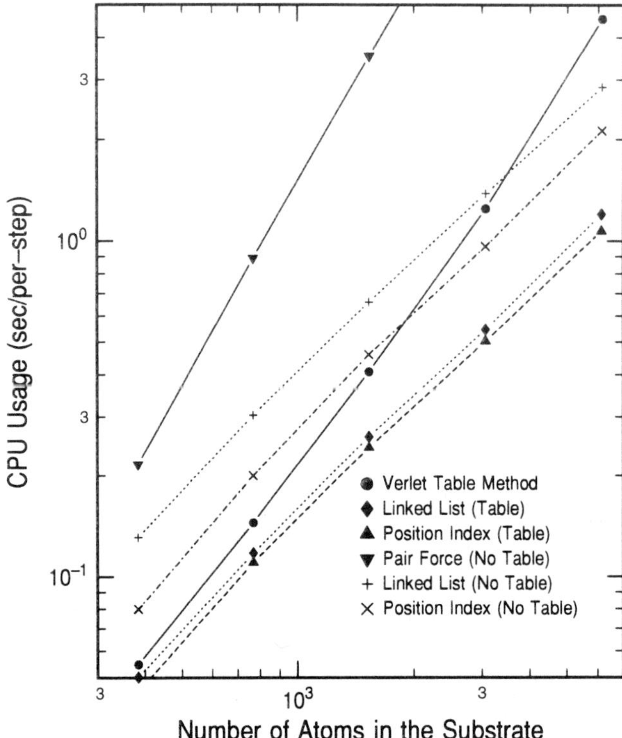

FIG. 2. Performance of various sorting algorithms. In each case, a 1-eV atom is deposited on the substrate. When the neighbor table is used, one sorting process is required to update the table approximately every 16 time steps (*63*).

The effectiveness of the linked list and position index methods compared to the neighbor tables is also shown in Fig. 2. The improved performance of these methods is primarily because sorting is no longer needed at each time step and is performed only when the movement of an atom is larger than the skin depth. The performance is even better in the case of the position index method since the neighbor information is stored within a space array, and there is no need to reconstruct the entire neighbor table whenever the atoms change their positions. The neighbor table can be updated locally by scanning the space array partially whenever the atoms change their cell indices. The position index method therefore provides an efficient sorting algorithm whenever a large number of atoms need to be considered in the MD simulation.

C. Discretization and Integration Scheme

The choice of an algorithm to integrate the equation of motion is generally based on the accuracy of the solution versus the efficiency. A reduced effort is desired for force evaluation during each time step as well as a faster approach to the equilibrium condition during the simulation. Verlet (*40*), Hockney and Eastwood (*44*), and Beeman (*45*) have presented second-order algorithms that are commonly used in MD simulations. They require only one force evaluation at each time step. Higher order methods are obviously more accurate, but more expensive. The Gear 4 algorithm (*46*) is a better choice for high resolution, but for reasonable accuracy Verlet-type algorithms are more efficient and more stable, even with large time steps. The Verlet, leap-frog (*44*), and Beeman schemes have been reported to produce similar orbits (*47,48*) and the leap-frog algorithm is considered to be more appropriate for MD calculations since it provides the best energy description and has lower memory requirements.

Since the incident energy of incoming particles in sputter deposition varies from 0.01 eV (thermal energy of sputtered atoms at a high working pressure) to several hundred electron-volts (energetic ions), it is advantageous to employ a nonuniform time interval scheme such that the time step is not limited by the velocity of the incident atom at the beginning of the deposition process, and can be increased whenever the incident atom or the collided atoms are less energetic. By using nonuniform time steps, the energetic collisions can be computed efficiently and the thermal relaxation can be completed faster.

In the valley region of the potential function, the second derivative of the Lennard–Jones potential with respect to the radial coordinate is positive, which requires that at least a third-order discretization be used in any numerical integration scheme. Using the forward and backward differences, we obtain

$$x_{n+1} = -\gamma^3 x_{n-1} + (1+\gamma^3)x_n + (1-\gamma^2)\,\Delta t_{\text{new}}\dot{x}_n + 0.5(1+\gamma)\,\Delta t_{\text{new}}^2 \ddot{x}_n + O(x_n^{iv}), \tag{9}$$

where $\gamma = \Delta t_{\text{new}}/\Delta t_{\text{old}}$, and the velocities can be approximated by

$$v_n = \frac{x_{n+1} - x_{n-1}}{\Delta t_{\text{new}} + \Delta t_{\text{old}}} + \frac{\ddot{x}_n}{2}(\Delta t_{\text{old}} - \Delta t_{\text{new}}). \tag{10}$$

This allows the use of nonuniform time intervals to improve the calculation speed by employing larger time steps whenever the largest velocities and forces become small. The time interval can be selected based on the criterion that the maximum displacement is never more than 0.02 lattice unit, L_U:

$$\Delta t = \frac{0.02 L_U}{V_{\max}}, \tag{11}$$

for which the energy fluctuation between a pair of particles interacting through the potential at the bottom is less than 1%. During the thermal relaxation period, one sometimes deals with a system for which $\langle F_i \rangle / m$ is large. Under such conditions, the time interval can be determined by

$$\Delta t = \left(\frac{2 * 0.002 L_U * m}{F_{\max}} \right)^{1/2}, \tag{12}$$

where $F_{\max}$ is the largest force acting on the particles. However, the use of this equation may cause large errors if the cutoff range of the truncated potential function is too small.

Five cases were considered to evaluate the effectiveness of the nonuniform time interval integration scheme. The average time step for each of these cases as a function of incident energy is given in Table I. In the case of a l-eV Ni atom deposited above the substrate, the time interval for the variable time scheme is determined by the temperature control algorithm

TABLE I

AVERAGE TIMES FOR THE CONSTANT AND VARIABLE TIME STEP SCHEMES FOR DIFFERENT INCIDENT ENERGIES (*43*)

Scheme	Type	Incident energy (eV)	Average time step (ps)
Constant	Ni	1	0.001861
Variable	Ni	1	0.001861
Constant	Ni	5	0.001712
Variable	Ni	5	0.001841
Constant	Ion	10	0.001013
Variable	Ion	10	0.001735
Constant	Ion	50	0.000435
Variable	Ion	50	0.001689
Constant	Ion	100	0.000320
Variable	Ion	100	0.001677

[Eq. (22)], to be discussed later, such that the temperature of the system can be held steady, and is actually fixed. When high-energy particles (5 to 100 eV) are deposited, the average time step in a constant interval algorithm is significantly smaller than that in the variable time scheme. In the case of energetic particles, the use of a constant time step generally results in a smaller time step and puts a heavy demand on computation if the evolution of the system requires a long time. The variable time scheme, on the other hand, allows larger time intervals to be used whenever the energetic collision process is completed and thermal relaxation has been achieved.

D. Constant Temperature Control Algorithms

Historically it was Andersen (*49*) who first proposed the idea of MD calculations using the constraints of constant temperature and constant pressure. In his constant temperature algorithm, the velocities of the randomly selected atoms are assigned using the Maxwell–Boltzmann distribution that corresponds to the collision with an imaginary external heat bath. Andersen's theory deals only with an isotropic system, for example, a fluid. A simple method to fix the temperature of a system is used that rescales the velocity of each atom at each time step by a factor $(T_{set}/T_{now})^{1/2}$ where T_{set} is the desired thermodynamic temperature and T_{now} is the current kinetic temperature (*50*). Velocity rescaling has proven to be a coarse method for solving a set of equations of motion that differ from the Newtonian ones even though the positions can be corrected accordingly (*51*). Several other schemes using constraint mechanics have been proposed. Nosé (*52,53*) used the extended system method to reformulate the Andersen theory in an elegant way. Further refinement of velocity rescaling was proposed by Berendsen *et al.* (*54*). In this method, the algorithm forces the system toward the desired temperature at a rate determined by t_T while perturbing the force at each molecular level only slightly (t_T is the time constant for coupling with the external heat bath). Wu and Friaut (*55*) later developed a conduction method to describe the isothermal boundary condition. Several other techniques to perform constant temperature MD calculations have been proposed and a number of them use the Brownian model to solve the many-particle Langevin equation (*56*).

To identify the fundamental processes in film growth, the energy relaxation and dynamics of incoming particles corresponding to the substrate temperature need to be understood and modeled properly. In laboratory experiments, it has been observed that the temperature of the film experiences a thermal spike whenever an incoming atom collides with the substrate particles. However the heat is conducted away very fast through the substrate and the temperature drops exponentially (*57*). To incorporate this phenomenon in the model, several different temperature control algorithms have been proposed. We have considered the generalized Langevin equation and extended system method to examine their strengths and weaknesses.

1. Generalized Langevin Equation. In the generalized Langevin equation (*58*) (GLE), the random force and friction terms are approximated in a way that permits a realistic description of the motion of particles:

$$\dot{x} = v(t), \tag{13}$$

$$\dot{v}(t) = -\int_0^t \zeta(t-t')v(t')\,dt' - \frac{\partial U(r)}{\partial r} + F_R(t), \tag{14}$$

where the coordinate of unit mass $x(t)$ moves in the Lennard–Jones potential well $U(r)$ and experiences a memory friction kernel $\zeta(t)$ and a random force $F_R(t)$. The random force originates from thermal motions of the particles inside the heat bath and is assumed to vanish in the mean, $\langle F_R(t)\rangle = 0$ and have a correlation, $\langle F_R(t')F_R(t)\rangle = \zeta(t-t')(\langle v^2\rangle)$. The mean velocity of the GLE particle (*59*) is then given by

$$\langle v^2(t)\rangle = \frac{K_B T_s}{m} + \left(\langle v^2(t')\rangle - \frac{K_B T_s}{m}\right)X^2(t-t'), \tag{15}$$

where

$$X(t) = L^{-1}\frac{1}{s+\hat{\zeta}(s)}. \tag{16}$$

Here, $\hat{\zeta}(s)$ is the friction in frequency space and L^{-1} denotes the inverse Laplace transformation. Equation (15) describes how the particles reach the equilibrium state. The time dependence of temperature T can be obtained from the derivative of the kinetic energy E_k:

$$\begin{aligned}\frac{dE_k}{dt} &= \frac{d\sum_i^N m_i v_i^2(t)/2}{dt}\\ &= \frac{d}{dt}\left(\frac{3K_B T_s}{2}\right) + \frac{d}{dt}\left\{\left(\sum_i^N \frac{m_i\langle v_i^2(t')\rangle}{2} - \frac{3K_B T_s}{2}\right)X^2(t-t')\right\}\\ &= \frac{3K_B}{2}\frac{d}{dt}\{(T_{t'} - T_s)X(t-t')\}.\end{aligned} \tag{17}$$

As usual, the memory function is taken as

$$X(t) = \frac{\beta_0}{\tau} e^{-t/\tau}. \tag{18}$$

For a small time step, Eq. (17) can be rewritten as

$$\frac{dT}{dt} = -2\beta_0(T - T_s), \tag{19}$$

which is similar to the expression obtained by Berendsen *et al.* (*54*). The global temperature coupling is then accomplished by

$$\dot{v} = -\frac{\partial U(x)}{\partial x} + \beta_0\left(\frac{T_s}{T} - 1\right)v, \tag{20}$$

which leads to an additional constraint force acting on the atom through a control variable ξ:

$$\xi = \beta_0\left(\frac{T_s}{T} - 1\right). \tag{21}$$

Equation (20) is used to control temperature and ξ is chosen to be proportional to the difference between the current temperature and the set value of the temperature bath. Using this temperature coupling condition, the GLE can be solved without implementing the random force, $F_R(t)$, distribution.

Using the difference method, Eq. (19) can be rewritten as:

$$T_{n+1} = (1 - \alpha)T_n + \alpha T_s, \tag{22}$$

where $\alpha = 2\beta_0\,\Delta t$. The stability condition for Eq. (22) requires that α be less than 1.0. Therefore, the time step Δt should be less than $1/2\beta_0$. Using the Debye model, Adelman and Doll (*60*) found the friction coefficient which leads to

$$\beta_0 = \tfrac{1}{6}\pi\omega_{\mathrm{D}}, \tag{23}$$

where ω_{D} is Debye frequency. Obviously, the GLE is a proportional control algorithm that changes temperature exponentially (increasing or decreasing) as a function of β_0 and the initial condition.

To examine the effect of constant temperature algorithm on deposition processes, the GLE is employed to calculate the temperature history of the incoming particle and the substrate with 768 atoms. Figure 3 presents the time history of the temperature for a 2-eV atom deposited above a cradle as well as an atomic site (Fig. 1). It is observed that the atom deposited

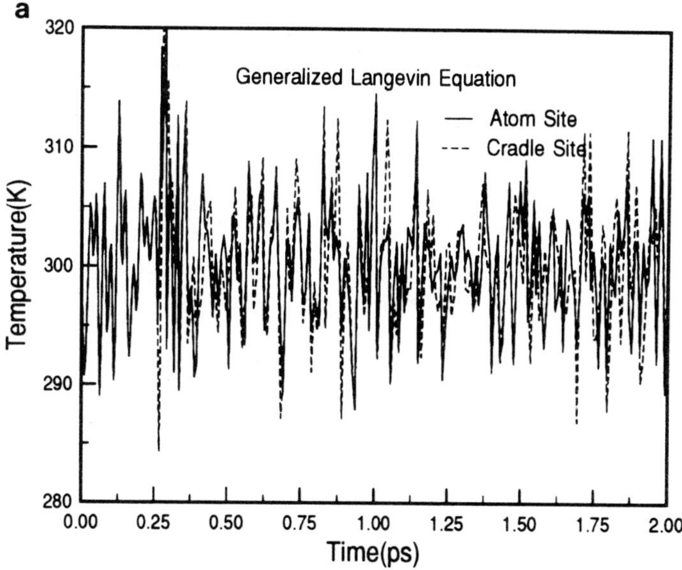

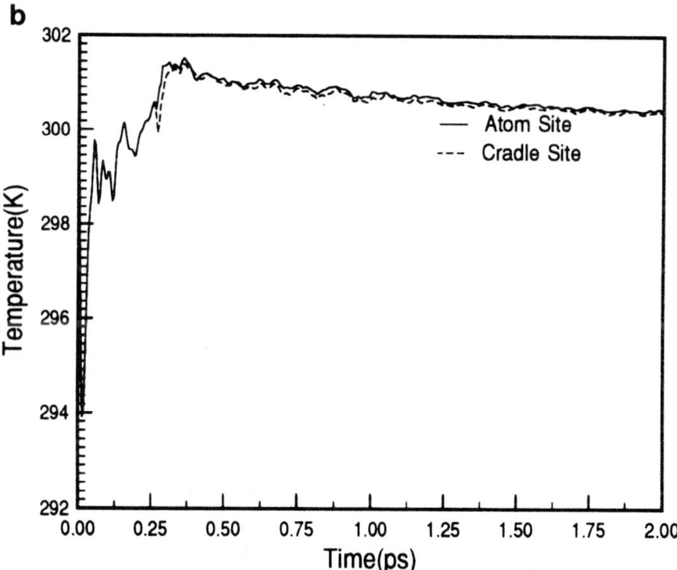

FIG. 3. (a) Instantaneous temperature of the 768 atoms and (b) time-averaged temperature of the 768 atoms (T_s = 300 K) (*63*).

above a cradle loses most of its kinetic energy much faster than the atom deposited above an atomic site. The temperature of the incoming atom deposited above an atomic site decays exponentially and diffuses into the stable cradle at a relatively slower rate. It is evident that the energy transfer process is very different in these two cases. Nevertheless, the average temperatures of both systems are almost identical (Fig. 3b).

In both cases, the temperature of the system increases after the energetic particle has collided with the substrate. The instantaneous temperature increases sharply and oscillates around the set temperature. The average temperature in each case shows a small deviation from the set temperature, T_s. Obviously, a proportional control algorithm forces the temperature of the system to decay fast to the set temperature. However, temperature deviations may exist in a small period of time due to the exponential decay behavior.

The GLE uses a proportional control algorithm to maintain the kinetic energy of the particle system constant. It follows from the heat transfer theory that the temperature of the system having a thermal spike will change exponentially even though the instantaneous temperature may fluctuate around the heat bath temperature due to the size effect of the anharmonic lattice. The average temperature of the simulated system is, however, higher than the set temperature and decays exponentially.

2. *Extended System Method.* In this method, the classical Hamilton equation is combined with the isothermal condition of the substrate and an additional parameter s' is introduced to account for the degree of freedom of the cooling system which is coupled to the physical system of N particles, with coordinate q_i', mass m_i', momentum p_i' and time t'. The equations of motion are then obtained as:

$$\frac{dq_i'}{dt'} = \frac{p_i'}{m_i}, \tag{24}$$

$$\frac{dp_i'}{dt} = -\frac{\partial U(q)}{\partial q_i'} - \frac{s' p_s' p_i'}{Q}, \tag{25}$$

$$\frac{ds'}{dt'} = s'^2 \frac{P_s'}{Q}, \tag{26}$$

$$\frac{dp_s'}{dt'} = \frac{\Sigma\, p_i'^2/m_i - gkT}{s'} - s' \frac{p_s'^2}{Q}, \tag{27}$$

where g is the number of degrees of freedom for the system, Q is the mass of the heat bath s', and p_s' is the momentum of s'. Finally, the motion of the particle can be obtained as

$$\frac{dp_i'}{dt'} = -\frac{\partial U(q)}{\partial q_i'} + \frac{p_i'}{Q}\int_0^t (T_s - T)\, d\tau. \tag{28}$$

Nosé's extended system method differs from the GLE and the heat conduction temperature control algorithms in many ways. In this model, the control variable ξ is given by (*61*)

$$\xi = \frac{1}{Q}\int_0^t (T_s - T)\, d\tau, \tag{29}$$

and is no longer dependent on the difference between the system and the set temperatures. In fact, it depends on the history of the temperature difference between the system and the heat sink. By integrating Eq. (28), the temperature of the system is found as

$$\ddot{T} = \left(\frac{\dot{T}^2}{T}\right) - \frac{1}{Q}T(T - T_s). \tag{30}$$

It seems that the Nosé extended system method uses the integral control mechanism to maintain the system temperature by constructing a stable fixed point T_s in the phase space (*62*).

By performing a fixed-point analysis of Eq. (30) at point $(T = T_s, \dot{T} = 0)$, the related temperature equation is obtained as:

$$\begin{aligned} \dot{T} &= y, \\ \dot{y} &= -\frac{T_s}{Q}T. \end{aligned} \tag{31}$$

The temperature around T_s is now dependent on the initial condition of the system, because the temperature torus in the phase space describes a harmonic oscillator having a frequency of

$$\omega = \sqrt{\frac{T_s}{Q}}\,. \tag{32}$$

To investigate the characteristics of the integral control algorithms, a 2-eV atom is deposited above a cradle site of the substrate. The instantaneous temperature and time-averaged temperature plotted in Fig. 4 show oscillations of instantaneous temperature after the incoming particle has

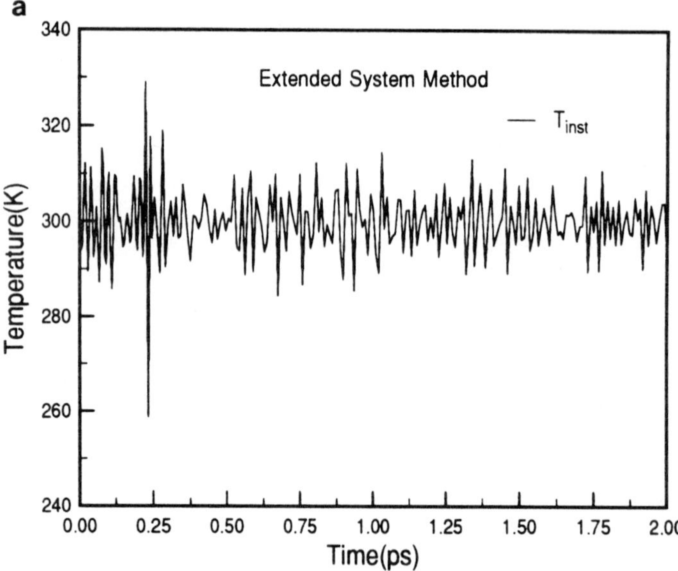

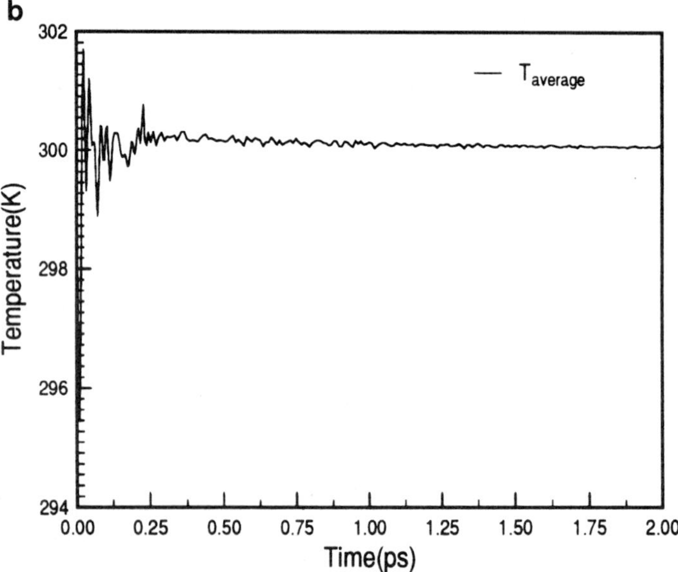

FIG. 4. (a) Instantaneous temperature of the 768 atoms and (b) time-averaged temperature of 768 atoms (case 1 uses a 2-eV atom); T_s = 300 K) (*63*).

collided with the substrate surface. Indeed, the average temperature also shows an oscillatory pattern instead of an exponential decay as observed in the experiments. These extraordinarily large variations in the instantaneous and average temperatures of the system indicate that the amplitude of oscillations is a function of the energy of the incoming particle.

In the extended system method, the temperature of the system increases initially as the particle is deposited, but it drops below the set temperature due to an energy exchange between the system and the heat sink. Even though a large fluctuation in the system temperature has been produced, the temperature of the deposited atom still decreases exponentially. However, the average temperature of the system is almost equal to the set value, T_s. Obviously, the integral control mechanism tries to compensate for the high-temperature disturbance by lowering the temperature of the system in order to minimize the accumulation error. It also shows that a large value of heat sink, Q, represents a slow response by the heat bath to the temperature disturbance. Theoretically, the final temperature of the system should not depend on mass Q. The amplitude of oscillation is, however, a strong function of the temperature of the heat bath, T_s, and its mass, Q.

3. Selection of an Appropriate Algorithm. It is evident that the generalized Langevin equation gives a reasonable description of the deposition process. The GLE can be derived from the proportional control algorithm using suitable values for the control variable. Nosé's extended system method, on the other hand, uses an integral control method to maintain a constant ensemble temperature. The proportional control algorithms are best suited for thin film deposition processes, particularly in the case of thermally evaporated (small perturbation) and sputter-deposited (large perturbation) thin films because the final temperature of the system is independent of the energy of the incoming particle. In contrast, the path of the temperature within the system, using Nosé's method, is determined by the energy of the deposited atoms above the substrate, even though the final temperature of the system is kept constant. The GLE is also found to be better than the traditional heat conduction method (*63*). It is therefore concluded that the generalized Langevin equation is more appropriate for simulating the deposition processes. In this scheme, the random force $R(t)$ and memory kernel $\zeta(t)$ can be selected appropriately to represent the physical condition of the system. Although the accuracy of and difficulty in implementing the random force $R(t)$ and defining $\zeta(t)$ may pose problems, the robustness of the algorithm and its application to ion bombardment, chemical reactions, and other energetic phenomena make it a better choice for a generalized MD model.

E. Periodic Boundary Conditions: Fixed versus Moving

To make accurate predictions of the stresses and deformations of the deposited film, the thermal expansion of the substrate with respect to its set temperature, T_s, must be considered. The zero stress state of the substrate avoids the preload deformation and stresses. Our preliminary calculations with fixed periodic boundary conditions showed that a perfectly deposited film might have large stresses because of an improper lattice space of the substrate (*43*). To overcome this artifact of the fixed periodic boundary condition, a movable boundary condition needs to be applied at the beginning of the simulation. This allows the boundary to expand or contract depending on the stresses within the substrate layer. After the substrate has achieved a zero stress state, the deposition can be carried out to study the mechanical properties of the film without any strain in the substrate.

Andersen's constant-pressure method (*51*) allows for isotropic changes in the volume of the simulation cell by coupling the system to an external variable. The constant-pressure condition therefore requires that the equation of motion be solved in terms of a scale factor which represents the size of the simulation box, allowing only an isotropic change in the volume. Parrinello and Rahman (*64*) extended this method to deal with the atomic solids that can change shape whenever the system experiences varying external stresses in different planes. Since this scheme allows for phase transitions that may involve variations in the simulation cell's dimension and orientation, the technique is particularly useful in describing the deformation of the solid. However, these constant pressure schemes do not account for the deformation between the film structure and the substrate due to the bilayer mode (film and substrate) of the growth. The scheme therefore needs to be modified to allow for stress relaxation during the deposition process, which requires that the relationship between the stresses in the film and the substrate be considered. This can be achieved by employing a further modified method of Parrinello and Rahman (*65*) where the simulation cell is allowed to change according to the variation in the external stresses. In this method, the stress caused by the force at the interface of the film and the substrate is described by (*66*):

$$\frac{\dot{P}_1}{P_2} = 3\frac{d_f}{(d_s + 3d_f)}, \tag{33}$$

where d_f is the thickness of the film, d_s is the thickness of the substrate, and P_1 and P_2 are the uniform external stresses applied to the film and the substrate, respectively. Introducing a matrix $h = (a, b, c)$ whose column vectors represent the 3-D simulation space, the real coordinates of particle i can be described by scaled coordinates s_i as

$$r_i = h \cdot s_i. \tag{34}$$

A homogeneous distortion of either the film or the substrate changes h_0 to h, and r_0 to r with

$$r = hs_0 = hh_0^{-1}r_0 \tag{35}$$

producing the strain,

$$\varepsilon_s = \tfrac{1}{2}\left(h_0^{T^{-1}}Gh_0^{-1} - 1\right), \tag{36}$$

where the tensor $G = h^T - h$. After the strain is identified, the elastic energy can be obtained as

$$V_{el} = p(\Omega - \Omega_0) - \Omega_0 Tr(S + p)\varepsilon_s, \tag{37}$$

where S is the external stress (either P_1 or P_2) and p is the isotropic hydraulic stress, which can be approximated as zero in the low-pressure chamber. In the limit of small strains, this leads to a Hamiltonian (*65*) from which the equation of motion can be obtained as

$$\frac{dv_i}{dt} = \frac{F_i}{m} - h^{T^{-1}}\dot{h}^T v_i, \tag{38}$$

$$W\ddot{h} = \Pi A_c - hh_0^{-1}Sh_0^{T^{-1}}\Omega_0, \tag{39}$$

where Π_i is the internal stress, A_c is the face across which the external and internal stresses act, Ω_0 is the original volume, and W is the virtual mass associated with the momentum of the scale matrix h.

During the simulation, the internal stresses of the film and the substrate are calculated at each time step. The internal and external stresses of the film are assumed to be equal. The external stress of the substrate is calculated by using Eq. (33). The transformation matrix h is then obtained from Eq. (39). This method allows the periodic boundary to either expand or shrink according to the internal stresses in the substrate and the force caused by these stresses in the film, and results in more accurate predictions of the stress.

F. Average Film Stress

The local stress in a specific plane of the simulation cell can be calculated from

$$\Pi_{A_i} = \Pi_{A_{i1}} + \Pi_{A_{i2}}. \tag{40}$$

Here, $\Pi_{A_{i1}}$ is the stress due to interatomic forces acting on the boundary plane A_i and is given by

$$\Pi_{A_{i1}} = \frac{1}{A_i} \sum_{R_s} F(N_r, N_l), \tag{41}$$

where R_s is the truncation range, and N_r and N_l are the atoms on right- and left-hand sides of the plane A_i, respectively. The stress $\Pi_{A_{i2}}$ is due to the momentum flow across the plane A_i, and can be described as

$$\Pi_{A_{i2}} = \frac{1}{A_i} \sum_{i=1}^{N_{\text{cross}}} \frac{m_i v_i}{\Delta t}, \tag{42}$$

where N_{cross} denotes the atoms which cross A_i during Δt. It is evident that this method (*67*) takes into account the interatomic forces and the momentum flux intercepted by an arbitrary plane, and the stresses are the time and spatial averages per unit area. Also, it can be applied to a nonhomogeneous system because the stress can be calculated for a local region averaged over short time intervals. Finally, the intrinsic stress in a simulation cell is given by

$$\Pi_s = \frac{1}{M} \sum_{i=1}^{M} \Pi_{A_i}, \tag{43}$$

where M is the number of planes within the simulation cell.

The external stress applied on the boundary is defined as:

$$P_b = \Pi_{A_b}, \tag{44}$$

where b is the boundary plane. However, in the case of a large number of planes M and uniform cross-sections, the intrinsic and external stresses become

$$\Pi_s = \frac{1}{V}\left\{\sum_{i=1}^{n} m_i v_i \cdot v_i + 0.5 \sum_{i=1}^{n}\sum_{j=1}^{n} F(N_i, N_j)\cdot r_{ij}\right\}, \tag{45}$$

$$P_b = \frac{0.5}{V}\sum_{i=1}^{n_L}\sum_{j=1}^{n_R} F(N_i, N_j)\cdot R_{\mathrm{Box}}, \tag{46}$$

where Box is the length of the simulation cell and n_L and n_R are the atoms on the left- and right-hand sides of the boundary, respectively. Since this method provides more information than the virial method, particularly when the stress at a specified cross-section is to be calculated, it should yield a better average (*43*).

G. Local Stress Function

To identify the origin of the stresses in the film, it is necessary to inspect the local stress around the impurity, point defects, grain boundary, etc. To accomplish this, the sum of the scalar product of virial, $F_{ij} \cdot r_j$ of particle i is calculated. A local stress function is then defined as the time average of the scalar virial over the interactive volume V_i of particle i:

$$\chi_i = \frac{1}{V_i\,\Delta t}\int_0^{\Delta t} \sum_{j=1,\, i\neq j,\, r_{ij}<R_c}^{N} F_{ij}\cdot r_{ij}. \tag{47}$$

The polygon (2-D) closed by the nearest surrounding atoms is used to represent the interactive volume for the particle and is obtained by triangulating the space of the atoms using the Voronoi diagram (*68*).

By examining the value of χ_i, the stress around the point defects such as the voids and impurities can be identified. The individual contributions of different types of point defects and grain boundaries can then be separated from the average stress of the film. This makes it possible to study the creation mechanism of the intrinsic stresses in the film. It is important to note that the stress resulting from either the mismatched thermal expansion between the film and the substrate or due to the artifact of the fixed boundary condition can be identified only by studying the local stress behavior. The local stress function is very helpful in studying the evolving film microstructure and stress distribution.

III. Transport of Sputtered Atoms from Target to Substrate

Since the properties of a thin film strongly depend on its microstructure, which is directly related to the incident angle and energy of the adatom, the effect of process parameters on sputtered atoms arriving at the substrate needs to be evaluated for more accurate predictions. It is very difficult to measure the characteristics of the adatoms (particularly the energy distribution) experimentally. A particle method based on Monte Carlo calculations (*69*) has therefore been developed to study the transport of sputtered atoms in a sputtering chamber and to investigate the effect of pressure on adatoms arriving at the substrate surface. Both the kinetic energy and incident angle of the atoms reaching the substrate after passing through the background gas are calculated during the thermalization process.

We assume that the initial energy distribution of sputtered atoms follows Thompson distribution (*70*) with cosine emission at the target:

$$\Xi(E, \phi)\, d\theta\, dE = Y \cos\phi \frac{1 - [(E_b + E)/\Lambda E_l]^{1/2}}{E^2(1 + E_b/E)^3}\, d\theta\, dE \qquad (48)$$

with

$$Y = \frac{\pi a^2 \Lambda E_a \kappa D \varphi_l}{16}, \qquad \Lambda = \frac{4M_1M_2}{(M_1 + M_2)^2}, \qquad (49)$$

and

$$E_a = \frac{2E_R(Z_1Z_2)^{7/6}(M_1 + M_2)}{eM_2}, \qquad (50)$$

where ϕ is the ejection angle, E_l is the energy of the incident ion, M_1 and M_2 are the respective masses of the ion and targets, ΛE_l is the maximum recoil energy, $a = a_0/(Z_1Z_2)^{1/6}$ is the screening radius of the interatomic potential ($a_0 = 0.53$ Å), E_a is the value of E_l that gives the distance of closest approach in a head-on collision, $\kappa = 0.52$, D is the nearest neighbor spacing in the target, E_R is the Rydberg energy, and φ_l is the flux of incident ions perpendicular to the surface of the target.

During the simulation, a particle is first ejected from the target with a predefined erosion profile (*71*) given by the Thompson energy distribution. An energy-dependent mean free path λ is then calculated to determine the position where the collision of this atom with the background gas will occur. The collision between the ejected atoms and the background gas is

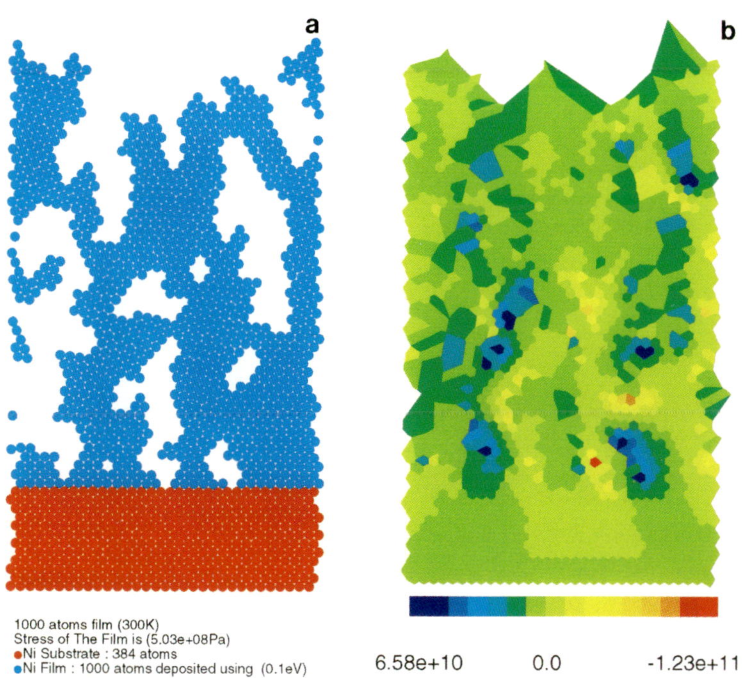

FIG. 5. Thin film (a) structure and (b) stress field χ for 0.1-eV nickel atoms deposited on a substrate maintained at 300 K.

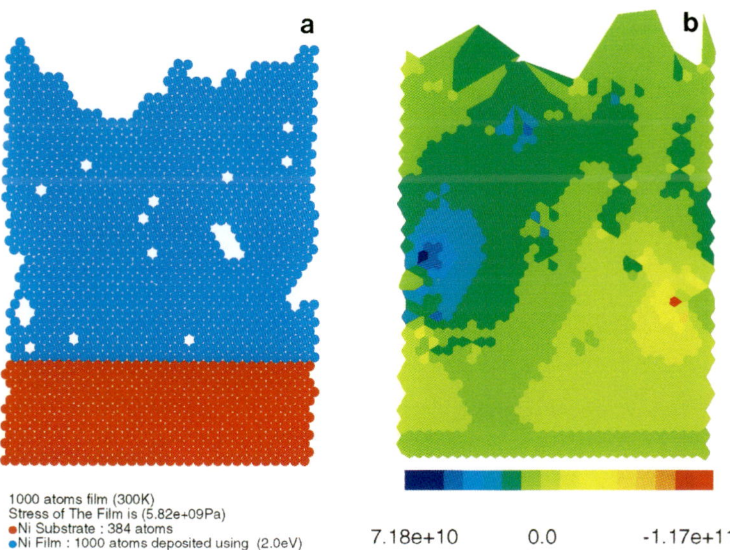

FIG. 6. Thin film (a) structure and (b) stress field χ for 2.0-eV nickel atoms deposited on a substrate maintained at 300 K.

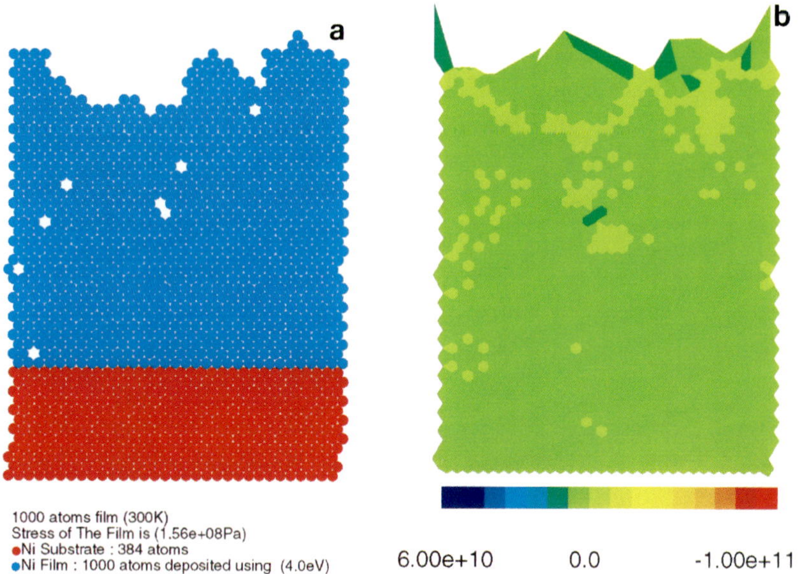

FIG. 7. Thin film (a) structure and (b) stress field χ for 4.0-eV nickel atoms deposited on a substrate maintained at 300 K.

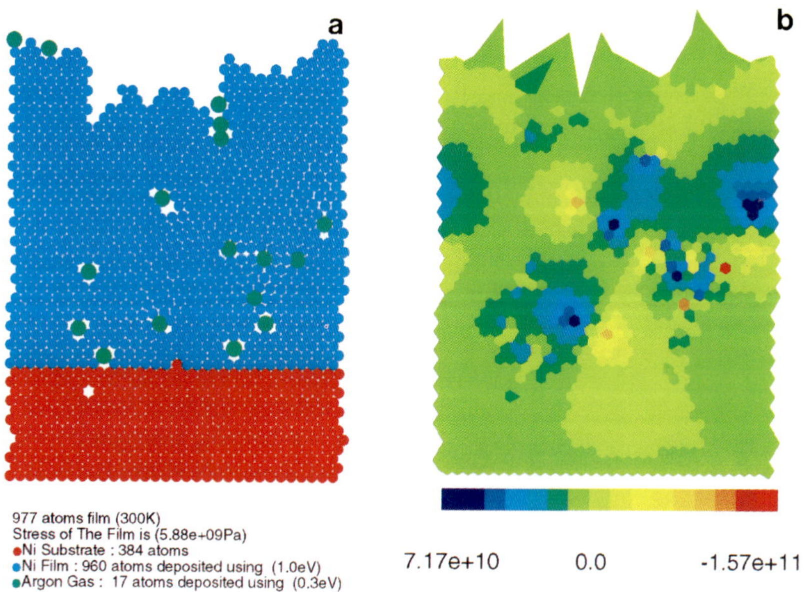

FIG. 9. Thin film (a) structure and (b) stress field χ for 0.1-eV nickel atoms deposited on a substrate maintained at 300 K, in the presence of 0.3-eV argon gas atoms and bombarded by 50-eV argon ions.

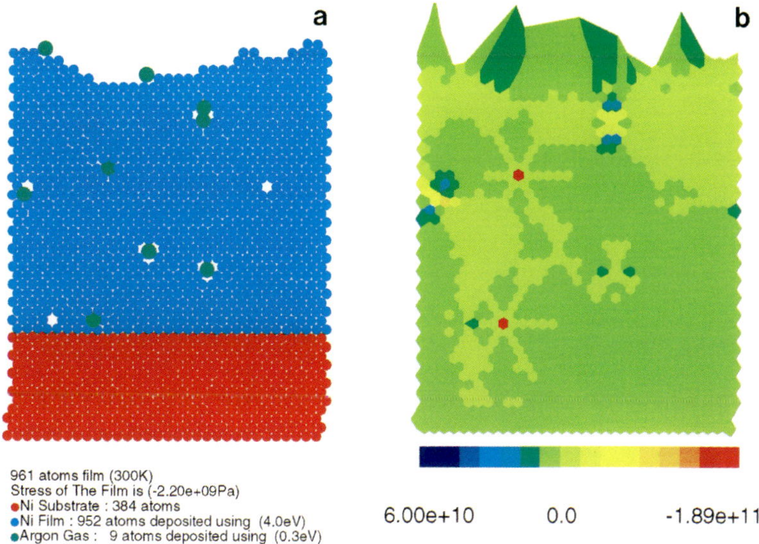

FIG. 10. Thin film (a) structure and (b) stress field χ for 4.0-eV nickel atoms deposited on a substrate maintained at 300 K, in the presence of 0.3-eV argon gas atoms and bombarded by 50-eV argon ions.

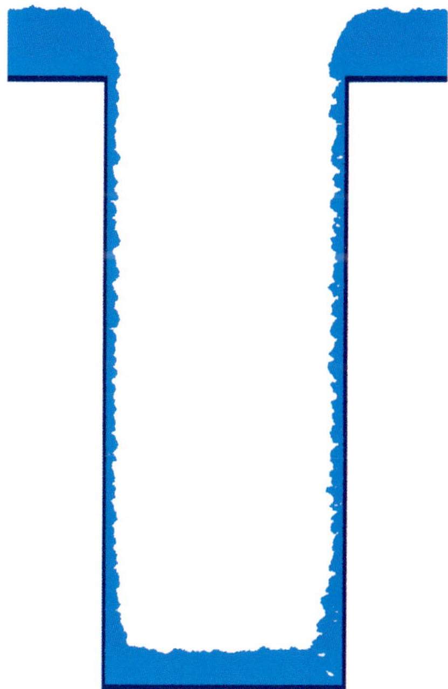

FIG. 22. Simulation of Ti deposition with a throw distance of 150 mm.

assumed to be a random process described by the Poisson distribution (*69*). The distance, l, between each collision is obtained from

$$l = \lambda \ln|R_n|, \tag{51}$$

where R_n is the random number between 0 and 1.0. From the kinetic theory of gases and considering that the ejected atoms collide only with the background gas atoms, it is known that λ will increase with the increasing relative speed V of the colliding atoms:

$$\lambda = \frac{M_g V^2}{2\sqrt{\pi}\, P p_{\max}\, \psi(x)}, \tag{52}$$

where P is the background gas pressure, and $p_{\max}$ is the maximum impact parameter:

$$\psi(x) = x \exp(-x^2) + \frac{(2x^2+1)}{2} \operatorname{erf}(x), \tag{53}$$

$$x = V\sqrt{\frac{M_g}{2K_B T_{Ar}}}. \tag{54}$$

Here, T_{Ar} is the temperature of the background gas, M_g is the atomic mass of the gas atom, K_B is Boltzmann's constant, and erf(x) is the error function. To obtain the maximum energy-dependent impact parameter, $p_{\max}$, two different approaches can be adopted, depending on the requirement of the precision and computational efficiency.

In the first scheme, the relationship between the impact parameter, the vertical distance between the trajectories of two molecules, and the reflection angle (in the center of mass frame) during the collision can be calculated from (*72*)

$$\theta = \pi - 2p \int_{r_0}^{\infty} \frac{dr}{r^2}\left(1 - \frac{p^2}{r^2} - \frac{U(r)}{E_c}\right)^{-1/2}, \tag{55}$$

where $U(r)$ is the atomic potential, E_c is the energy available in the center of mass system, p is the impact parameter, and r_0 is the distance of the closest approach obtained from

$$1 - \frac{U(r_0)}{E_c} - \left(\frac{p}{r_0}\right)^2 = 0. \tag{56}$$

It is obvious that there is no limit on the value of the maximum impact parameter, p_{max}. However, we define the maximum impact parameter at each energy level when the scattering angle is equal to 2°, and obtain it from Eq. (55). Because of the complexity of the potential function, $U(r)$, the solution of Eq. (55) is carried out numerically. The integration of Eq. (55) to obtain p_{max} is considered to be time consuming, especially when hundreds of thousands of collisions need to be taken into account. Calculation of the deflected angle during the collision course as a function of energy and impact parameter is also a time-consuming task. One of the solutions to reduce the CPU requirement is to create a table for p_{max} as a function of energy, and a table for reflected angles as a function of the energy of incoming atoms and the impact parameter during binary collisions. Tables with arbitrary intervals (smaller interval at the low-energy axis) are used to account for the nature of the Thompson distribution (high-energy tail) to obtain a high precision.

Somekh (*73*), on the other hand, has proposed a different approach to improve the computational efficiency with reasonable accuracy. Instead of the integration of Eq. (55), he assumes a complete linearity between the reflected angle, θ, and the impact parameter, p, and prescribes the maximum impact parameter as 2.2 times its value when the deflected angle is $\pi/2$. The scattering angle in his method is then chosen randomly, assuming that the impact occurs uniformly over the entire cross-section:

$$\theta = \pi\sqrt{R_n}\,. \tag{57}$$

After the scatter angle is determined either from Eqs. (55) or (57), the energy transfer ratio, K_t, and the direction of the scattered target atom, ϑ, can be obtained from

$$\begin{aligned} K_t &= \frac{4M_sM_g}{(M_s+M_g)^2}\sin^2\left(\frac{\theta}{2}\right), \\ \vartheta &= \tan^{-1}\left(\frac{M_s\sin(\theta)}{\left(M_g+M_s\cos(\theta)\right)}\right). \end{aligned} \tag{58}$$

The scattering procedure is carried out repeatedly until the particles arrive at the substrate (a cumulative total of $l\cos\vartheta$ gives the total distance traveled toward the substrate). The final energy and incident angle of each sputtered atom are then recorded.

IV. Deposition on Plane Substrates: Microstructure and Stresses

The MD model presented in Sec. II has been employed to study the microstructure and intrinsic stresses of sputter-deposited thin films for a wide range of parameters. The film is deposited with and without *in situ* argon ion bombardment and the background gas. The average adatom energy that is a function of the working pressure is allowed to vary in order to investigate its influence on film structure, followed by the simulation for a wide range of voltage bias.

The growth of sputter-deposited thin films is studied by performing calculations in two dimensions. A periodic boundary (either fixed or movable) is applied in the planar direction to minimize the surface effect and thereby to simulate more closely the behavior of an infinite system. An open surface is allowed in the growth direction so that new particles can be introduced (Fig. 1). Initially, 12 layers of substrate containing 12×32 atoms are arranged in a close-packed structure using the lattice constant at 0 K. The bottom 4 layers are fixed such that the interaction with the atoms in the bottom portion of the movable layers is more than complete. A perfect lattice representation of the moving layers at the substrate is then assured. The velocity of each movable atom is assigned using the Boltzmann distribution as a function of the substrate temperature.

Initially, the computations are performed with quasi-moving and periodic boundary conditions to obtain the zero stress state at 300 K. A fixed periodic condition is then applied and maintained during the entire deposition process. Once equilibrium is achieved for the substrate, an atom is introduced from the top of the substrate in each run. The initial horizontal position of the incident atom is generated randomly and the location of the atom above the previously deposited film and/or the substrate atoms is chosen in such a way that there is no interaction between the new incoming particle and the deposited film or the substrate atoms. The impinging angle of the incoming atoms is uniformly distributed within 0 to 10° measured from the direction normal to the substrate. The incident energy of the adatom is assigned from a Boltzmann distribution function having a desired mean value. During the deposition of an incident particle, we need to determine whether the incoming atom is trapped or desorbed from the surface in order to decide whether a new particle should be deposited or the force calculation should be carried out continuously. This is done by tracing the normal velocity of the incoming particle. It the sign of the velocity has been switched twice or more, it is assumed that the atom has been trapped. Otherwise, the calculation of the ejected incident atom has been trapped. Otherwise, the calculation of the ejected incident atom or the desorbed atoms in the film is continued until there is no interaction between the deposited (and/or substrate) atom(s) and the escaping atom. Even though this provision in our model increases the computational cost, it is essential to minimize the effect of a high deposition rate.

A. Effect of Adatom Energy

First, the effect of the atom's incident energy on film structure and intrinsic stresses is studied by depositing 1000 atoms in the absence of any background gas and the bombardment of ions. Computations are performed for mean incident energies in the range of 0.1 to 6 eV. The range of adatom energy considered here represents the working pressures generally employed in sputter deposition processes. Figures 5 through 7 present the film structure and the distribution of local stress function in the film for mean incident energies of 0.1, 2, and 4 eV, respectively. The density and average stress of the film as functions of incident energy are plotted in Fig. 8. Due to relatively small number of atoms in the film, the data show some scattering but the trend is very clear. Figure 5 demonstrates that the less energetic atoms (0.1 eV) deposited on the substrate produce a columnar structure and create large open voids within the film. This kind of

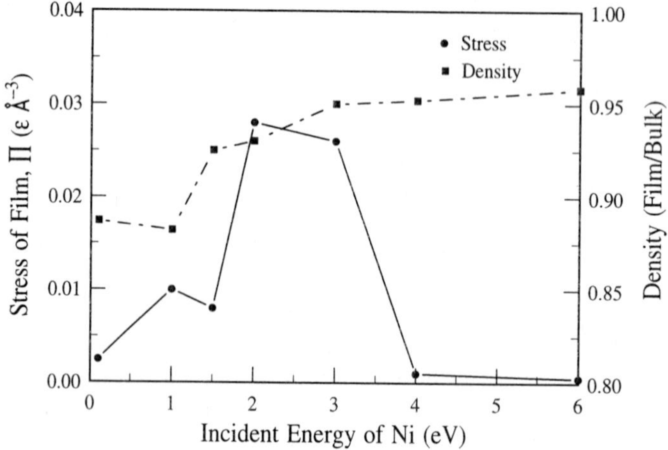

FIG. 8. Effect of incident energy on density and intrinsic stress of the Ni film.

structure is a result of the low thermal mobility as well as the shadowing effects of the previously deposited atoms. The density of the film is low and the intrinsic stress produced in the film is tensile (Fig. 8). The distribution of local stress (Fig. 5) indicates that there are small areas within the film where the stress is compressive but most of the film is under tension. Within the film (either at the top of the film or in the skirt of the columnar structure), the atoms can relax into the minimum stress configuration because of the weaker constraints, resulting in zero or low tensile stress in the film. On the other hand, the compressive stress is induced in the region where the film structure is dense. As shown in Fig. 8, the average intrinsic stress in the film is, however, tensile and very low.

As the adatom energy is increased, the columnar structure disappears and the large voids are eliminated. However, small defects such as vacancies still exist in the film at moderate incident energies (Fig. 6 for 2-eV adatoms). Note that the local stress around the vacancies is tensile. This is primarily due to the fact that the force interaction is more effective at this distance (the attractive force of the two-body potential is large before passing through the equilibrium position). The average stress in the film with these vacancies is tensile and large. At high incident energies (≥ 4 eV), the defects disappear to a large extent and a nearly perfect crystalline structure is produced due to the good thermal mobility of the incoming particles (Fig. 7). Consistent with the microstructure, the density of the film increases with the incident energy at a faster rate at a low energy level and much slower when the adatom energy is high (>3 eV). The density curve seems to exhibit an asymptotic behavior at high incident energies (Fig 8). Between 0.1 and 2 eV, the tensile stress in the film increases with the incident energy. It reaches a maximum between 2 and 3 eV and then starts decreasing with any further increase in the incident energy. The 4- to 6-eV data in Fig. 8 indicate that the stress becomes constant, almost an order of magnitude smaller than the maximum value when a perfect crystalline structure is obtained. The stress value in the 4- to 6-eV incident energy range suggests that a saturated state of the film structure has been achieved.

A similar study was performed by Müller (*30*) but only in the incident energy range of 0.01 to 1.5ε (0.01 to 2.00 eV) for 500 atoms impinging normal to the surface. His microstructure and average stress plots demonstrate similar behavior as observed in Figs. 5, 6, and 8. Experimental observations (*18*), however, indicate that the sputter-deposited films have compressive residual stress whenever either the working pressure is low or a large enough negative voltage bias is applied across the plasma and the substrate. To investigate this contradiction between the experimental data and the predictions, calculations were also performed to simulate the trappings of the background gas and the effect of ion bombardment. Impurity incorporation is introduced because it is well known that the films sputtered in an argon environment often contain several atomic percent argon. We postulate that since the argon cross section is significantly larger than that of the nickel, an introduction of the larger argon impurity atoms may distort the local arrangement of the nickel atoms in such a way that significant compressive stresses may result.

B. Combined Effect of Ion Bombardment and Gas Entrapment on Ni Film

The simulations are carried out with the concurrent ion bombardment using 50-eV argon ions as well as the presence of the background argon gas atoms. The ion-to-atom and gas-to-atom flux ratios are taken as 0.76 and 0.143, respectively. One thousand film atoms with 143 argon gas atoms and 760 ions are deposited on the substrate during this simulation. Here, we present the results for the case when both the gas impurity and ion bombardment are simultaneously incorporated into the process. Their independent effects were also studied and can be found in Ref. *74*.

Figures 9 and 10 show the resulting microstructure and local stress distribution for the incident energies of 1.0 and 4 eV, respectively. These plots demonstrate that a small number of argon atoms have been trapped in the film. This is because the ion bombardment inhibits the formation of columnar structure and the creation of large voids into which the argon impurities can be readily incorporated. Also, the energetic argon ions clean the weakly bonded argon gas from the film surface. At 1.0 eV, the average tensile stress is a maximum (Fig. 11) and the intrinsic stress now decreases whenever the structure becomes more compact (increase in incident energy). The stress is compressive when the average incident energy of Ni atom is larger than 1 eV. Under these conditions the film is quite dense (Fig. 11a). The amount of argon impurity in the film decreases with an increase in incident energy of Ni atoms (Fig. 11b). An inspection of the local stress field (Fig. 10) shows that the stress around the argon atom is compressive whenever it is tightly surrounded by Ni atoms. The resulting compressive stress in the film seems to be a combined result of the strong compressive stresses around the argon impurities and the zero stress of the perfect-like atomic structure in other regions. The magnitude of the compressive stress appears to be constant when the amount of argon is

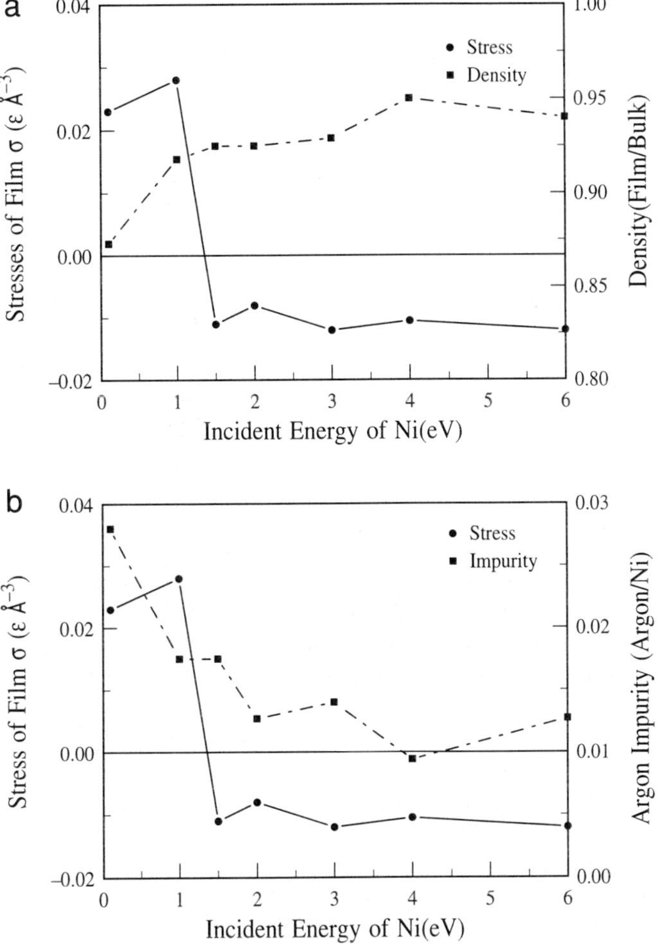

FIG. 11. Combined effect of ion bombardment (50-eV Ar ions) and gas impurity on (a) stress and density and (b) stress and impurity. The ion-to-atom and gas-to-atom flux ratios are 0.76 and 0.134, respectively (*74*).

kept constant (2 to 6 eV in Fig. 11). The transition from tensile stress at 1 eV to compressive stress at 1.5 eV is not associated with any significant change in the argon impurity.

By examining the deposition processes *in situ*, we observe that forward sputtering plays an important role in creating the tightly surrounded gas impurities. First of all, forward sputtering of either the film adatom or the trapped gas impurity into a void does not need as much energy as that required by complete sputtering. The gas impurity, for example, can then be easily trapped in a vacancy by forward sputtering. Secondly, the gas impurity can be tightly surrounded by the forward sputtered atoms, which fill the vacancies around the gas atom much more easily and provide a stable configuration than a direct implantation of the substitutional interstitials into the atomic site of the film. Note also that the low-energy ion bombardment is required to decrease the void contents and incorporation of the gas impurity, and pack the gas impurity firmly by forward sputtering. On the other hand, a high-energy ion bombardment, say 500 eV, may clean the gas impurity to a large extent, such that the film is almost free of stress, particularly compressive stress. However, such high levels of ion energy are known to damage the film surface.

Experimental results for the deposition of tantalum above the glass and tantalum substrates (*75*) and of Cu above the polyimide substrate (*11*) demonstrate similar trends (Figs. 12 and 13). As shown by the stress curves in Fig. 11, the stress rises and reaches a peek, and then drops sharply with a reduction in the working pressure. A transition from tensile to compressive stress has also been observed in these experiments. Our predictions are quite consistent with these experimental observations, and also agree with the results reported by Hoffman and Thornton (*8*). This is the first time that a theoretical model has predicted compressive stresses in the sputter-deposited thin films and the mechanism of the creation of tensile and compressive stresses can be explained in terms of the governing parameters and physical phenomena.

To further characterize the effect of the ion bombardment on gas impurity incorporation, calculations are also performed for 100 eV ions with adatom energy varying between 0.1 to 6 eV. An increase in ion energy results in a denser structure, fewer impurities, and larger compressive stress (Fig. 14). Here, the tensile stress is only produced at 0.1 eV instead of below 1.0 eV in the case of 50-eV ion bombardment. Note that under 100 eV ion bombardment, a larger compressive stress is associated with a decrease in the gas impurity (see the 0.1- and 1-eV data in Fig. 14b).

In conclusion, the magnitude of the compressive stress strongly depends on the amount of tightly trapped gas impurities within the film instead of the total amount of gas impurity. The well-known steep transition pressure observed in the experiments by Hoffman and Thornton (*19*) seems to be the result of a significant change in the amount of tightly trapped argon atoms in the film. Our study on the creation mechanism of intrinsic stresses confirms the theory of the atomic peening mechanism and the effect of the gas impurity (*21*). A columnar film structure produced at a

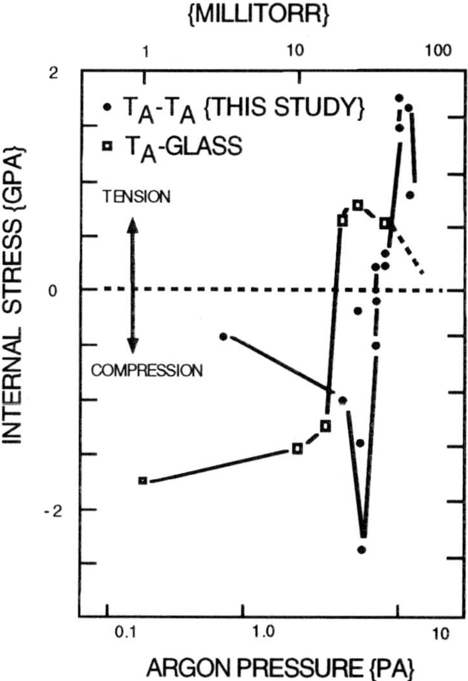

FIG. 12. Internal stress as a function of working pressure for tantalum sputtered in an argon environment and deposited on a tantalum foil using a planar magnetron and on glass wafers using a cylindrical magnetron (*75*). Note that the adatom energy decreases as the working pressure is increased.

high working pressure, with a large number of vacancies, has tensile stresses because of the attractive atomic forces acting across the voids. In the case of either the ion bombardment or low working pressure, the atomic peening of the energetic molecules on the film surface drives the film atoms closer and reduces the void contents. The gas atoms can then be bonded firmly with the concurrent ion bombardment, resulting in compressive stress in the film. In the absence of ion bombardment, it is unlikely that the gas atoms will be trapped into the film tightly. It is, however, believed that both the entrapment of gas impurity and the atomic peening mechanism strongly depend on the power input, working pressure, and mass ratio between the adatom and background gas, and play an important role in producing the intrinsic stresses in the sputter-deposited thin films.

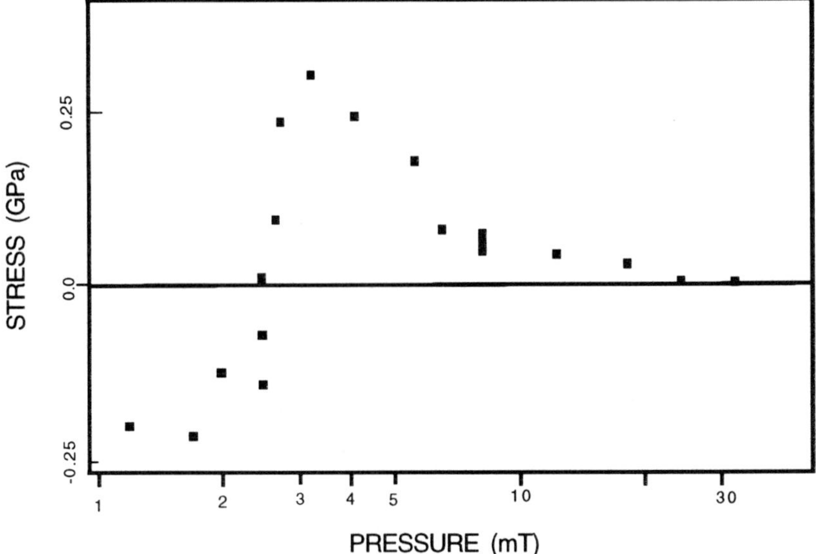

FIG. 13. Internal stress as a function of argon pressure for 0.25-μm Cu films deposited at 2 Å/s onto flexible polyimide sheets using a dc planar magnetron (*11*). Note that the adatom energy decreases as the working pressure is increased.

C. Effect of Voltage Bias on Film Stresses

In sputter depositions, the columnar morphology may be eliminated by applying a voltage bias to the substrate. It has been observed in the previous section that the density and stress of the film can be altered significantly by concurrent bombardment of ions. To study the effect of the voltage bias, the incident energy of Ni atoms is kept constant at several different values, between 0.1 and 2.0 eV. The ion-to-atom and gas-to-atom flux ratios are taken as 0.76 and 0.143, respectively. To characterize the effect of the voltage bias, the calculations are carried out for the ion energy varying between 10 and 400 eV.

At low ion energy (0 to 20 eV) with adatom energy of 0.1 eV, the film has a columnar structure and the density is low. The stress of the film is tensile and small (Fig. 15a). Also a large amount of impurity is found within the film (Fig. 15b). With a further increase in the ion energy (up to 200 eV), the columnar structure is destroyed by the energetic ion and the film becomes denser. The film is still under tension. It is believed that the resultant tensile stress as shown in earlier study (Fig. 5), is primarily due to

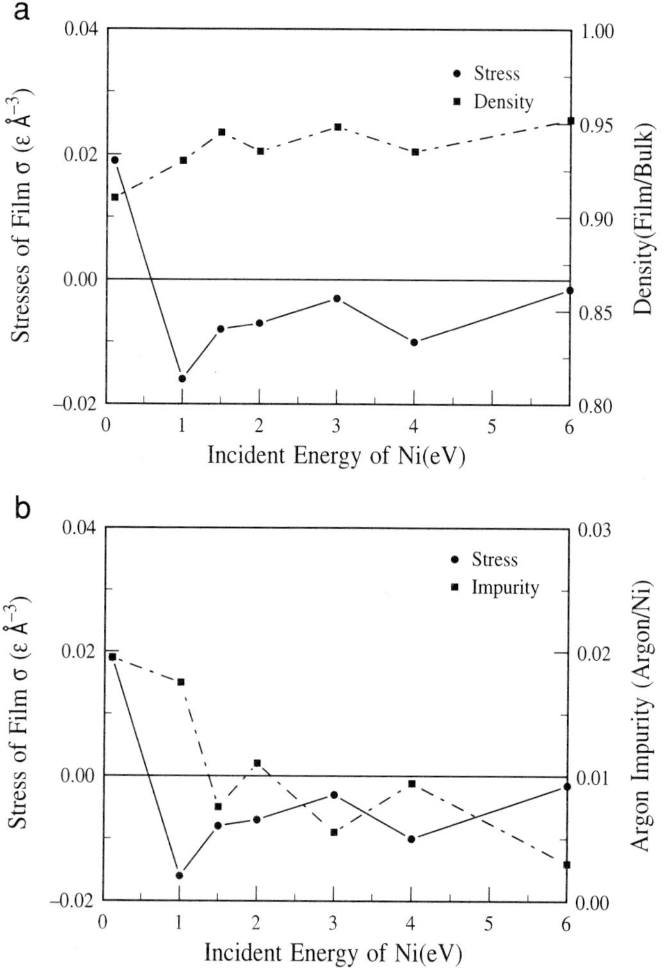

FIG. 14. Combined effect of ion bombardment (100-eV Ar ions) and gas impurity on (a) stress and density and (b) stress and impurity. The ion-to-atom and gas-to-atom flux ratios are 0.76 and 0.134, respectively (*74*).

small voids and vacancies in the film. An increase in the ion energy has, however, increased the density of the film structure, as a result of the cleaning of the gas on the surface and the compacted film structure. The ion energy when increased up to 400 eV causes a transition from tensile to compressive stress, even though this increase in bombardment energy (100

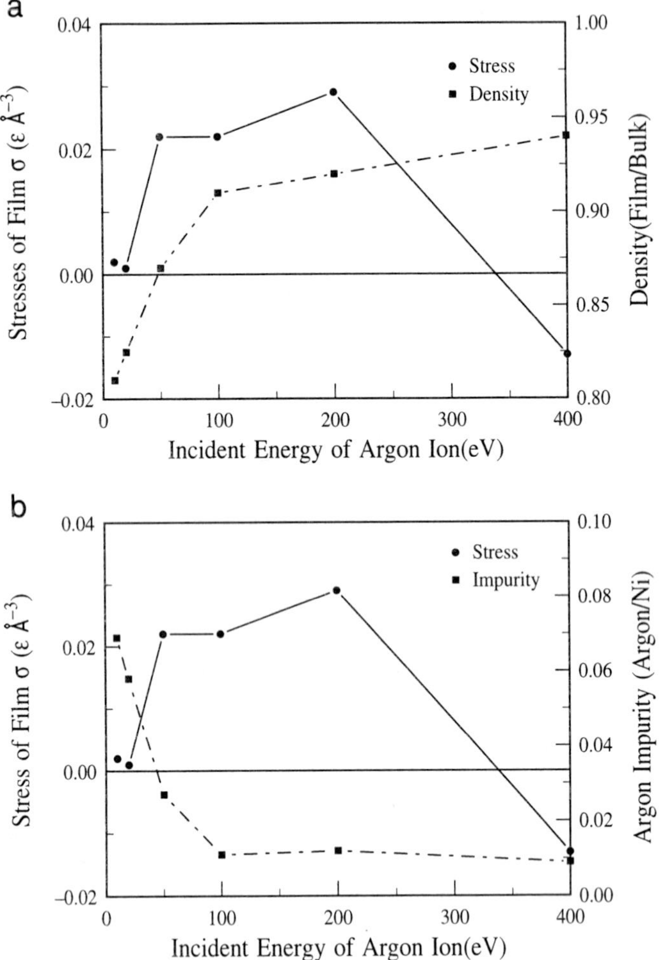

FIG. 15. Effect of ion bombardment energy (Ni adatom energy is 0.1 eV) on (a) stress and density and (b) stress and impurity. The ion-to-atom and gas-to-atom flux ratios are 0.76 and 0.134, respectively.

to 400 eV) does not reduce the impurity level significantly. As noted earlier, the film stress seems to depend strongly on the amount of the tightly packed gas impurity. At this Ni adatom energy (0.1 eV), the ion transition energy, which changes the film stress from tensile to compressive, is very high (>350 eV).

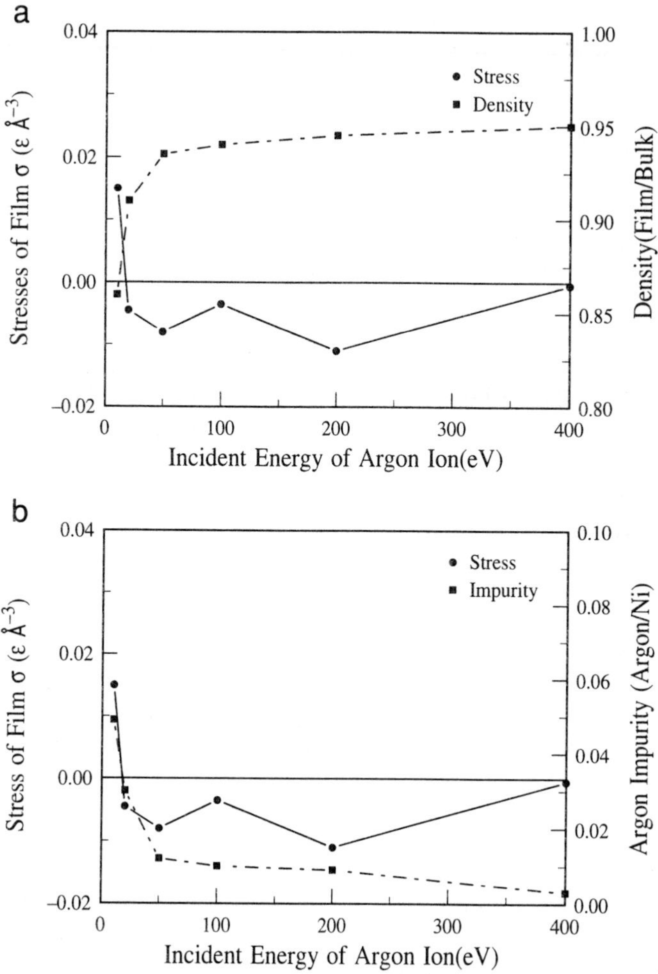

FIG. 16. Effect of ion bombardment energy (Ni adatom energy is 2.0 eV) on (a) stress and density and (b) stress and impurity. The ion-to-atom and gas-to-atom flux ratios are 0.76 and 0.134, respectively.

The simulation results for 2.0 eV demonstrate that the ion energy where the maximum tensile intrinsic stress occurs decreases from 200 eV at 0.1-eV adatom energy to 10 eV (Fig. 16). Also, the ion energy at which the transition from tensile to compressive stress takes place reduces from 350 to 20 eV, respectively (Figs. 15 and 16). In the case of adatom energy of

2.0 eV, the film density approaches the saturation value much more rapidly and the impurity level drops faster with an increase in the ion energy. The film has compressive stress whenever the incident energy of ion is large enough to pack the gas impurity tightly. However, note that when the ion energy is very high such that the gas atoms can be relaxed from a tightly packed site to a more stable deposition site or desorbed from the surface due to the cleaning effect of the energetic bombardment, the stress of the film no longer remains compressive. Finally, a perfect-like film structure is obtained with zero stress. This is shown by the upward moving portions of the stress curves in Fig. 16. These results therefore demonstrate that the intrinsic stress of the films can be controlled successfully by the use of ion energy, with depends on the voltage bias across the substrate. As shown in Fig. 17, the predicted behavior is quite consistent with the experimental observation of Bland *et al.* (*76*) and Kaminsky (*77*). A similar agreement between the experimental result and numerical simulation of Tungsten film has also been demonstrated by Fang *et al.* (*78*).

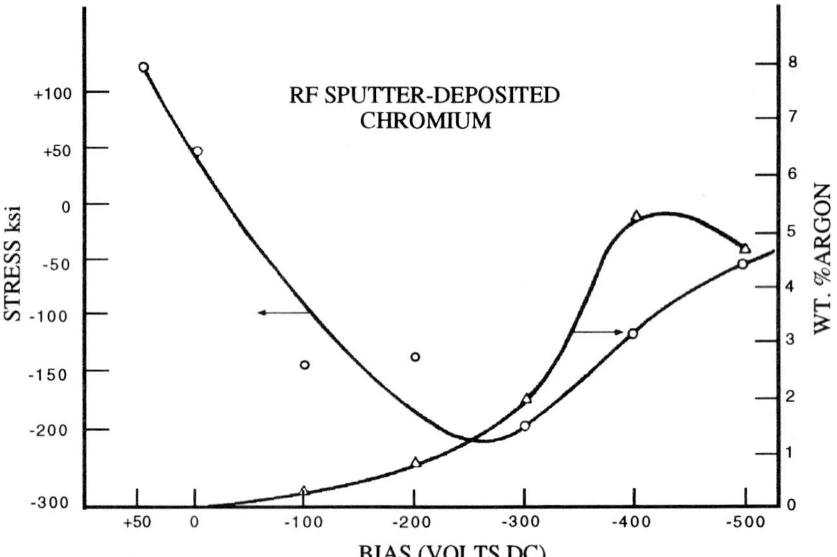

FIG. 17. Stress in a thick rf sputter-deposited chromium film as a function of the bias deposition (from Ref. *76*).

V. Step Coverage of a Submicron Liner

During the metallization of the IC fabrication process, Ti liners are widely used to improve the contact resistance with the underlying substrate in the contact or via-filling applications. TiN layers are then used to prevent the attack of WF_6 on the Ti layer. As the critical dimensions of electronic devices fall below 0.5 μm and the aspect ratio of these contact holes/liners is increased, it becomes extremely difficult to provide a sufficient coverage for the liner or to fill the contact hole completely. The use of collimated sputter deposition (*79*) and low-pressure/long-throw sputtering that can yield sufficient step coverage at both the bottom and the sidewall was demonstrated by Joshi and Brodsky (*12*) and Bulter *et al.* (*13*), respectively. However, the basic physics of these processes are not clearly understood. This poses a challenge to their potential applications. To develop a better understanding of these processes, we have combined the feature scale model with the tool-scale model (*80–82*). This extended model accounts for the important parameters governing the process, such as pressure, power input, etc., and describes the ballistics and energetics of the sputtered atoms during their transport from the target to the substrate, as well as the collimated sputter deposition of the film. The Monte Carlo method (Sec. III) is used to calculate the incident energy and arrival angle of the sputtered adatoms and neutrals above the substrate, as functions of pressure, target-to-substrate distance, collimator aspect ratio, collimator location, etc. A simplified MD model described in Sec. V.A.2 is then employed to determine the position, velocities, and forces of incident adatoms and energetic neutrals and to predict the evolution of the film structure and step coverage as the deposition takes place.

In these calculations, the temperature of the background gas is kept constant at 350 K and the distance between the target and the substrate is varied for various processes. The incident energy of the ion approaching the target is taken as 410 eV for Ti. The incident angle and energy of the sputtered atom approaching the substrate are then calculated as functions of the background gas pressure and throw distance. To obtain a good statistical average, 2×10^5 atoms are typically emitted from the target during the simulation.

The model is also used to examine the effect of the collimating filter on angular and energy distributions of the adatoms. A cylindrical filter is centrally located between the target and the substrate, and the particles sticking to the collimator surface are removed from the scattering process.

The final energy, position, and incident angles of each sputtered atom passing through the collimator are then recorded. Four different aspect ratios, 0, 1, 1.5 and 2, are used to examine the effect of the collimator on step coverage and to compare with the experimental results of Joshi and Bradsky (*12*).

A. Collimated Sputtering

1. Process Characteristics of Collimated Deposition. With the introduction of a collimator which is generally located between the target and the substrate, only the atoms with nearly normal incident angles can reach the substrate. For the current simulations, a 20-cm-diameter target is considered with a 0.5-cm-diameter collimator. First, the throughput of the collimated sputtering is calculated as a function of the working pressure and aspect ratio of the collimator. To study the effect of the collimating filter, an operating pressure of 2.8 m Torr and target-to-wafer distance of 8.2 cm are used first. At 2.8 m Torr, the use of a collimator of Ar = 1.0 results in an 80% decrease in the throughput. An increase in collimator aspect ratio from 1.0 to 2.0 reduces the Ti deposition rate (normalized with the noncollimated deposition rate) by 44%. This agrees well with the experiments as well as with the theoretical results of Hsieh and Joshi (*80*) (Fig. 18). Note that at this pressure, the mean free path of the incident particle is shorter than the throw distance (8.2 cm in this case) and the

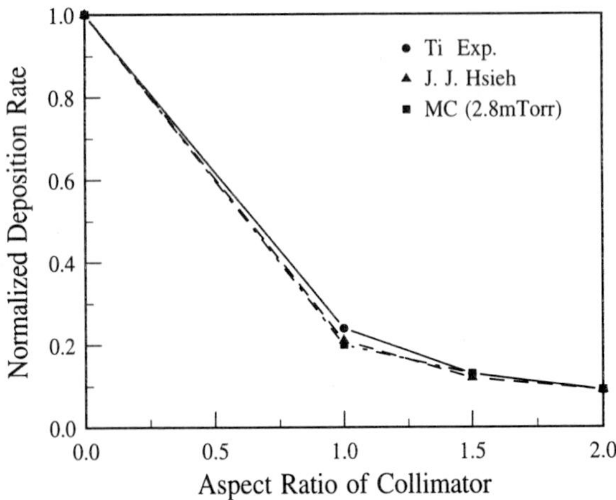

FIG. 18. Normalized deposition rate as a function of the aspect ratio of the collimator (*81*).

adatom experiences collisions with the background Ar gas before reaching the substrate.

The effect of collimator aspect ratio of average incident angle and energy is also investigated for a pressure of 2.8 m Torr. Without the collimator, the average incident angle of the adatom above the substrate is about 30° (Fig. 19). The use of a collimator results in an increase in incident energy because the collimator screens out low-energy adatoms that have oblique incident angles due to collisions with the background Ar atoms. Figure 19 also demonstrates a reduction in incident energy when the aspect ratio of the collimator is increased beyond 1.0. This reduction in average incident energy with an increase in collimator aspect ratio at both the high and low pressures seems to have been caused by the blockage of the directional atoms.

2. Step Coverage of Ti Thin Film in a Trench. To study the formation of a thin film at the submicron level, simulations of a large number of atoms need to be carried out, which makes the MD computations prohibitively expensive and difficult. To overcome this difficulty, a simplified MC approach has been adopted following the procedures of Xia *et al.* (*83*) and Müller (*84*). First, a Ti substrate is constructed based on the aspect ratio of the trench. Then, a full MD calculation is performed to obtain the proper lattice space at 300 K. The Ti adatoms, generated by Monte Carlo

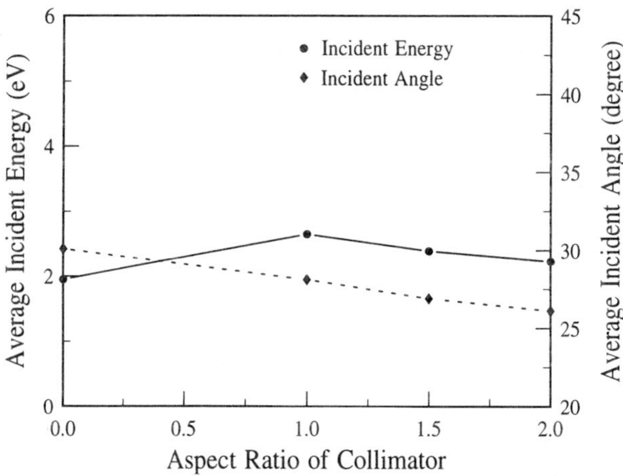

FIG. 19. Average incident energy of Ti atoms arriving at the substrate surface as a function of the aspect ratio of the collimator.

calculations and passed through the collimating filter, are then deposited on the substrate. The nuclear scattering theory (*85*) is used to determine the energy loss and the deflection angle of the incident atom when the adatom approaches the trench surface. The energy loss/transfer to the collided atom in the substrate/deposited film is obtained as a function of the deflection angle. If the remaining energy of the adatom is large enough to overcome the local activation barrier (*84*), the atom may move and can be treated as a newly arriving atom. When the remaining kinetic energy is less than the energy barrier, the mobility parameter M, which measures the maximum number of hopping for the adatom at current temperature, is calculated to determine the number of hopping to the lowest energy neighbor site. During the present simulations, 60,000 particles are deposited as a 512×400 lattice unit substrate ($0.15 \times 0.11\ \mu$m) to study the film profile and its characteristics.

Figure 20 compares the experimental data with predictions of bottom step coverage as a function of the collimator aspect ratio for a nontapered trench with an aspect ratio of 2.5. Two methods of predictions were used for comparison. The first method uses unity sticking probability together with the flux distribution obtained from the Monte Carlo calculations at 2.8 m Torr. The predicted trend for trench bottom coverage is qualitatively similar to that obtained from the experiments, that is, an increase in collimator aspect ratio improves the coverage at the bottom due to a strong directional atomic flux. However, the predicted bottom coverage is

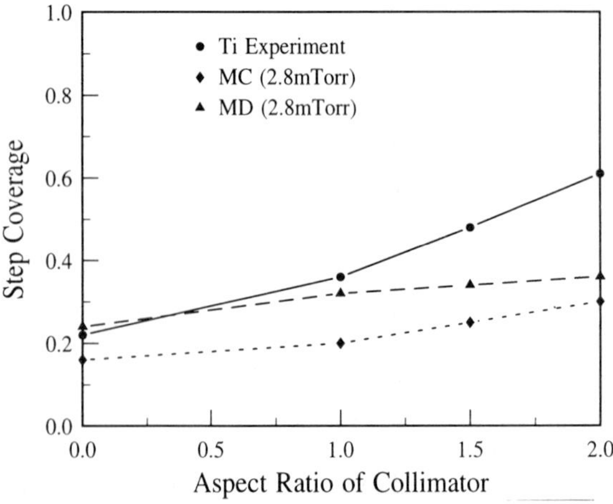

FIG. 20. Step coverage as a function of the collimator aspect ratio (*81*).

too low compared to the experiments. The second method uses the MD model to perform this simulation. It is found that the prediction of bottom coverage improves and is significantly better than that predicted by the Monte Carlo calculation with sticking probability of unity. These improved predictions may be related to the nonunity value of the sticking probability of sputtered Ti adatoms in the MD model. However, the coverage predicted by the MD model is still lower than the experimental data. Although this quantitative disagreement may be due to a difference in the dimension of collimator used for the experiments and the model needs to be further investigated. A qualitative agreement between the predicted and experimental trends, however, demonstrates that the combined MC/MD modeling approach has great potential for simulating thin film formation phenomena.

By examining the film profiles (*82*), we also note that the overhang usually observed in the case of noncollimated sputtering is much less when a high-aspect-ratio collimator is used. This is again believed to be a result of the higher adatom energy which causes the cascade collision and an almost-directional flux distribution during the deposition process when the collimator is present.

B. Low-Pressure/Long-Throw Sputtering

Because collimated sputtering is a filtering process, it is slow and inefficient. As demonstrated in the previous section, the use of a collimator with an aspect ratio of 2 can result in an 80% reduction in the throughput. The collimators can also get partially clogged with the depositing material, which can further reduce the deposition rate, and usually need to be removed for cleaning. Particle performance such as flaking from the collimator is also a major concern. In this section, we demonstrate the use of our model to investigate the effectiveness of low-pressure sputter deposition with a hollow cathode to achieve both the bottom and sidewall coverage required for a high-aspect-ratio feature with a high throughput. This process is again modeled for the deposition of a Ti liner and is compared with experiments (*86*).

In the magnetron sputtering deposition experiments, a hollow cathode is used to inject electrons and sustain a plasma at low pressures. This allows the operation of the cathode at full power in a low pressure range (below 1 m Torr). To evaluate the quality of the step coverage, the thickness of the Ti layer is varied between 0.05 and 0.15 μm. Sixty thousand Ti adatoms with or without 40,000 ions are deposited on a 512 × 400 lattice unit substrate to simulate this process.

1. The Effect of Pressure and Distance on the Step Coverage. Numerical results for low-pressure/long-throw sputtering are presented in Fig. 21 together with the experimental data for collimated sputtering of a bottom coverage of Ti liners with an aspect ratio of 2.5. In the case of collimated sputtering at 2.8 m Torr with a throw distance of 82 mm, the bottom coverage changes from 22 to 61% when the collimator aspect ratio is increased from 0.0 to 2.0. The improvement in bottom coverage is because only the directional flux is deposited on the substrate after the collimator has blocked most of the oblique angle sputtered atoms ejected from the target. Though an increase in bottom coverage can be obtained with the use of a collimator, the throughput decreases significantly, making the collimated sputtering a slow process. Also, the particle performance and the need to clean the collimator on a regular basis are major concerns. In the case of low-pressure/long-throw sputtering with a throw distance of 82 mm, 50% bottom coverage is predicted by the simulation. It seems that at 0.3 m Torr, the incident Ti atoms not only have strong directional angular distribution but also large incident energies, which result in a low sticking coefficient at the substrate. As the throw distance is increased from 82 to 300 mm, the bottom coverage increases from 50 to 63%. The Ti liner deposited with a throw distance of 150 mm at 0.3 m Torr is shown in

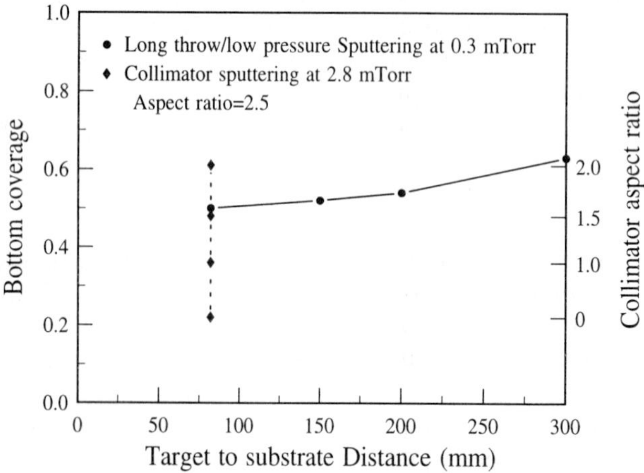

FIG. 21. Bottom coverage of Ti sputtered at 0.3 m Torr as a function of throw distance (*82*).

Fig 22. The bottom coverage is considered to be excellent (more than 50%). This kind of desirable coverage at the bottom can only be achieved with a collimator aspect ratio of 2.0 at 2.8 m Torr (*12*) in the collimated sputtering. An increase in bottom coverage in the present case may be attributed to the fact that the sputtered Ti atoms only with a normal angle can reach the substrate when the throw distance is long. Certainly, this line-of-sight type of deposition requires that the adatoms have fewer collisions with the background gas during their travel from the target to the substrate. A low working pressure, then, is essential in this process. However, an increase in throw distance also results in a significant reduction of the throughput. For example, when the throw distance is increased from 82 to 150 mm, the throughput decreases by 85%.

An increase in throw distance R generally results in a decrease in the throughput proportional to R^2. Interestingly, an increase in throw distance does not necessarily provide a significant improvement in the bottom coverage. A small improvement can be attributed to a reduction in the incident energy of the adatom (associated with an increase in the sticking coefficient), even though an increased directional angular distribution is obtained as the throw distance is increased. It seems that an optimal value for the throw distance exists as a function of the process parameters for the best throughput and bottom coverage.

2. Step Coverage and Ion Bombardment. The experimental results and numerical calculations for bottom and sidewall coverage of Ti liners with an aspect ratio of 4.5 are presented in Figs. 23 and 24, respectively. At 2.0 m Torr with throw distance of 15 cm, the bottom and sidewall coverage are 15 and 14% respectively. As the working pressure is decreased from 2.0 to 0.2 m Torr, the bottom coverage increases to 53%. The SEM photograph of a Ti liner deposited at 0.2 m Torr, using low-pressure sputter deposition, is shown in Fig. 25. The bottom coverage for Ti is considered to be excellent (more than 50%), while good coverage (29%) is maintained on the sidewall. This kind of desirable coverage on both the bottom and the sidewall can only be achieved by using collimated sputtering with a collimator aspect ratio of 2.0 at 2.8 m Torr (*12*). An increase in bottom coverage in the present case may be attributed to the increase of incident energy and a strong directional flux of adatoms (Fig. 23) when the pressure is decreased.

To understand this technique better, MD calculations are carried out with the energy and angular distributions obtained from the MC thermalization model. The MD results of bottom coverage are found to agree well with the experimental observation (Fig. 23). However, the sidewall coverage is much less than the experimental data (Fig. 24). To examine this contradiction in more detail, the effect of energetic bombardment is introduced in the model. A case where 50 eV ion bombardment normal to the substrate is applied during the film growth is simulated. The results for both the bottom and the sidewall coverage now agree well with the experimental data. By examining the predicted film structures for cases

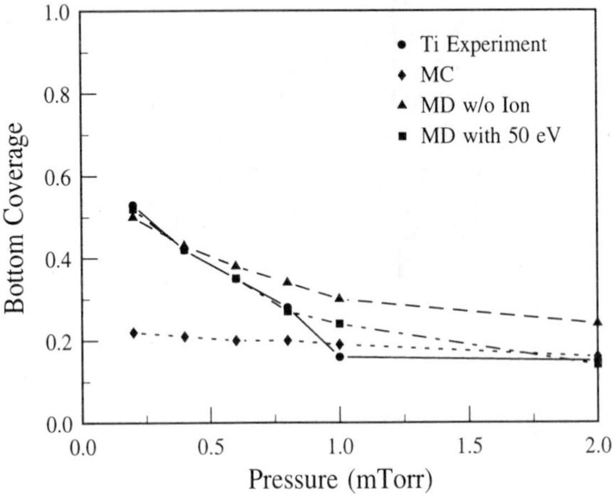

FIG. 23. Bottom coverage as a function of the working pressure for Ti.

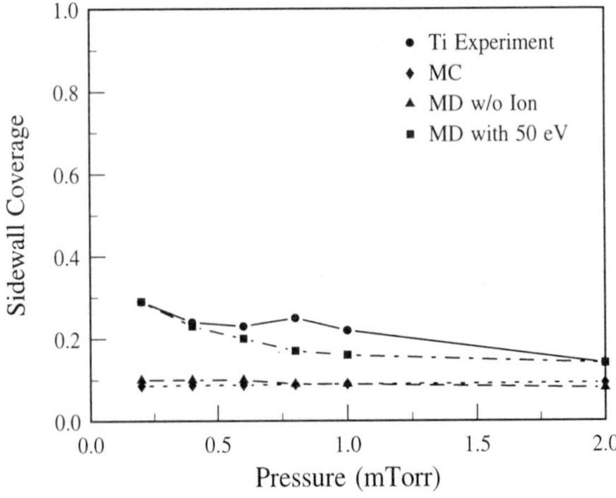

FIG. 24. Sidewall coverage as a function of the working pressure for Ti (*86*).

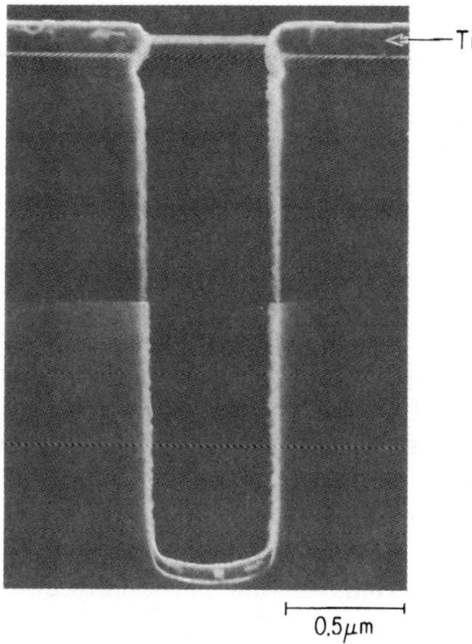

FIG. 25. SEM photograph showing step coverage for Ti (*86*).

with and without ion bombardment, it is found that the energetic bombardment helps in resputtering of the deposited atoms at the substrate. Because the resputtering occurs at the bottom, most of these adatoms are deposited onto the sidewall because the resputtered atoms cannot reach the opening of a high-aspect-ratio feature easily. This results in better sidewall coverage in the case of ion bombardment.

VI. Conclusions

A molecular dynamics model is developed to study the microstructure and intrinsic stresses of sputter-deposited thin films and the step coverage as a function of the process conditions. To perform efficient integration of the equation of motion and force calculations, a third-order discretization scheme with nonuniform time step is employed. Special attention is also given to development of schemes for fast search of the neighbor atoms. A "position index method" that updates the Verlet neighbor table locally instead of reconstructing the entire neighbor table is proposed and implemented in order to perform faster calculations of the forces. The generalized Langevin equation (GLE) has been used to simulate the constant temperature condition during the deposition processes. This can allow the study of the effects of ion bombardment, chemical reactions, and many other processing conditions on thin film formation phenomena.

A movable periodic boundary condition based on a constant external pressure technique is employed that considers the relationship between the stresses in the film and permits stress relaxation during the deposition process. The present algorithm can therefore demonstrate successfully the effect of substrate conditions, such as external stresses and temperature, on the deposition phenomena. The global average intrinsic stresses of the film are calculated using an improved method and it is possible to study the evolution of this stress with the film. A novel scheme to study the effect of the microstructure on local stresses with the help of a local stress parameter is proposed and implemented. This helps in developing an understanding of the creation mechanism of tensile and compressive stresses in the film. The MD model developed here is very general and can be used to study many kinds of thin film deposition phenomena at the atomic level.

Two-dimensional simulations are performed by using this MD model to predict the microstructure and show that the intrinsic stresses depend on the defect contents such as voids, columnar structure, interstitials, and grain boundary of the film. The tensile stress induced in the film increases when a porous film, such as a columnar structure, transforms into a more compact film. When the voids become small and are less frequent at low pressures (high incident energies) or under energetic ion bombardment, the tensile stress decreases sharply. The existence of the trapped gas impurity with concurrent ion bombardment, however, causes the gas atoms to be replaced by the vacancies and form a substitutional impurity within the deposited film. This results in compressive intrinsic stresses in the film when it is dense. It is therefore concluded that the compressive stress is related to the trapped gases and the concurrent ion bombardment during the growth. However, the self interstitial, which is expected to be observed when the film is dense, is not predicted by this two-dimensional model. This may be due to a smaller space provided by the 2-D model in the (111) plane or a symmetric character of the pair potential function. In spite of this shortcoming, the formation mechanisms of compressive stresses in sputter-deposited films are successfully explained by this two-dimensional molecular dynamics model. The simulations for a wide range of voltage bias have demonstrated that the stress in sputter-deposited films can be controlled successfully by the use of the ion energy. Further computations with a large simulation cell and a much larger number of deposited atoms, particularly in three dimensions, may demonstrate some other interesting aspects of the deposition phenomena as well as the film characteristics. Although the MD model presented here can be easily used for three-dimensional simulations, computing power is still a constraint for such calculations, particularly when the complex physics of the deposition process are included in the model.

To study further the effect of the tool-level phenomena on adatoms arriving at the substrate for deposition, a model based on Monte Carlo calculations is developed. This model accounts for the energy distribution at the target, ejection angle of the sputtered atoms, distance between the target and the substrate, collisions between the energetic atoms and the background gas atoms, and scattering. It then predicts the energy and directional angle of the adatoms arriving at the substrate.

The tool-level MC model is then combined with the feature-scale MD model to investigate the step coverage and show the effectiveness of the present algorithm. Two different processes, currently under development, collimated sputter deposition and low-pressure/long-throw magnetron sputtering, are considered to examine their suitability in producing desirable bottom and sidewall coverage of Ti liners. Interesting effects that result from the collimator aspect ratio, throw distance, working pressure, and ion bombardment on step coverage in trenches of various aspect ratios are exhibited by these simulations. Predicted coverage and their trends agree well with the experimental observations, and demonstrate that a particle method is the most appropriate approach for modeling thin film deposition particularly at the submicron level. Such methods can be successfully used to study sputter deposition phenomena and process design.

References

1. F. Kurger, British Patent 157,909 (1919).
2. P. J. Martin, *J. Mater. Sci.* **21**, 1–25 (1986).
3. R. S. Nowichi, W. D. Buckley, W. D. Mackintosh, and I. V. Mitchell, *J. Vac. Sci. Technol.* **11**, 675–679 (1974)
4. J. L. Vossen, *J. Vac. Sci. Technol.* **8**, S12–S30 (1971).
5. J. A. Thornton, *J. Vac. Sci. Technol.*, **11**, 666–670 (1974).
6. R. C. Sun, T. C. Tisone, and P. D. Cruzan, *J. Appl. Phys.* **46**, 112–117 (1975).
7. J. A. Thornton, *J. Vac. Sci. Technol.*, **12**, 830–835 (1975).
8. D. W. Hoffman and J. A. Thornton, *J. Vac. Sci. Technol.* **17**, 380–383 (1980).
9. D. W. Hoffman and J. A. Thornton, *J. Vac. Sci. Technol.* **16**, 134–137 (1979).
10. A. G. Blachman, *J. Vac. Sci. Technol.* **10**, 299–302 (1973).
11. A. Entenberg, V. Lindberg, and K. Fletch, *J. Vac. Sci. Technol. A* **5**, 3373–3377 (1987).
12. R. V. Joshi and S. Brodsky, *Appl. Phys. Lett.* **61**, 2613–2615 (1992).
13. D. C. Bulter, P. J. Holverson, P. Rich, J. Hems, G. A. Dixit, and S. Poarch, *in* "Advance

Metallization for ULSI Applications in 1994" (R. Blumenthal and G. Janssen, eds). pp. 277–284. *Mater. Res. Soc. Proc.*, Pittsburgh, Pennsylvania, 1994.
14. D. E. Harrison, *Radiat. Eff.* **70**, 1–64 (1983).
15. H. P. Murbach and H. Wilman, *Proc. Phys. Soc.* **B66**, 905–910 (1953).
16. R. W. Hoffman, R. D. Daniels, and E. C. Crittenden, Jr., *Proc. Phys. Soc.* **B67**, 497–499 (1954).
17. E. Klokholm and B. S. Berry, *J. Electrochem. Soc.* **115**, 823–826 (1968).
18. D. W. Hoffman and J. A. Thornton, *Thin Solid Films* **40**, 355–363 (1977).
19. D. W. Hoffman and J. A. Thornton, *Thin Solid Films* **45**, 387–396 (1977).
20. R. S. Wanger, A. K. Sinai, T. T. Shoeing, H. J. Levinstein, and F. B. Alexander, *J. Vac. Sci. Technol.* **11**, 582–590 (1974).
21. W. W. Lee and D. Oblas, *J. Vac. Sci. Technol.* **7**, 129–133 (1970).
22. A. Pan and J. E. Greene, *Metallurgical and Protective Coatings*, 25–34 (1980).
23. B. A. Movchan and A. V. Demchishin, *Fiz. Met. Metalloved.* **28**, 653 (1969).
24. G. Ciccotti and W. G. Hoover, "Molecular-Dynamics Simulation of Statistical–Mechanical System." North-Holland, Amsterdam, 1986.
25. D. Henderson, M. H. Bradsky, and P. Chaudhari, *Appl. Phys. Lett.* **25**, 641–645 (1974).
26. S. Kim, D. Henderson, and P. Chaudhari, *Thin Solid Films* **47**, 155–158 (1977).
27. M. J. Brett, K. L. Westra, and T. Smy, *Can. J. Phys.* **67**, 347–350 (1989).
28. A. G. Dirks and H. J. Leamy, *Thin Solid Films* **47**, 219–233 (1977).
29. H. J. Leamy, G. H. Gilmer, and A. G. Dirks, *Curr. Top. Mater. Sci.* **6**, 309–344 (1980).
30. K. H. Müller, *J. Appl. Phys.* **62**, 1796–1799 (1987).
31. C. Kittel, "Introduction to Solid State Physics," 6th Ed. Wiley, New York, 1986.
32. M. P. Allen and D. J. Tildesley, "Computer Simulation Liquids." Oxford Univ. Press (Clarendon), Oxford, 1987.
33. A. S. Nandedkar, G. R. Srinivasan, and C. S. Murthy, *Phys. Rev.* **B 43**, 7308–7311 (1991).
34. G. Moliére, *Z. Naturforsh.* **A2**, 133 (1947).
35. J. Tersoff, *Phy. Rev.* **B37**, 6991–7000 (1988).
36. M. S. Daw and M. I. Baskes, *Phys. Rev. Lett.* **50**, 1285–1288 (1983).
37. R. A. Joshnson and D. J. Oh, *J. Mater. Res.* **4**, 1195–1201 (1989).
38. A. M. Guellil and J. B. Adams, *J. Mater. Res.* **7**, 639–652 (1992).
39. F. Sullivan, R. D. Mountain, and J. O'Connell, *J. Comp. Phys.* **61**, 138–153 (1985).
40. L. Verlet, *Phys. Rev.* **159**, 98–103 (1967).
41. A. Chialvo and P. G. Debenedetti, *Comput. Phys. Commun.* **64**, 15–18 (1991).
42. A. Chialvo and P. G. Debenedetti, *Comput. Phys. Commun.* **60**, 215–224 (1990).
43. C. C. Fang, F. Jones, and V. Pradad, *J. Vac. Sci. Technol. A* **11**, 2778–2789 (1993).
44. R. W. Hockney and J. W. Eastwood, "Computer Simulation Using Particles," p. 28. McGraw–Hill, Advanced Publications Series, New York, 1981.
45. D. Beeman, *J. Comput. Phys.* **20**, 130–139 (1976).
46. C. W. Gear, "Numerical Initial Value Problems in Ordinary Differential Equations." Prentice-Hall, Englewood Cliffs, New Jersey, 1971.
47. A. Amini and D. Fincham, *Comput. Phys. Commun.* **56**, 313–324 (1990).
48. H. J. C. Berendsen and W. F. van Gunsteren, *in* "Molecular-Dynamics Simulation of Statistical–Mechanical Systems" (G. Ciccotti and W. G. Hoover, eds.), pp. 43–65, North-Holland, Amsterdam, 1986.
49. H. C. Andersen, *J. Chem. Phys.* **72**, 2384–2393 (1980).
50. L. V. Woodcock, *Phys. Lett.* **10**, 257–261 (1971).
51. H. C. Andersen, M. P. Allen, A. Bellemans, J. Board, J. H. R. Clarke, M. Ferrario, J. M. Haile, S. Nos, J. V. Opheusden, and J. P. Ryckaert, *Rapport d'Activité, Scientifique du CECAM*, 82–115 (1984).

52. S. Nosé, *Mol. Phys.* **52**, 255–268 (1984).
53. S. Nosé, *J. Chem. Phys.* **81**, 511–519 (1984).
54. H. J. C. Berendsen, J. P. M. Postma, W. F. Van Gunsteren, A. DiNola, and J. R. Haak, *J. Chem. Phys.* **81**, 3684–3690 (1984).
55. Y. Wu and R. J. Friaut, *J. Appl. Phys.* **65**, 4714–4718 (1987).
56. T. Schneider and E. Stoll, *Phys. Rev.* **B17**, 1302–1322 (1978).
57. C. Kittel, "Thermal Physics," p. 427. Wiley, New York, 1969.
58. M. P. Allen, *Mol. Phys.* **40**, 1073–1087 (1980).
59. S. Okuyama and D. W. Oxtoby, *J. Chem. Phys.* **84**, 5824–5829 (1986).
60. S. A. Adelman and J. D. Doll, *J. Chem. Phys.* **64**, 2375–2388 (1976).
61. W. G. Hoover, "Molecular Dynamics." Springer-Verlag, New York, 1986.
62. M. Tabor, "Chaos and Integrability in Nonlinear Dynamics." Wiley, New York, 1989.
63. C. C. Fang, F. Jones, and V. Prasad, "Thin Film Heat Transfer: Properties and Processing," Vol. 184, pp. 63–71. American Society of Mechanical Engineers, New York, 1991.
64. M. Parrinello and A. Rahman, *Phys. Rev. Lett.* **45**, 1196–1199 (1980).
65. M. Parrinello and A. Rahman, *J. Appl. Phys.* **52**, 7182–7190 (1981).
66. J. D. Wilcock, Stress-Thin Film, Ph.D. Thesis, Electrical Engineering Department, Imperial College, London University, U.K., June 1967.
67. D. H. Tsai, *J. Chem. Phys.* **70**, 1375–1382 (1979).
68. M. Tanemura, Y. Hiwatari, H. Matsuda, T. Ogawa, N. Ogita, and A. Ueda, *Prog. Theor. Phys.* **58**, 1079–1095 (1977).
69. G. M. Turner, I. S. Falconer, B. W. James, and D. R. McKenzie, *J. Vac. Sci. Technol. A* **10**, 455–467 (1992).
70. M. W. Thompson, *Philos. Mag.* **18**, 377–414 (1968).
71. T. Motohiro, *J. Vac. Sci. Technol. A* **4**, 189–195 (1986).
72. L. D. Landau and E. M. Lifshitz, "Mechanics," p. 48. Pergamon, London, 1976.
73. R. E. Somekh, *J. Vac. Sci. Technol. A* **2**, 1285–1291 (1984).
74. C. C. Fang, F. Jones, and V. Prasad, *J. Appl. Phys.* **74**, 4472–4482 (1993).
75. A. F. Jankowski, R. M. Bionta, and P. C. Gabriele, *J. Vac. Sci. Technol. A* **7**, 210–213 (1989).
76. R. D. Bland, G. J. Kominiak, and D. M. Mattox, *J. Vac. Sci. Technol.* **11**, 671–674 (1974).
77. M. Kaminsky, "Atomic and Ionic Impact Phenomena on Metal Surface," p. 292. Academic Press, New York, 1965.
78. C. C. Fang, F. Jones, R. R. Kola, and G. K. Celler, *J. Vac. Sci. Technol. B* **11**, 2947–2952 (1993).
79. S. M. Rossnagel, D. Mikalsen, H. Kinoshita, and J. J. Cuomo, *J. Vac. Sci. Technol. A* **9**, 261–265 (1991).
80. J. J. Hsieh and V. Joshi, *in* "Advanced Metallization for ULSI Applications in 1993" (D. F. Favreau, Y. Shacham-Diamond, and Y. Horiike, eds.), p. 207. Mater. Res. Soc. Proc., Pittsburgh, Pennsylvania, 1993.
81. C. C. Fang, J. J. Hshieh, R. V. Joshi, F. Jones, and V. Prasad, *in* "Advance Metallization for ULSI Applications in 1994" (R. Blumenthal and G. Janssen, eds.), pp. 315–319. Mater. Res. Soc. Proc., Pittsburgh, Pennsylvania, 1994.
82. C. C. Fang and V. Prasad, *in* "Modeling and Simulation of Thin-Film Processing," p. 131. Mater. Res. Soc. Proc., Pittsburgh, Pennsylvania, 1995.
83. M. Xia, X. Liu, and P. Gu, *Appl. Opt.* **32**, 5443–5446 (1993).
84. K. H. Müller, *J. Appl. Phys.* **58**, 2573–2576 (1985).
85. J. P. Biersack and L. G. Haggmark, *Nucl. Instrum. Methods* **174**, 257–269 (1980).
86. C. C. Fang, R. V. Joshi, V. Prasad, and C. Ouyang, Presented at AVS 1995.

Feature Scale Transport and Reaction during Low-Pressure Deposition Processes

TIMOTHY S. CALE AND VADALI MAHADEV

Department of Chemical, Biological, and Materials Engineering, Center for Solid State Electronics, Arizona State University, Tempe, Arizona

I. Introduction

The focus of this chapter is on feature scale transport and reaction in features on patterned wafers during low-pressure processing. The theory for transport and reaction is valid for both deposition and etch processes; however, the examples focus on deposition processes as is appropriate for this text. The goal of transport and reaction modeling and simulation is to develop and validate models which in turn provide a unified understanding of film profile evolution. Step coverage or film conformality is a well-known production-level constraint in processes used to deposit conducting, semiconducting and insulating films onto patterned wafers in microelectronics fabrication. A less well-known constraint is film composition, which can be important in the deposition of multicomponent films. These constraints must be met in addition to satisfying a number of other film-specific property constraints. Specifications on all of these properties must be met at each location across a wafer (intrawafer uniformity) as well as between wafers (interwafer uniformity). The process must also be production worthy; that is, the film must be deposited at acceptable deposition rates with acceptable source gas utilization to meet throughput and cost targets. An understanding of the origins of step coverage limitations and composition variations for a given process may be important in identifying reactor designs and reactor operating windows, in guiding experimental efforts to optimize the process, and/or in guiding the search for different deposition precursors and/or processes for the same or similar films.

Our understanding of the physics and chemistry which govern deposition processes has improved significantly over the last few years. This can be attributed largely to (1) the development of "first principles," "physically based" models for deposition processes, and (2) the use of computer programs (simulators) based on those models. *First principles* means that the governing equations which provide the model framework are well-accepted balance equations, for example, species balances. *Physically based* means that these models include, and the simulation programs rely on, submodels for species, momentum and energy transport, and reaction kinetics. The majority of studies have focused on predicting film profile evolution; however, a few have considered the possibility of composition profiles. In this chapter, we present examples of combined simulation and experimental studies which were designed to elucidate critical physical and chemical aspects of topography evolution. Although the majority of references are taken from our work, many unreferenced studies on similar systems have been performed by other groups. The references to many of these studies can be accessed by reading the referenced material as well as the other chapters in this book.

The focus of this chapter is on "continuum" models of film profile evolution, as opposed to models which maintain film granularity, which might be important to the performance of polycrystalline films. Our work on atomistic modeling of deposition processes is reported in Refs. *1–3*. Other work in this area is referenced in those publications and in Chapter 1 in this volume.

In addition to conformality and composition of deposited films, planarization of deposited films by flow or reflow is important in current and some future process flows. It will certainly play an important role for years to come, even with the increased interest and acceptance of chemical–mechanical polishing and the more common deposition and etchback schemes. These topics are beyond the scope of this chapter; however, our recent work on thin film flow is covered in Refs. *4–7*.

A. Transport during Deposition Processes

Gas phase transport can be characterized by a ratio of the mean free path of species in the gas phase to a characteristic length scale for transport. This ratio is called the Knudsen number. A high Knudsen number ($\gg 1$) implies that collisions between species in the gas phase are much less frequent than collisions between gas phase species and the surfaces which define the flow of interest. The source volume adjacent to the feature does not strictly need to be included as part of the simulation domain. It is critical to know the angular distribution of the flux of each species arriving at the surface. This transport regime is termed *free molecular* or *ballistic* transport. To date, modeling and simulation efforts for feature scale transport and reaction have focused on solving the Clausing-like integral equations developed by Cale and introduced in Refs. *8–10*, which represent species (material) balances. The examples in this chapter deal with processes which occur in this transport regime.

A small Knudsen number ($\ll 1$) implies that collisions between species in the gas phase occur more frequently than collisions between gas phase species and feature surfaces. Thus, transport in this regime is governed by differential equations which are part of every engineer's education. This is called the continuum transport regime. We do not deal with deposition in this transport regime, and our work in the area can be found in Refs. *7, 11, 12*.

The Boltzmann equation is the starting point for modeling transport. In the derivation of the governing equations in the ballistic and continuum transport regimes, the Boltzmann equation is replaced by equations which are much easier to solve. To date, the focus has been on solving the appropriate mass balance equations; that is, the local temperature is specified. It may vary with time, but is assumed to be spatially uniform on the scale of microns. More generally, an energy balance equation needs to be solved in order to predict temperature nonuniformities in features.

The intermediate regime of Kn (~ 1) is termed the *transition* regime, where the full Boltzmann equation is usually assumed to govern the transport. Very little work has been done in the transition regime; however, Monte Carlo simulations have been used to advantage (for example, see *13–15*).

Central to the choice of transport regime (and hence the form of the governing equations) is the choice of characteristic length scale for transport. Often, the choice is clear, or relatively unimportant, for example, because all reasonable choices are of the same order. When the choice of length scale is not clear, the decision is validated by comparing simulation and experimental results. For example, consider the situation in which the feature width is used as the characteristic length scale, and that the transport regime at the start of the decomposition is well within the continuum regime. As deposition proceeds toward feature closure, transport will pass through the transition regime and into the free molecular regime. The choice of transport regime(s) to use for modeling a deposition process should be validated by comparison with experiments. The same can be said of cases where equations which govern either ballistic transport or continuum transport are used when Kn ~ 1: Do the predictions compare well with experiments? These concerns are not present for deposition processes which occur in the free molecular flow regime, except perhaps right at feature closure.

In this chapter, we first review basics of chemistry and free molecular transport and the ballistic transport and reaction model (BTRM) valid for low-pressure deposition (and etch) processes inside micron scale features on patterned wafers. We then discuss how studies which combine experiments and simulations are being used to help develop constitutive properties for selected LPCVD, PECVD, and PVD systems. For PVD processes, we also discuss a study of curvature-driven diffusion. We then present the results of a combined simulation and experimental study of feature scale composition nonuniformities in the sputter deposition (PVD) of titanium–tungsten films. We finish with an example of how feature scale simulations can be used to guide process design.

B. Chemical Reaction Kinetics

Homogeneous as well as heterogeneous kinetic processes can be important in deposition processes. For feature scale simulations, heterogeneous processes are most important. In feature scale transport and reaction simulations of low-pressure (high Kn) processes, only heterogeneous reactions are explicitly considered. Homogeneous reactions can occur in the reactor volume, which is almost always in the continuum transport regime or transition regime, and might well play critical roles in the deposition processes. For example, there are systems for which homogeneous reactions apparently produce reactive precursors, which then participate in deposition reactions. For transport and reaction simulations of high-pressure deposition processes, homogeneous reactions can occur in features, because collisions between species can occur inside features. In this article, we focus on heterogeneous reactions.

It is now accepted that chemical reaction mechanisms should be used whenever possible, at least for thermal CVD processes and for selected reactions in PECVD processes, rather than "sticking factors" or even overall rate expressions. Unfortunately, mechanisms have been proposed for few deposition processes of industrial interest. Overall rate expressions, in which deposition (reaction) rates are expressed in terms of local temperature and local fluxes, are becoming more common. They can be based on proposed mechanisms or be used simply to represent kinetic data. The concept of "sticking factor" can be useful; and is still widely used, particularly for PVD and PECVD processes. However, it can also be misleading. The sticking factor of a given species, when defined as the fraction of arriving reactant species which contributes to film deposition, is in general a function of temperature and local flux conditions; that is, it can vary as a function of position inside features. In addition, a low sticking factor of a critical reactant does not necessarily lead to high conformality, contrary to the conformality predictions of simplistic sticking factor based models. Poor conformality for systems in which key reactants have low sticking factors requires more complete heterogeneous kinetics such as Langmuir–Hinshelwood kinetics. These arise naturally from reaction mechanistic considerations.

The energetics of reactions and incoming species have not been treated explicitly in the vast majority of studies; however, it is clear that species energetics are critical in certain processes, for example, plasma processes. Work at Sandia National Laboratories has included the explicit treatment of energy-dependent reaction rates (*16*).

C. Transport and Reaction Simulations

We have focused on developing transport and reaction simulators, which solve the governing equations appropriate for a given transport regime, rather than developing modules for specific processes. EVOLVE (*17*), dealt with in the following sections, is an example of a transport and reaction simulator. Simulations for a specific process are accomplished by introducing the transport and reaction submodels which are the current best estimates for, or are being proposed for, that process.

The topography simulations discussed here are done using EVOLVE on features which can be represented in two dimensions: long lines or trenches and features of circular horizontal cross-section. Thus, line segments in the plane of the paper represent either infinitely long strips for lines or trenches and sections of cones for circular features. Circular features are inherently symmetric, whereas trenches can either be symmetric or of fairly arbitrary cross-section. More importantly, surface evolution can be done in two dimensions (with certain assumptions about the flux distributions). Topography simulations on completely three-dimensional surfaces require considerably more computational resources, but are being developed to meet the need for realistic simulations of device fabrication in "virtual fabs" (e.g., see *18–20*). On the other hand, process understanding has been developed using simulations on 2-D surfaces. This will continue to be the case, in general, simply because a method for comparing simulator predictions of surface evolution with experimental surfaces has not been developed for evolving three-dimensional surfaces. Even after methods are established, which is an important goal in order to validate 3-D simulation packages, working with 3-D surfaces experimentally will be tedious. We refer to simulators like EVOLVE, in which the transport is done in 3-D and surfaces are represented in 2-D, as 3-D/2-D topography simulators. 3-D/2-D simulation packages will continue to play a significant role in process development, particularly in establishing the transport and kinetic models for particular processes. Transport is done in three dimensions in both cases, of course. Note that not all flux distributions are appropriate for the evolution of 2-D surfaces, general flux distributions can convert a 2-D surface into a 3-D surface. Our work with EVOLVE-3d, a 3-D/3-D topography simulator, is discussed in Refs. *7, 21*, and *22*.

II. Chemistry

In our deposition modeling and simulation efforts, we emphasize the use of chemistry, as well as we know it, for each particular deposition process. The chemistry of a deposition system is defined by the stoichiometries and reaction kinetics of all the reactions which occur in the process (*23*). These reactions occur in the gas phase (homogeneous) or at the evolving surface (heterogeneous). In general, both homogeneous and heterogeneous reactions can be important. Homogeneous reactions occur in the volume of the reactor and in features, in cases where there are intermolecular collisions within features, that is, at low Kn. In the following, we discuss how reactions are handled inside EVOLVE.

A given reaction is specified by its stoichiometry. Ideally, the reaction rates are given in terms of the detailed mechanisms for the chemical reactions which occur. A "detailed reaction mechanism" for the majority of deposition processes is one which expresses the rate of formation of intermediate species as functions of reactant fluxes and surface concentrations (heterogeneous) or concentrations (homogeneous), at a given temperature. For the purposes of film profile evolution, it is usually not necessary to consider the details of how the molecules adsorb, react, and desorb; the geometry of bond formation and breakage is not needed.

Unfortunately, reaction mechanisms are not available for the majority of the deposition processes of interest to the microelectronics industry. Even the stoichiometries of the reactions are often unknown. We can expect more reaction mechanisms to be presented during the next few years.

More common than reaction mechanisms are *global* or overall reaction kinetics for a given reaction. A global rate expression expresses the molar rate of reaction as a function of local reactant concentrations or partial pressures. A global rate expression for a specific reaction might be based on some assumed mechanism, or it may simply be used to correlate kinetic data. Whenever possible, the complete mechanism should be used in deposition simulations, rather than a global rate expression. One danger of using a global rate expression derived from a detailed mechanism for kinetics is that a rate limiting step is often assumed. The rate limiting step may actually be different for different reaction conditions; species concentrations, and temperatures for homogeneous reactions and species fluxes and temperatures for heterogeneous reactions. For example, the rate limiting step for a given reaction may change due to the changes in the reactant fluxes which occur during a deposition process, even at the same temperature. The species fluxes can change during a deposition process used to fill a feature; that is, as closure is approached, fluxes can change dramatically.

A reaction rate expressed in terms of sticking factors of reactants is the simplest approach to modeling deposition processes. This approach is used when little is understood about a system, but when engineering estimates of film profile evolution are needed, for example, for plasma systems. The reaction kinetics using this approach are linear, and many of the issues which arise due to nonlinear kinetics cannot be represented. On the other hand, this approach can provide useful trends in topography evolution by solving equations which are often much easier to solve than those which include nonlinear kinetics. Monte Carlo simulations have been used to simulate depositions using sticking factors (*24–26*).

As demonstrated in the following sections, we use global rate expressions or reaction mechanisms for thermal CVD processes. For PVD processes, sticking factors usually provide at least semi quantitative predictions. We use that approach. For PECVD, we use a combination of sticking factor based rates and global reaction rate expressions.

The reactions which occur as well as the kinetics for a specific reaction can depend on the materials in the substrate, in addition to the deposition conditions (e.g., see *27*). Thus, if the same reaction (same stoichiometry) occurs on more than one material, the rate expression might well be different for each material, at least during the initial stages of deposition.

To account for different reactions which occur during a deposition process, the materials in the substrate, at least those exposed to the reactant fluxes, need to be known at any given time in the deposition process.

A. Reaction Stoichiometry

For modeling and simulation purposes, the stoichiometries of the reactions which occur in a given deposition process are specified by the "generalized" stoichiometric coefficient matrices, which can be written as

$$\mathbf{S} = \begin{bmatrix} \alpha_{1,1} & \cdots & \alpha_{1,ns} \\ \vdots & & \vdots \\ \alpha_{nr,1} & \cdots & \alpha_{nr,ns} \end{bmatrix}, \tag{1}$$

where the α_{ij} are generalized stoichiometric coefficients, which are negative for reactants and positive for products in a given reaction. The first subscript on α is the reaction index and the second subscript is the species index. There are *nr* reactions and *ns* species involved in the reactions. Separate matrices can be written for homogeneous and heterogeneous reactions, or they may be included in a single matrix. Stoichiometric coefficient matrices can be written for the overall reactions which lead to product formation, or can represent the mechanistic reactions which occur.

For example, consider the deposition of tungsten silicide from dichlorosilane and tungsten hexafluoride. For demonstration purposes, we present plausible overall stoichiometries and reaction kinetics for three reactions (*28,29*). Each reaction forms a different solid product species; however, they all contribute to the material called tungsten silicide. The reactions are

$$\begin{aligned} SiH_2Cl_2 &\rightarrow Si + 2HCl, \\ 4SiH_2Cl_2 + WF_6 &\rightarrow WSi_2 + 8HCl + SiF_2 + SiF_4, \\ 11SiH_2Cl_2 + 5WF_6 &\rightarrow W_5Si_3 + 22HCl + SiF_2 + 7SiF_4. \end{aligned} \tag{2}$$

If SiH_2Cl_2 is species 1, WF_6 is species 2, Si is species 3, WSi_2 is species 4, W_5Si_3 is species 5, HCl is species 6, SiF_2 is species 7, and SiF_4 is species 8, the generalized stoichiometric coefficient matrix is

$$\mathbf{S} = \begin{bmatrix} -1. & 0. & 1. & 0. & 0. & 2. & 0. & 0. \\ -4. & -1. & 0. & 1. & 0. & 8. & 1. & 1. \\ -11. & -5. & 0. & 0. & 1. & 22. & 1. & 7. \end{bmatrix}. \tag{3}$$

It is also important to specify which materials each reaction occurs on, as well as which material any solid products help to form. Thus, it is assumed that the products of a given reaction contribute to a single material. One way to accomplish this is by indexing the materials present in a given system and introducing another matrix, called the reaction-material matrix:

$$\mathbf{M} = \begin{bmatrix} m_{1,1} & \cdots & m_{1,nm+1} \\ \vdots & & \vdots \\ m_{nr,1} & \cdots & m_{nr,nm+1} \end{bmatrix}, \tag{4}$$

where each row corresponds to the reaction specified by the same row in **S**, and each of the first nm columns corresponds to a material of the problem, of which there are nm, and the last column contains the material index of the material which the solid products of reaction help to produce. An entry of one for m_{ij} ($j \leq nm$) means the i-th reaction occurs on the j-th material, while an entry of zero means that it does not. In general, if column $nm + 1$ contains the index which represents a material on which the reaction occurs, then there is a one in the column, which indicates that the reaction occurs on that material. If the rates are different on different materials, then a reaction stoichiometry can be entered for each material.

For the tungsten silicide example, assume that there are three materials in the problem: (1) silicon in the original substrate, (2) silicon dioxide in the original substrate, and (3) tungsten silicide formed by all three reactions. Further assume that the reaction stoichiometries and rates are the same on all materials. Then **M** is

$$\mathbf{M} = \begin{bmatrix} 1 & 1 & 1 & I_{\mathrm{WSi}} \\ 1 & 1 & 1 & I_{\mathrm{WSi}} \\ 1 & 1 & 1 & I_{\mathrm{WSi}} \end{bmatrix}, \tag{5}$$

where column 1 corresponds to silicon, column 2 corresponds to silicon dioxide, and I_{WSi} is the material index for tungsten silicide. If any of the reactions did not occur on silicon dioxide, for example, then column 2 would contain a zero for that reaction. If the reaction kinetics (discussed later) were different for one or more reactions on each material, then additional rows would have to be introduced for those new reactions.

To keep track of the atomic composition of the deposited films, the stoichiometry of each of the solid products of reaction needs to be specified. One way to accomplish this is to assign a number to each element in the system and build the *atomic stoichiometric* matrix:

$$\mathbf{A} = \begin{bmatrix} a_{1,1} & \cdots & a_{1,ne} \\ \vdots & & \vdots \\ a_{nns,1} & \cdots & a_{nns,ne} \end{bmatrix}, \tag{6}$$

where there are *nns* solid species and *ne* elements in the simulation. For the tungsten silicide example, **A** is

$$\mathbf{A} = \begin{bmatrix} 1 & 0 \\ 2 & 1 \\ 3 & 5 \end{bmatrix}, \tag{7}$$

where column 1 corresponds to silicon and column 2 corresponds to tungsten, and each row corresponds to one of the solid products of reaction.

B. Global Rate Expressions

Global rate expressions are often available after kinetic studies have been performed. For the tungsten silicide example introduced earlier, the reaction kinetics have been expressed in terms of global rate expressions as (*28,29*)

$$\begin{aligned} r_1 &= \frac{6.6 \times 10^{22} e^{-45.300/T} P_{\mathrm{DCS}}^2 P_{\mathrm{WF_6}}}{1 + 10000 P_{\mathrm{WF_6}}}, \\ r_2 &= \frac{1.8 \times 10^{30} e^{-60,4000/T} P_{\mathrm{DCS}} P_{\mathrm{WF_5}}}{1 + 1000 P_{\mathrm{WF_6}}}, \\ r_3 &= 9.5 \times 10^4 e^{-20,100/T} P_{\mathrm{DCS}}^{0.5} P_{\mathrm{WF_6}}. \end{aligned} \tag{8}$$

Once again, we emphasize that these rate expressions are intended for demonstration purposes only. These rate expressions were developed to represent the limited amount of kinetic data available; they were not derived from proposed mechanisms for the reactions. Notice that the reaction rate expressions cannot be derived from the reaction stoichiometry, because the stoichiometries presented are overall and not mechanistic. The use of global rate expressions derived from simple reaction mechanisms is discussed in Sec. V.

C. Reaction Mechanisms

Examples of how we use simple mechanisms are presented in Secs. V, VI, and VII. The approach currently used for reaction mechanisms in EVOLVE differs in the details from the approach used in CHEMKIN (*30*) and Surface CHEMKIN (*31*), but most of the same features are present. As an example, workers at Sandia National Laboratory recently proposed a detailed reaction mechanism for the deposition of SiO_2 from TEOS (tetraethyoxysilane, $Si(OCH_2CH_3)_4$) (*32–34*). Our use of this mechanism is summarized in Sec. V; the values used in the mechanism were supplied by Sandia personnel. The proposed mechanism includes a homogeneous (gas phase) reaction and eight heterogeneous (surface) reactions. The gas phase reaction proposed is

$$Si(OC_2H_5)_4 \leftrightarrow SiOH(OC_2H_5)_3 + C_2H_4 \tag{9}$$

or, letting E represent an ethoxy group for convenience,

$$SiE_4 \leftrightarrow Si(OH)E_3 + C_2H_4.$$

The rate of this reaction is (subscripts 1, 2, and 5 represent TEOS, $Si(OH)E_3$ and C_2H_4, respectively)

$$R = 2.459 * 10^{10} * \exp\left(\frac{-30861}{T}\right) * c_1 - 8.34 * \exp\left(\frac{-24417}{T}\right) * c_2 * c_5. \tag{10}$$

The relative fluxes of the species to the surface depends on the extent of this reaction, which in turn depends on reactor configuration and operating conditions. The surface reactions are given by

$$\begin{aligned}
SiE_4 + SiG_3(OH) &\rightarrow SiO_2(D) + SiGE_3 + C_2H_5OH,\\
SiG_3E &\leftrightarrow SiG_3(OH) + C_2H_4,\\
SiG(OH)E_2 &\leftrightarrow SiG(OH)2E + C_2H_4,\\
SiGE_3 &\leftrightarrow SiG(OH)E_2 + C_2H_4,\\
SiG(OH)2E &\leftrightarrow SiG_3(OH) + C_2H_5OH,\\
SiG(OH)E_2 &\leftrightarrow SiG_3E + C_2H_5OH,\\
SiG(OH)2E &\leftrightarrow SiG_3E + H_2O,\\
Si(OH)E_3 + SiG_3(OH) &\leftrightarrow SiO_2(D) + H_2O + SiGE_3,
\end{aligned} \tag{11}$$

where G stands for the oxygen-glass bond and D represents deposited

$$\begin{bmatrix} 1 \sim \text{SiE}_4 \\ 2 \sim \text{Si(OH)E}_3 \\ 3 \sim \text{H}_2\text{O} \\ 4 \sim \text{C}_2\text{H}_5\text{OH} \\ 5 \sim \text{C}_2\text{H}_4 \\ 6 \sim \text{vacancy} \\ 7 \sim \text{SiG(OH)E}_2 \\ 8 \sim \text{SiGE}_3 \\ 9 \sim \text{SiG}_3\text{E} \\ 10 \sim \text{SiG}_3\text{(OH)} \\ 11 \sim \text{SiG(OH)}_2\text{E} \end{bmatrix}, \tag{12}$$

the stoichiometric coefficient matrix for the eight heterogeneous reactions of Eq. (11) is

$$\mathbf{S} = \begin{bmatrix} -1 & 0 & 0 & 1 & 0 & 1 & 0 & 1 & 0 & -1 & 0 \\ 0 & 0 & 0 & 0 & 1 & 0 & 0 & 0 & -1 & 1 & 0 \\ 0 & 0 & 0 & 0 & 1 & 0 & -1 & 0 & 0 & 0 & 1 \\ 0 & 0 & 0 & 0 & 1 & 0 & 1 & -1 & 0 & 0 & 0 \\ 0 & 0 & 0 & 1 & 0 & 0 & 0 & 0 & 0 & 1 & -1 \\ 0 & 0 & 0 & 1 & 0 & 0 & -1 & 0 & 1 & 0 & 0 \\ 0 & 0 & 1 & 0 & 0 & 0 & 0 & 0 & 1 & 0 & -1 \\ 0 & -1 & 1 & 0 & 0 & 1 & 0 & 1 & 0 & -1 & 0 \end{bmatrix}. \tag{13}$$

Because this is a detailed reaction mechanism, these reactions are assumed to occur "as written" in terms of the appropriate "activity," that is, surface concentration (υ) or flux (η), as

$$R_1 = 2.2 * 10^{14} * \exp\left(\frac{-22900}{T}\right) * \eta_1 * \upsilon_{10},$$

$$R_2 = 1.2 * 10^{12} * \exp\left(\frac{-13400}{T}\right) * \upsilon_9 - 1.26 * 10^6 * \exp\left(\frac{-19240}{T}\right) * \eta_5 * \upsilon_{10},$$

$$R_3 = 2.4 * 10^{12} * \exp\left(\frac{-23650}{T}\right) * \upsilon_7 - 3.96 * 10^6 * \exp\left(\frac{-19180}{T}\right) * \eta_5 * \upsilon_{11},$$

$$R_4 = 3.6 * 10^{12} * \exp\left(\frac{-23650}{T}\right) * \upsilon_8 - 4.45 * 10^7 * \exp\left(\frac{-19310}{T}\right) * \eta_5 * \upsilon_7,$$

$$R_5 = 1.4 * 10^{12} * \exp\left(\frac{-22140}{T}\right) * \upsilon_{11}$$

$$- 1.76 * 10^{6} * \exp\left(\frac{-25970}{T}\right) * \eta_4 * \upsilon_{10},$$

$$R_6 = 1.4 * 10^{12} * \exp\left(\frac{-22140}{T}\right) * \upsilon_7 - 2.76 * 10^{6} * \exp\left(\frac{-25970}{T}\right) * \eta_4 * \upsilon_9,$$

$$R_7 = 1.4 * 10^{12} * \exp\left(\frac{-22650}{T}\right) * \upsilon_{11} - 9.78 * 10^{6} * \exp\left(\frac{-26440}{T}\right) * \eta_3 * \upsilon_9,$$

$$R_8 = 6.68 * 10^{11} * \exp\left(\frac{-6587}{T}\right) * \eta_2 * \upsilon_{10}$$

$$- 1.42 * 10^{14} * \exp\left(\frac{-9199}{T}\right) * \eta_3 * \upsilon_6 * \upsilon_8. \quad (14)$$

III. Low-Pressure Transport

In this section, some of the basics of free molecular transport and reaction are discussed. Further discussions can be found in Refs. *8–10,25,35–40*.

A. Coordinate Systems

In the following, we consider both global and local coordinate systems and both rectangular and spherical coordinates. In the global coordinate system, the z-axis points upward, the x-axis points to the right along the page, and the y-axis points into the page. In the local coordinate system, the z-axis is specified by some other attribute of the problem, such as the direction of a normal vector to a surface. In either the global or local spherical coordinate system, the angle θ is measured relative to the z-axis and ranges from 0 to π. The angle ϕ is measured relative to the x-axis and ranges from 0 to 2π. Vectors in these coordinates, when written relative to the origin, are

$$\mathbf{r} = x\mathbf{i} + y\mathbf{j} + z\mathbf{k} \quad (15)$$

in rectangular coordinates and

$$\mathbf{r} = r\mathbf{e}_r = r[\sin\theta\cos\phi\mathbf{i} + \sin\theta\sin\phi\mathbf{j} + \cos\theta\mathbf{k}] \tag{16}$$

in spherical coordinates, where $\mathbf{e}_\mathbf{r}$ is the unit vector in the direction specified by θ and ϕ. Integrations can be written in terms of either coordinate system.

In the following a feature is defined by a surface, which is given symbolically by ∂R and a model or simulation domain given symbolically by R; ∂R refers only to the solid surface, and does not include the boundary in the source volume ∂V. Usually, there is no need to define the top of the simulation domain (or to completely define R), although that is sometimes done.

B. Single Particle Distribution Function

To treat transport, it is helpful to introduce the *single particle distribution function* or pdf. We bypass the details for brevity, and focus on the essentials. For a more complete discussion of the Boltzmann equation and its implications and limitations for the problem at hand, see Refs. *41, 42*. Let $f(\mathbf{x}, \mathbf{v}, t)\,\mathbf{dx}\,\mathbf{du}$, where f represents the pdf, be the number of "particles" in physical volume $\mathbf{dx}$ about $\mathbf{x}$ having velocities within $\mathbf{dv}$ about $\mathbf{v}$ at time t, where $\mathbf{dx} = dx\,dy\,dz$ and $\mathbf{dv} = dx_x\,dv_y\,dv_z$ in rectangular coordinates. In the following, we assume that the time dependence of the distribution function can be ignored, except perhaps for some prescribed function of time. Integrating over all velocities in rectangular coordinates $(-\infty < v_i < \infty)$, we obtain the number of particles per unit volume or number density at position $\mathbf{x}$ as

$$n(\mathbf{x}) = \int_{-\infty}^{\infty} f(\mathbf{x}, \mathbf{v})\,\mathbf{dv} = \int_{-\infty}^{\infty}\int_{-\infty}^{\infty}\int_{-\infty}^{\infty} f(\mathbf{x}, v_x, v_y, v_z)\,dv_x\,dv_y\,dv_z, \tag{17}$$

or in spherical coordinates,

$$n(\mathbf{x}) = \int_0^{\infty}\int_0^{2\pi}\int_0^{\pi} f(\mathbf{x}, v, \theta, \phi)v^2 \sin\theta\,d\theta\,d\phi\,dv. \tag{18}$$

C. Fluxes

Consider a differential area $d\mathbf{A}$ about $\mathbf{x}$, specified by $dA\mathbf{a}$, where $\mathbf{a}$ is its normal vector. This area can be in space or on a surface. Given f at the point $\mathbf{x}$, the flux away from it in the direction of $\mathbf{a}$ (out of the same side of the area which the normal points) is

$$\eta^1(\mathbf{x}) = \int_0^\infty \int_0^{2\pi} \int_0^{\pi/2} f(\mathbf{x}; v, \theta, \phi)(\mathbf{e}_v \cdot \mathbf{a}) \xi v^3 \sin\theta \, d\theta \, d\phi \, dv, \quad (19)$$

where the superscript 1 signifies leaving and

$$\xi = u(\mathbf{e}_v \cdot \mathbf{a}), \quad (20)$$

where u is the unit step function; that is, $\xi = 1$ if its argument is non-negative and zero if its argument is negative. This is introduced for convenience as well as to emphasize that only particles traveling in the direction of the normal are considered; this is not net flux. The integration is done in the local coordinate system with its origin at $\mathbf{x}$, and its z $(\theta = 0)$ axis in the direction of $\mathbf{a}$, and the upper limit on the theta integration is used to emphasize this fact. The subscript v on the unit vector in the radial direction in spherical coordinates is used to emphasize that the integration is performed over velocity space. Note, however, that specifying theta and phi specifies a direction in physical space. For a differential area in space, the flux leaving the area from the same side as $\mathbf{a}$ is the same as the flux arriving in that direction from the other side of the area. In an analogous manner, the flux of particles to a point $\mathbf{x}$ (in the direction opposite $\mathbf{a}$) is

$$\eta^a(\mathbf{x}) = -\int_{-\infty}^\infty \int_0^{2\pi} \int_0^{\pi/2} f(\mathbf{x}, v, \theta, \phi)(\mathbf{e}_v \cdot \mathbf{a}) \psi v^2 \sin\theta \, d\theta \, d\phi \, dv, \quad (21)$$

where superscript a signifies arriving, and

$$\psi = 1 - u(\mathbf{e}_v \cdot \mathbf{a}) \quad (22)$$

so that ψ is zero when the argument of the step function is non-negative. Thus, specifying the pdf at a point in space, and perhaps at a given time, specifies the flux distribution. In other words, specifying the pdf arriving at the surface specifies the flux distribution to the surface.

The flux which leaves one point and arrives at another is central to free molecular transport. In the absence of collisions and under steady conditions, the Boltzmann equation indicates that f does not change along particle trajectories. Consider a differential area $d\mathbf{A}' = dA'\mathbf{a}'$ around point $\mathbf{x}'$ which is distance s from $\mathbf{x}$ and subtends solid angle

$$d\omega = \sin\theta \, d\theta \, d\phi = -\frac{dA'(\mathbf{e}_s \cdot \mathbf{a})}{s^2} \quad (23)$$

as measured at $\mathbf{x}$. The flux leaving a point $\mathbf{x}$ and arriving at $\mathbf{x}'$ can be arrived at by substituting for $d\omega$ in Eq. (21) and dividing by dA':

$$\eta(\mathbf{x} \to \mathbf{x}') = -\left[\frac{(\mathbf{e}_s \cdot \mathbf{a})(\mathbf{e}_s \cdot \mathbf{a}')\xi\psi'\kappa}{s^2}\right]\int_0^\infty f(\mathbf{x}, v, \theta, \phi)v^3\, dv. \quad (24)$$

The factors ξ and ψ' have the same meaning as in Eqs. (20) and (22), and κ is a *visibility factor* which is 1 if the two points are in line of sight and 0 if their mutual view is blocked. Thus, f does not change from $\mathbf{x}$ to $\mathbf{x}'$, unless transport in that direction is blocked. Note that the geometric variables are specified by the relative positions of the two points $\mathbf{x}$ and $\mathbf{x}'$ and their normal vectors. If f is known at $\mathbf{x}$, then the integral can be evaluated. For convenience, we introduce the *differential transmission probability* as

$$q(\mathbf{x}; \mathbf{x}') = -\left[\frac{(\mathbf{e}_s \cdot \mathbf{a})(\mathbf{e}_s \cdot \mathbf{a}')\xi\psi'\kappa}{s^2}\right] = -\left[\frac{\cos\Omega \cos\Omega'\xi\psi'\kappa}{s^2}\right], \quad (25)$$

which is purely geometric, and Ω is the angle which $\mathbf{e}_s$ makes relative to the normal at $\mathbf{x}$, and Ω' is the angle which $\mathbf{e}_s$ makes relative to the normal at $\mathbf{x}'$. More generally, the unit vector $\mathbf{e}_s$ in the local coordinate system at $\mathbf{x}$ can be replaced by the unit normal pointing from $\mathbf{x}$ to $\mathbf{x}'$.

D. Flux Distributions

Some specific flux models can be arrived at by considering the flux from a differential area $d\mathbf{A}$, as in Sec. III.C. We assume that the pdf for a species can be written as

$$f(\mathbf{x}, \mathbf{v}) = n(\mathbf{x})p(\mathbf{v}), \quad (26)$$

where the velocity distribution function is normalized so that the integration indicated in Eq. (17) or (18) yields the number density. If the gas is in collisional equilibrium, then Eq. (26) would be the usual Maxwellian distribution function. If we assume that the function $p(\mathbf{v})$ can be written as a product of a function $p'(v)$ of the speed and a function $g(\theta)$ of the angle measured relative to the normal to $d\mathbf{A}$, the flux of the i-th species passing through $d\mathbf{A}$ in the solid angle $d\omega$ [see Eq. (23)] can be written

$$\eta(\theta, \phi) = n(\mathbf{x})g(\theta)\cos\theta\, d\omega\int_0^\infty p'(v)\, dv = n(\mathbf{x})g(\theta)\cos\theta\, d\omega P, \quad (27)$$

where P has been defined implicitly and is a constant for a specified speed distribution function. This assumes cylindrical symmetry and that speed is not a function of angle. The total flux through $d\mathbf{A}$ is

$$\eta^t = 2\pi n(\mathbf{x}) P \int_0^{\pi/2} g(\theta) \cos\theta \sin\theta \, d\theta \tag{28}$$

and Eq. (27) can be written as

$$\eta(\theta, \phi) = \frac{\eta^t g(\theta) \cos\theta \, d\omega}{G} = \frac{\eta^t g(\theta) \cos\theta \sin\theta \, d\theta \, d\phi}{2\pi \int_0^{\pi/2} g(\theta) \cos\theta \sin\theta \, d\theta}, \tag{29}$$

where G has been defined implicitly, and is the normalization constant for a specified g. The flux distribution is said to be "cosine" if $g(\theta) = 1$ and $G = \pi$. If $g = 1$, and the distribution of velocities is Maxwellian, then $P = \langle v \rangle / 4\pi$, where $\langle v \rangle$ is the mean speed.

A reasonable assumption for CVD processes is that $g = 1$ for all species in the source volume and that the total flux of each species is given by $\langle v_i \rangle n_i / 4$, where $\langle v_i \rangle$ is the average speed of the i-th species and n_i is its number density. For systems which have more directional species fluxes, other simple models have been used. For sputtered atoms in PVD systems, we have used

$$g(\theta) = \frac{1}{\rho \sin^2\theta + \cos^2\theta}. \tag{30}$$

If $\rho = 1$, then $g = 1$ and the source is isotropic (cosine). If $\rho > 1$, then the source is termed over-cosine and there is some "beaming" of the atoms. On the other hand, if $\rho < 1$, then the source is termed under-cosine. The more collisions which occur between the target and the substrate, the more cosine the distribution becomes. A function which has been used for ions in plasma deposition systems is

$$g(\theta) = \frac{1}{\sqrt{2\pi}\sigma} \exp\left(-\frac{\theta^2}{2\sigma^2} \right), \tag{31}$$

where the width of the distribution is specified by σ and depends on the number of collisions of ions with other particles in the sheath, which in turn depends on operating conditions.

The preceding analysis for the angular dependence of flux from the source to the features can be extended to emission from interior surfaces. There is not much information available regarding the angular distribution of species emitted from surfaces typically found in semiconductor materials processing. The angular distribution of flux may in general depend on the species considered and on the composition and structure of the emitting surface. In the absence of such information, a reasonable assumption is that species which equilibrate with the surface are emitted diffusely with a cosine distribution (all directions equally likely). This assumption has provided excellent comparisons between simulated and experimental films profiles (e.g., see *43*).

These simple models are not adequate for all simulations, and more complete flux distributions can be obtained. Our Monte Carlo based simulations used to estimate the flux distributions for species traveling through plasma sheaths are discussed in Refs. *44–47*. Our simulations for transport through collimators are discussed in Refs. *48–51*.

IV. Ballistic Transport and Reaction Model

More complete descriptions of the BTRM can be found in publications by Cale *et al.* (*8–10*), IslamRaja *et al.* (*25*), Singh *et al.* (*37*), and Hseih and Joshi (*38*). The BTRM is based on species balances for the various species which exist in the deposition system, both in the gas phase and on the surface of the evolving film.

A. Model Statements

The basic assumptions of the BRTM are as follows:

1. The frequency of collisions between gas phase species is small relative to the frequency of collisions between gas phase species and surfaces; that is, intrafeature transport is by free molecular flow.
2. Deposition occurs by heterogeneous reactions between gas phase species and the evolving film surface.
3. The film grows slowly relative to the redistribution of fluxes to the feature surfaces caused by film evolution.
4. The open end of the feature is exposed to a gas, from which species travel to the surface, arriving with well-defined flux distributions.

The third assumption is supported by comparing molecular speeds with the speed of the evolving film profile. The second assumption does not imply that homogeneous reactions are not important to film deposition. Homogeneous reactions are not significant inside features in this transport regime; however, they might be important in the reactor volume. Homogeneous reactions can indeed lead to precursors which then react on the surface. The fourth assumption implies that the total and the angular dependence of species fluxes arriving at the surface need to be known, ideally from reactor scale simulations.

B. GAS/SOLID INTERACTIONS

This is a critical point in the development of the BTRM. To go further, assumptions about how molecules interact with the evolving surfaces determine whether the equations which govern transport can be represented in closed form. The details of how species interact with the solid are not considered in the topography evolution models or simulators to date. In the following, we consider two simple limiting cases: diffuse and specular re-emission. Diffuse means incoming molecules can be considered to reach thermal equilibrium with the surface, before being re-emitted. In that case, a re-emitted molecule carries forward no knowledge of its velocity before hitting the surface. Specular means that the molecule does not equilibrate with the surface. It is reflected, leaving with a velocity which is determined by its incoming velocity and the local surface normal. Interactions in reality can be expected to be somewhere between these two extremes. Whether these simple limiting cases are satisfactory depends on the accuracy of the topography evolution predictions; that is, how well the simulators based on these models reflect experimental information. In the vast majority of cases studied to date, diffuse re-emission was apparently a reasonable assumption. In a few cases, specular re-emission has been used to advantage (*50,52,53*). Although these idealized re-emission models may not be realistic, they form important limiting cases from which to build more complete models, if they are necessary. During the next few years, situations that use more realistic re-emission modes will be introduced, perhaps beginning with combinations of the two simple modes of re-emission; that is, situations will probably arise for which the limiting assumptions will fail to predict deposited film profiles. More general discussions of modeling gas/solid interactions are available (*41,54*).

C. DIFFUSE RE-EMISSION

Diffuse re-emission as used here means that the species which are being re-emitted have reached thermal equilibrium with the re-emitting surface. Although not necessary, a Maxwellian velocity distribution is usually also assumed. Since each species travels through the vacuum, independent of all other species, the following equations are valid for any species. The flux of the i-th species arriving at point $\mathbf{x}$ on a surface of a feature directly from the source volume is written as an integral over the source, using a modification of Eq. (21), once the distribution function of the arriving species and surface topography are specified, as

$$\eta_i^{a,1}(\mathbf{x}) = -\int_{-\infty}^{\infty}\int_0^{2\pi}\int_0^{\pi} f_i^s(v,\theta,\phi)(\mathbf{e}_v\cdot\mathbf{a})\psi\chi(\mathbf{x},\mathbf{e}_v)v^3\sin\theta\,d\theta\,d\phi\,dv \qquad \text{for } \mathbf{x}\in\partial R, \quad (32)$$

where χ is 1 if the source volume can be seen from $\mathbf{x}$ in the direction $\mathbf{e}_v$, and 0 if its view to the source volume is blocked by material (shadowing). The superscript s on f signifies that this is the distribution function of the i-th species arriving at the surface from the source volume. Again, note that specifying f specifies a flux distribution.

The flux of the i-th species arriving at point $\mathbf{x}$ from point $\mathbf{x}'$ after striking the surface once and only once can be written as [see Eqs. (24) and (25)]:

$$\eta_i^{a,2}(\mathbf{x}'\to\mathbf{x}) = q(\mathbf{x}';\mathbf{x})\int_0^{\infty} f_i^1(\mathbf{x}',v,\theta,\phi)v^2\,dv \qquad \text{for } \mathbf{x}',\mathbf{x}\in\partial R, \quad (33)$$

where the superscript $a,2$ means that this is the flux of the i-th species arriving at $\mathbf{x}$ for their second impact and the argument means that it is the incremental flux from $\mathbf{x}'$. In general, there will be "particles" arriving at $\mathbf{x}$ for their second impact from other points on the surface. Similarly, the flux of the i-th species which arrive at $\mathbf{x}$ from $\mathbf{x}'$ after having their second impact is

$$\eta_i^{a,3}(\mathbf{x}'\to\mathbf{x}) = q(\mathbf{x}';\mathbf{x})\int_0^{\infty} f_i^2(\mathbf{x}',v,\theta,\phi)v^3\,dv \qquad \text{for } \mathbf{x}',\mathbf{x}\in\partial R. \quad (34)$$

More generally, the flux of the i-th species arriving at $\mathbf{x}$ from $\mathbf{x}'$ after hitting the surface k times is

$$\eta_i^{a,k+1}(\mathbf{x}'\to\mathbf{x}) = q(x';\mathbf{x})\int_0^{\infty} f_i^k(\mathbf{x}',v,\theta,\phi)v^3\,dv \qquad \text{for } \mathbf{x}',\mathbf{x}\in\partial R. \quad (35)$$

The total flux of the i-th species to $\mathbf{x}$ from $\mathbf{x}'$, independent of how many times they have struck the surface is naturally the sum over the impacts

$$\eta_i^{a,t}(\mathbf{x}'\to\mathbf{x}) = q(\mathbf{x}';\mathbf{x})\sum_{k=1}^{\infty}\int_0^{\infty} f_i^k(\mathbf{x}',v,\theta,\phi)v^3\,dv \qquad \text{for } \mathbf{x}',\mathbf{x}\in\partial R. \quad (36)$$

If the form of f_i at $\mathbf{x}'$ does not depend on the history of the emitted species, and it is proportional to the number of molecules which have struck k times, then this sum can be evaluated. For example, for diffuse re-emission and a Maxwellian velocity distribution, each integral is the flux leaving $\mathbf{x}'$ after their k-th impact, divided by π, and the sum is the total flux leaving $\mathbf{x}'$, independent of the number of impacts, divided by π. More generally, if the speed dependence of f_i is uncoupled from the angular dependencies, each integral can be written as the total flux of the i-th species leaving $\mathbf{x}'$, which have struck the surface k times, multiplied by a geometric factor $g_i(\mathbf{x}', \mathbf{x})$ (see Sec. III, and *35*). This is accomplished by integrating over speed, and $g_i(\mathbf{x}', \mathbf{x})$ depends on the form of the distribution function of the i-th species at $\mathbf{x}'$ and on the angles θ and ϕ set by the unit vector pointing from $\mathbf{x}'$ to $\mathbf{x}$ measured relative to the normal vector at $\mathbf{x}'$, namely,

$$\int_0^{\infty} f_i^k(\mathbf{x}', v, \theta, \phi) v^3 \, dv = g_i(\theta, \phi) \eta_i^{1,k}(\mathbf{x}'), \tag{37}$$

where $g = 1/\pi$ for Maxwellian re-emission. The total flux arriving at $\mathbf{x}$ from $\mathbf{x}'$, independent of the number of previous impacts, is

$$\begin{aligned} \eta_i^{a,t}(\mathbf{x}' \to \mathbf{x}) &= q(\mathbf{x}; \mathbf{x}') g_i(\theta, \phi) \sum_{k=1}^{\infty} \eta_i^{1,k}(\mathbf{x}') \\ &= q(\mathbf{x}; \mathbf{x}') g_i(\theta, \phi) \eta_i^{1,t}(\mathbf{x}') \qquad \text{for } \mathbf{x}', \mathbf{x} \in \partial R. \end{aligned} \tag{38}$$

The total flux of the i-th species to $\mathbf{x}$, from the source as well as from all other points $\mathbf{x}'$, is, for the case of diffuse re-emission with a Maxwellian velocity distribution,

$$\eta_i^{a,t}(\mathbf{x}) = \eta_i^{a,1}(\mathbf{x}) + \int_{\partial R} \frac{q(\mathbf{x}'; \mathbf{x}) \eta_i^{1,t}(\mathbf{x}')}{\pi} \, d\mathbf{x}' \qquad \text{for } \mathbf{x}', \mathbf{x} \in \partial R. \tag{39}$$

The total flux leaving $\mathbf{x}'$ needs to be evaluated. In the absence of reaction, the total flux leaving is equal to the total flux arriving, and Eq. (39) can be solved once a surface (feature) and the distribution function of the i-th species from the source are specified. In the presence of reactions, a species balance can be arrived at in a manner similar to that used earlier, by including the areal rate of generation R_i of the i-th species at $\mathbf{x}'$, which in general is a function of the fluxes $\boldsymbol{\eta}^{a,t}$ of each species arriving at $\mathbf{x}'$, the surface fractions $\boldsymbol{v}$ of all of the surface species at $\mathbf{x}'$, and the temperature, through the reaction kinetics. In the derivation, one must carefully consider the number of times the generated molecules of the i-th species have been re-emitted since being created. If the areal rate is not a function of the number of strikes for any species, and all molecules of a given species are indistinguishable, which is implied by the assumption of thermal equilibrium, then the total flux arriving at $\mathbf{x}$ can be written as

$$\eta_i^{a,t}(\mathbf{x}) = \eta_i^{a,1}(\mathbf{x}) + \int_{\partial R} q(\mathbf{x}',\mathbf{x}) g_i(\theta,\phi)\left[\eta_i^{a,t}(\mathbf{x}') + R_i(\boldsymbol{\eta}^{a,t},\boldsymbol{\vartheta},T)\right] d\mathbf{x}' \quad \text{for } \mathbf{x}',\mathbf{x} \in \partial R. \tag{40}$$

In some cases, it is convenient to introduce the *reactive sticking factor* of the i-th species σ_i which is defined as the fraction of arriving flux which is not re-emitted; that is, the fraction of incoming flux which adsorbs or reacts. When this definition can be used, the total flux arriving at $\mathbf{x}$ is written as

$$\eta_i^{a,t}(\mathbf{x}) = \eta_i^{a,1}(\mathbf{x}) + \int_{\partial R} q(\mathbf{x}',\mathbf{x}) g_i(\theta,\phi)\eta_i^{a,t}(\mathbf{x}')(1-\sigma_i)\, d\mathbf{x}'. \tag{41}$$

There is one of these equations [e.g., Eq. (39)] for each ballistic species in the deposition, etch (or catalytic) system. The equations are in general coupled through the reaction term, because reaction rates depend on the fluxes of more than one species in general. There are several ways to solve this equation, and direct quadrature works well in the majority of cases. For features which can be represented in two dimensions (long lines or trenches and features of circular horizontal cross-section) the surface is fully represented by a set of line segments, and the integrals are evaluated to estimate the total flux of all species to the center point on each segment from the source volume as well as the entire three-dimensional feature.

D. Specular Transport

For specular transport, the simplification which is possible for diffuse re-emission to get from equations like Eq. (35) to those like Eq. (38) is not viable; because the distribution functions of re-emitted molecules are specified by the distribution functions of the arriving molecules, rather than being independent of the history of the incident species. Thus, it is fruitless to proceed in the same way as for diffusely re-emitted species; nevertheless, the form of Eq. (39) is desirable in order to retain integration accuracy similar to that obtained for the arrival of diffusely re-emitted species to a point on the surface. This is done by tracking the fluxes to a point on the surface back to the source, where the distribution function for each species is specified.

For specularly reflected species, there is a simple relationship between f's for the incident and reflected particles:

$$f^1(\mathbf{x},\mathbf{v}) = f^a(\mathbf{x},\mathbf{v} - 2\mathbf{a}(\mathbf{a}\cdot\mathbf{v})) \qquad \text{for } \mathbf{a}\cdot\mathbf{v} > 0, \mathbf{x} \in \partial R, \tag{42}$$

where $\mathbf{a}$ is the outward normal to the surface at $\mathbf{x}$ and $\mathbf{v}$ is the velocity leaving $\mathbf{x}$. The second argument on the right-hand side is the velocity arriving at $\mathbf{x}$, which is designated $\mathbf{v}^1$ to indicate that it has been shifted once to arrive at the velocity of interest, that is,

$$\mathbf{v}^1 = \mathbf{v} - 2\mathbf{a}(\mathbf{a}\cdot\mathbf{v}). \tag{43}$$

Thus, the only difference between the two f's is that the velocities have shifted. Along with the constancy of f along particle trajectories, this is key to simulating specular transport. Generalizing somewhat, we introduce β, which is the fraction of the incoming flux of the i-th species which is re-emitted specularly:

$$f_i^1(\mathbf{x},\mathbf{v}) = \beta_i(\mathbf{x},\mathbf{v}^1) f_i^a(\mathbf{x},\mathbf{v}^1) \qquad \text{for } \mathbf{a}\cdot\mathbf{v} > 0, \mathbf{x} \in \partial R, \tag{44}$$

where β depends on species, position (perhaps through the material at $\mathbf{x}$), and the velocity vector.

Consider point $\mathbf{x}$ on the surface and the direction specified by velocity $\mathbf{v}$, such that species will arrive from another point of the surface $\mathbf{x}^{k-1}$ by traveling in the direction specified by $\mathbf{v}$. Because the distribution function of species which leave point $\mathbf{x}^{k-1}$ does not change along a trajectory, then

$$f_i^{a,k}(\mathbf{x}^k,\mathbf{v}) = f_i^{l,k-1}(\mathbf{x}^{k-1},\mathbf{v}) \qquad \text{for } \mathbf{a}^k\cdot\mathbf{v} < 0, a^{k-1}\cdot\mathbf{v} > 0, \mathbf{x}^k, \mathbf{x}^{k-1} \in \partial R, \tag{45}$$

where superscript a, k means arriving for the k-th impact, and superscript $l, k-1$ means leaving after the $k-1$-th impact. The distribution function for the i-th species leaving point $\mathbf{x}^{k-1}$ is given in terms of the distribution function for the i-th species arriving for their $k-1$-th impact along trajectory $\mathbf{v}^1$, where the superscript 1 has the meaning introduced above, by

$$f_i^{l,k-1}(\mathbf{x}^{k-1},\mathbf{v}) = \beta_i^{k-1}(\mathbf{x}^{k-1},\mathbf{v}^1) f_i^{a,k-1}(\mathbf{x}^{k-1},\mathbf{v}^1) \qquad \text{for } \mathbf{a}^{k-1}\cdot\mathbf{v}^1 < 0, \mathbf{a}^{k-1}\cdot\mathbf{v} > 0, \mathbf{x}^{k-1} \in \partial R, \tag{46}$$

similarly, $f_i^{a,k-1}(\mathbf{x}^{k-1},\mathbf{v}^1)$ is given by

$$f_i^{a,k-1}(\mathbf{x}^{k-1},\mathbf{v}^1) = f_i^{1,k-2}(\mathbf{x}^{k-2},\mathbf{v}^1) \qquad \text{for } \mathbf{a}^{k-1}\cdot\mathbf{v}^1 < 0, \mathbf{a}^{k-2}\cdot\mathbf{v}^1 > 0, \mathbf{x}^{k-1}, \mathbf{x}^{k-2} \in \partial R, \tag{47}$$

and $f_i^{l,k-2}(\mathbf{x}^{k-2}, \mathbf{v}^1)$ is given by

$$f_i^{l,k-2}(\mathbf{x}^{k-2}, \mathbf{v}^1) = \beta_i^{k-2}(\mathbf{x}^{k-2}, \mathbf{v}^2) f_i^{a,k-2}(\mathbf{x}^{k-2}, \mathbf{v}^2)$$
$$\text{for } \mathbf{a}^{k-2} \cdot \mathbf{v}^2 > 0, \mathbf{a}^{k-2} \cdot \mathbf{v}^1 > 0, \mathbf{x}^{k-2} \in \partial R, \quad (48)$$

where the velocity is shifted at $\mathbf{x}^{k-2}$ according to the local normal vector. More generally, for the j-th previous impact,

$$f_i^{a,k-j}(\mathbf{x}^{k-j}, \mathbf{v}^j) = f_i^{l,k-j-1}(\mathbf{x}^{k-j-1}, \mathbf{v}^j) \quad \text{for}$$
$$j = 0 \text{ to } k-2, \mathbf{a}^{k-j} \cdot \mathbf{v}^j < 0, \mathbf{a}^{k-j-1} \cdot \mathbf{v}^j > 0, \mathbf{x}^{k-j}, \mathbf{x}^{k-j-1} \in \partial R \quad (49)$$

and

$$f_i^{l,k-j-1}(\mathbf{x}^{k-j-1}, \mathbf{v}^j) = \beta_i^{k-j-1}(\mathbf{x}^{k-j-1}, \mathbf{v}^{j+1}) f_i^{a,k-j-1}(\mathbf{x}^{k-j-1}, \mathbf{v}^{j+1})$$
$$\text{for } j = 0 \text{ to } k-2, \mathbf{a}^{k-j-1} \cdot \mathbf{v}^{j+1} < 0, \mathbf{a}^{k-j-1} \cdot \mathbf{v}^j > 0, \mathbf{x}^{k-j-1} \in \partial R. \quad (50)$$

Let the first impact from the source volume be $j = k - 1$, then

$$f_i^{a,1}(\mathbf{x}, \mathbf{v}^{k-1}) = f_i^s(\mathbf{v}^{k-1}) \quad \text{for } \mathbf{a}^1 \cdot \mathbf{v}^{k-1} < 0, \mathbf{x}^1 \in \partial R, \quad (51)$$

which indicates that the flux arrives directly from the source, and the value of f^s to be used is the value for the $\mathbf{v}$ determined by the original $\mathbf{v}$ in the integration at $\mathbf{x}^k$. The distribution function for the i-th species which arrives at point $\mathbf{x}^k$ along the trajectory specified by $\mathbf{v}$ can be written as

$$f_i^{a,k}(\mathbf{x}^k, \mathbf{v}) = f_i^s(\mathbf{v}^{k-1}) \prod_{j=1}^{k-1} \beta_i^j(\mathbf{x}^j, \mathbf{v}^{k-j}) \quad \text{for } \mathbf{a}^j \cdot \mathbf{v}^{k-j} > 0, \mathbf{x}^j \in \partial R, \quad (52)$$

where it is understood that the distribution function of the i-th species in the source is evaluated at the velocity which has been shifted $k - 1$ times from the initial trajectory $\mathbf{v}$, and that these shifts depend on the normal vectors at each of the points on the surface along the path from $\mathbf{x}^k$ to the source.

The flux to point $\mathbf{x}$ can be written as

$$\eta_i^a(\mathbf{x}) = -\int_0^\infty \int_0^{2\pi} \int_0^\pi f_i^s(\mathbf{v}^{k-1})[\mathbf{e}_v \cdot \mathbf{a}]\psi \left[\prod_{j=1}^{k-1} \beta_i^j(\mathbf{x}^j, \mathbf{v}^{k-j})\right] \times v^3 \sin\theta \, d\theta \, d\phi \, dv. \quad (53)$$

The integration of Eq. (53) by quadrature, leads immediately to discrete values of v, as well as different trajectories. Note that k is in general different for each trajectory. It is determined by tracing the path which the molecules must have taken to arrive specularly at $\mathbf{x}$ along $\mathbf{v}$. The integration limits in Eq. (53) specify all trajectories, possible or not, and the factor ψ limits the trajectories to those which arrive at $\mathbf{x}$ from the vacuum. Equation (53) is of the same form as Eq. (40), except for the term for the flux directly from the source volume. If the quadratures used to evaluate these are the same, then the equations should provide about the same accuracy. As before, the integrations over speed can be performed once, if β does not depend on velocity and the form of the distribution function of the i-th species in the source is of appropriate form. On the other hand, there are some practical aspects of evaluating Eq. (53) which must be taken into account. One is the number of strikes at each increment of angle. Some reasonable upper limit on k is needed, above which the contribution to the flux is ignored. A surprisingly small number (say, 20) usually suffices, because a molecule will leave the feature after a few bounces, in general. Another point is that the points on the surface along the path back to the source must be computed. This involves a number of computations and introduces different errors than in the case of diffuse re-emission.

The next point of concern is what happens to the molecules which are not re-emitted specularly. In our simple model, they are converted to diffuse species. In that case, they either react, stick, or re-emit according to the reaction kinetics. Finally, the case in which specular species are created by reactions on the surface has not been considered to date.

E. Surface Species

We now consider surface species, mentioned in the preceding discussion, in more detail. We focus the discussion on features which can be represented by a cross-section, as described earlier. In that case, and within the overall assumptions used in feature scale transport and reaction simulations, gradients in the surface concentrations occur along the cross-section. A material balance on a differential length of the surface cross-section leads to

$$\frac{\partial \vartheta_i(s,t)}{\partial t} = R_i(\boldsymbol{\eta}, \boldsymbol{\vartheta}, T) + \frac{\partial}{\partial s}\left[D_i \frac{\partial \vartheta_i(s,t)}{\partial s}\right], \tag{54}$$

where s is measured relative to a point on the feature cross-section. Within the pseudo-steady-state approximation used throughout this chapter, this reduces to

$$R_i(\boldsymbol{\eta}, \boldsymbol{\vartheta}, T) = -\frac{\partial}{\partial s}\left[D_i \frac{\partial \vartheta_i(s,t)}{\partial s}\right]. \tag{55}$$

Surface diffusivities and particularly their dependencies on surface properties are not well known, and in practice they are taken outside the operator. Often, the surface diffusion of these species can be neglected for the purposes of profile evolution. In that case, the rate of generation of each surface species is zero within the pseudo-steady-state approximation.

The relationship between surface species and ballistic species should be clear. Consider a surface without spatial gradients in the arriving fluxes; for example, a flat surface exposed to spatially uniform fluxes. The arriving species fluxes and surface temperature determine the properties of the surface, that is, the surface concentrations of adsorbed species, the reactions which occur, and the rate of surface evolution. We distinguish between ballistic and surface species, even if they have the same chemical formula. The relationships between these species are represented by adsorption and desorption reactions. Although this increases the number of species in a problem, as well as the number of reactions in some cases, there is a significant advantage gained by solving these distinctly different governing equations in a segmented fashion.

Though Eq. (55) is valid, the role of such surface diffusion in film profile evolution is not clear. The lifetimes of migrating surface species are in general not large enough for significant migration distances. Because of this, none of the examples in this chapter deal explicitly with concentration-driven surface diffusion of intermediates. On the other hand, curvature-driven surface diffusion has been shown to play a role in the film profile evolution of PVD films (*55,56*). Although the mechanism of curvature-driven surface diffusion may be the same as concentration-driven surface diffusion, it is treated very differently. An example of curvature-driven surface diffusion is discussed in Sec. XI.

F. Solution of Governing Equations

As noted earlier, the integrals are discretized over angles (in general, the velocity integrations can be performed once). Spatially, the surface is represented by line segments. The integrals are evaluated to determine the species fluxes to the midpoint of each segment. Emission from each segment is over the entire segment and the flux leaving each point on a segment is the average flux from that segment. Integrations for trenches and for features of circular horizontal cross-section are written in terms of the line segments by fully accounting for the entire surface represented by each segment. The conversion from three-dimensional to two-dimensional integrations, valid for most cases of interest to date, is discussed for infinite features in Ref. *8*. The differential operators are discretized using second-order accurate finite differences.

The discretized equations are solved using a *nonlinear overrelaxation* (NLORM) technique (*57*) in general. The guesses for species fluxes and surface fractions used to initiate the NLORM are the values on flat surfaces under spatially uniform flux conditions. The surface fraction of each surface species on each material in the problem is determined for deposition conditions at the water surface. Using these initial values, the ballistic species fluxes leaving each segment are computed from the specified chemical kinetics. This leads immediately to an updated set of arriving ballistic species fluxes and surface fractions. Iteration continues until convergence is obtained. Not surprisingly, this is the most time-consuming part of the simulations. Fortunately, the method virtually always works (*17*). A second method, used in situations for which NLORM method does not work, is beyond the scope of this chapter.

V. LPCVD of Silicon Dioxide from TEOS

This section discusses the use of EVOLVE, along with film profiles obtained from carefully designed experiments, to discriminate between two reaction rate expressions which have been proposed for the deposition of SiO_2 from TEOS. The two rate expressions are based on "simple" mechanistic models. They are tested by comparing EVOLVE's predicted film profiles, after optimizing each model's parameters, with experimental film profiles obtained from films deposited in a "differential" reactor. Details of how such studies are conducted can be found in Refs. *43*, and *58–61*.

A. BTRM for LPCVD

For LPCVD, two additional assumptions are added to the four basic assumptions of the BTRM listed in Sec. IV.A:

5. Species (unreacted reactants, intermediates, or products) re-emit from surfaces diffusely, with Maxwellian distributions.

6. Reactive sticking factors do no depend on the incident angle or collision history of the impinging molecules.

There is little information regarding these two assumptions; however, all CVD process simulations to date have included them and good agreement between simulation and experiments can be obtained; that is, these assumptions do not appear to be a limiting factor in the agreement. For CVD processes, it is also assumed that the species arriving from the reactor volume to the surface obey a cosine distribution; that is, they are from a Maxwellian distribution. Although this assumption is difficult to validate quantitatively, it requires only that there be local equilibrium near the wafer surface. To complete the models used in EVOLVE for this LPCVD process, intrinsic reaction rate expressions need to be specified. These are discussed later.

In addition to the kinetic and transport models, the temperature, the total species fluxes to the surface, and the details of the surface features need to be specified. The experimental portion of this study was performed in a "differential" reactor, where conversion levels were kept below 1%, typically below 0.1%, in order to assume that the change in TEOS partial pressure is small. This allows the TEOS partial pressures at the small (about 1 cm^2) substrate to be estimated from the inlet flows and the total pressures. The deposition temperature is assumed to be that measured by a thermocouple placed against the small substrate. More complete descriptions of the experimental work can be found in Refs. *58* and *62*.

B. Kinetics Models

Each proposed rate expression is based on a simple proposed mechanism or model, which is intended to capture the important physical phenomena. The kinetic models are called (I) the homogeneous intermediate mediated deposition model and (II) the by-product inhibition model. We should emphasize that the purpose is not to test mechanisms, but rather to choose between rate forms to determine which provides better predictions of step coverage or film profile evolution. These are of central interest to process engineers, not the mechanism. Nevertheless, the result of such studies can help decide between the conceptual models represented by the rate expressions and help narrow the range of experiments needed to further test the most favorable mechanisms.

Model I is adapted from the work of Desu (*34*), who used a hot wall, volume loaded (batch) reactor to study SiO_2 deposition from TEOS. His proposed "simplified" mechanistic model can be represented by

$$TEOS_{(g)} \overset{K_g}{\longleftrightarrow} I_{(g)} + R_{(g)}, \tag{56}$$

$$I_{(g)} + * \overset{K_I}{\longleftrightarrow} I^*, \tag{57}$$

$$I^* \overset{k_d}{\longrightarrow} SiO_2 + * + \text{by-products}_{(g)}, \tag{58}$$

where $R_{(g)}$ is an unspecified by-product of the homogeneous partial dissociation step, and $I_{(g)}$ is an unspecified gas phase derivative of TEOS. The nature of the by-product $R_{(g)}$, as well as the gaseous by-products of the intermediate surface decomposition step, depend on the assumed nature of the intermediate. Desu speculates that the intermediate might be diethoxysilane, although the exact identity of the intermediate is not critical to the current analysis. For the mechanistic sequence given, and assuming that the surface decomposition step is rate limiting, one can derive the following rate expression:

$$R_{SiO_2} = \frac{k_d K_I P_I}{1 + K_I P_I} = \frac{k_d K_I (K_g P_{TEOS})^{1/2}}{1 + K_I (K_g P_{TEOS})^{1/2}}. \tag{59}$$

This expression yields a half-order rate dependence on TEOS partial pressure in the limit of low surface coverage by the intermediate. Under these conditions, the decomposition rate is first order in intermediate partial pressure. A first-order dependence on TEOS partial pressure can be realized if the reversible homogeneous decomposition reaction does not go to equilibrium. This expression is equivalent to that arrived at by Adams and Capio (*63*), who also studied this reaction in a volume loaded, hot wall reactor. A number of studies have shown that the mechanism proposed by Adams and Capio is not consistent with available data. For example, the sticking factor for TEOS is on the order of 10^{-5} under reaction conditions. Within the kinetic model proposed by Adams and Capio, this would lead to excellent conformality, which is at odds with experimental evidence. The unspecified intermediate in Desu's model has a much larger effective sticking factor (on the order of 10^{-2}) than TEOS.

In Model II, the deposition of SiO_2 proceeds through heterogeneous decomposition of adsorbed TEOS, and that gaseous by-products formed by the deposition reaction readsorb on the growing film surface, inhibiting deposition. This model is based on the observed characteristic behavior of TEOS sourced SiO_2 films. After decreasing in thickness as the feature is entered from the flat surface outside the feature, the film thickness is relatively uniform inside the feature. Indeed, this model was proposed by Schlote *et al.* (*64*) and Cale *et al.* (*43*) to explain the characteristic conformality behavior of TEOS sourced SiO_2. For kinetic modeling purposes, the mechanism is

$$\mathrm{TEOS} + * \overset{K_T}{\longleftrightarrow} \mathrm{TEOS}^*, \tag{60}$$

$$\mathrm{TEOS}^* \overset{k_d}{\longrightarrow} \mathrm{I}^* + \mathrm{R}_{(g)}, \tag{61}$$

$$\mathrm{I}^* \overset{k_s}{\longrightarrow} \mathrm{SiO}_2 + * + (\nu_\mathrm{R} - 1)\mathrm{R}_{(g)} + \text{by-products}_{(g)}, \tag{62}$$

$$\mathrm{R}_{(g)} + * \overset{K_R}{\longleftrightarrow} \mathrm{R}^*. \tag{63}$$

In this sequence, ν_R is the stoichiometric coefficient of the inhibiting by-product in the net deposition reaction. Assuming that the initial heterogeneous decomposition step is rate limiting, and that both adsorption processes are in equilibrium, the rate expression is given by

$$R_{\mathrm{SiO}_2} = \frac{kK_T P_{\mathrm{TEOS}}}{1 + K_T P_{\mathrm{TEOS}} + K_R P_R}. \tag{64}$$

In the limit of low surface coverage by TEOS, and low conversion of TEOS [and hence low production of $R_{(g)}$], this expression yields a first-order dependence on TEOS pressure. A fractional order dependence may be observed for conditions which yield moderate TEOS surface converges, or moderate converges by the by-product (R). The latter condition can reasonably only be achieved at higher conversion levels, which in turn will lead to finite values of P_R.

C. Results

Figures 1 through 3, selected from among the experimental kinetic and conformality data (*62*), are used to demonstrate that model II is superior to model I. These are features taken from the same sample, a piece of a trench wafer. Thus, it is reasonable to assume that the deposition conditions are the same for all of them; the same temperature and species fluxes. The conditions are 998 K and 1 Torr of TEOS. This reasonable assumption minimizes the number of model parameters which need to be estimated and eliminates some uncertainty in comparing models. The step converges at two different places on selected features are used to test the models. *Step coverage* in this context means the ratio of film thickness at

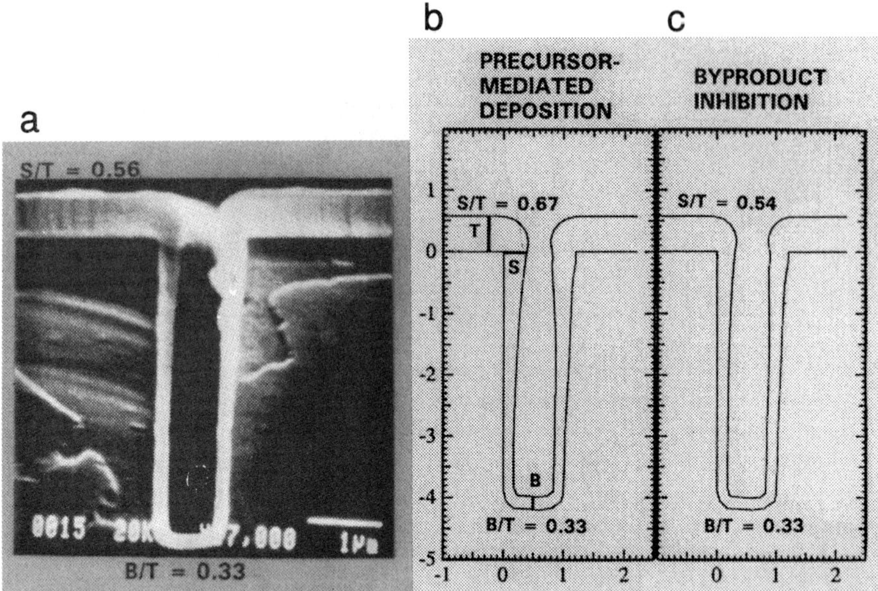

FIG. 1. Scanning electron micrograph of experimental silicon dioxide film profile in high-aspect-ratio trench, compared with the predicted profiles and values of step coverages using the precursor-mediated mechanism and the product inhibition mechanism in EVOLVE. Scales on the predicted film profiles are in micrometers (*58*).

some point in the feature to the film thickness deposited on the flat regions next to the feature. One of the two points is at the middle of the base of the feature (B/T) and the other is just inside the feature mouth—at the original substrate surface level (S/T). These points are marked on Fig. 1, as are the experimental and simulation results.

To ensure consistency with the observed first-order dependence of the deposition rate on TEOS pressure (see the rate data in *34*), the adsorption term in the denominator of Eq. (59) is neglected ($1 \gg K_I P_I$). Likewise, the TEOS adsorption term in the denominator of Eq. (64) is neglected ($1 \gg K_T P_{\mathrm{TEOS}}$). The partial pressure of the by-product in the source volume external to the feature is assumed negligible in light of the extremely low TEOS conversions experienced in the microreactor and the fact that by-products are readily swept out of the reactor. On the other hand, by-product fluxes within a feature are significant. The remaining parameters are the lumped rate parameter [coefficients in the numerators of Eqs. (59) and (64)], the homogeneous precursor's partial pressure at the

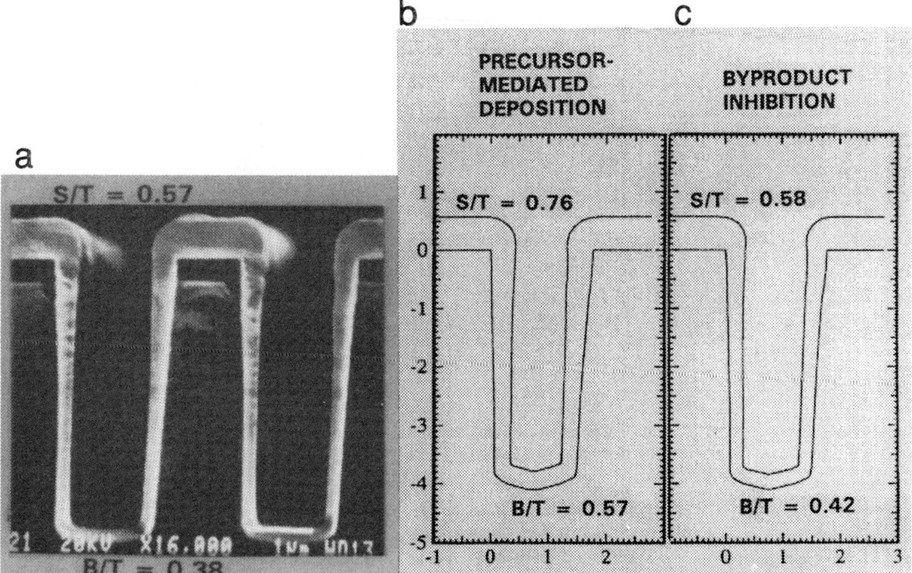

FIG. 2. Scanning electron micrograph of experimental silicon dioxide film profile in medium-aspect-ratio trench, compared with the predicted profiles and values of step coverages using the precursor-mediated mechanism and the product inhibition mechanism in EVOLVE. Scales on the predicted film profiles are in micrometers (*58*).

surface (model I), and the adsorption parameter for the readsorbing by-product (model II). The lumped rate parameter in each model is determined by the values of the other parameters and the observed deposition rate (6.3 nm/min). Thus, each model has one parameter which can be adjusted to improve the comparison between simulated and experimental step converges.

Consider Fig. 1, which shows the feature with the highest aspect ratio of the three. For simulations using model I, Eq. (59), the partial pressure of the homogeneous intermediate at the feature mouth was adjusted to a value of 0.000025 Torr to provide a good fit in step coverage at the feature base (B/T). For simulations using model-II, Eq. (64), the adsorption equilibrium constant K_R was adjusted to a value of 90,000 Torr^{-1} to fit the experimental step coverage at the base of this feature (33%). The value of S/T predicted by EVOLVE using model II is in better agreement with experiment. It seems clear that model II provides better agreement

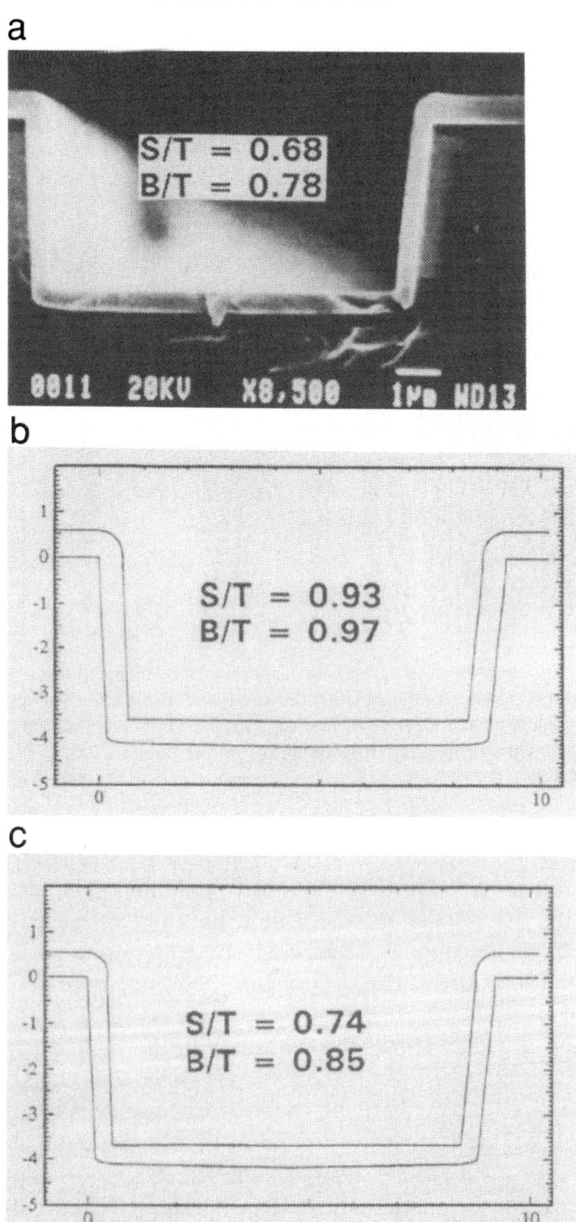

FIG. 3. Scanning electron micrograph of experimental silicon dioxide film profile in low-aspect-ratio trench, compared with the predicted profiles and values of step coverages using the precursor-mediated mechanism and the product inhibition mechanism in EVOLVE. Scales on the predicted film profiles are in micrometers (*58*).

with the experimental profile than model I. The same model parameters were used to simulate depositions in the features shown in Figs. 2 and 3. Model II provides better agreement with experimental results than model I.

We emphasize that this approach does not prove or disprove the mechanisms which led to the rate forms; there is no substitute for direct mechanistic studies. For example, the particular rate form used for the homogeneous precursor model likely does not represent the actual reaction steps. It is easy to visualize the formation of multiple precursors, each at a different rate and with different sticking factors. IslamRaja *et al.* (*65*) have used two precursors with different sticking factors to explain the conformality of TEOS sourced silicon dioxide films. On the other hand, there is evidence available which suggests that both heterogeneous and homogeneous paths are important (*32*). Some work has supported the by-product inhibition mechanism (*66*), while other work concludes that by-product inhibition does not play a role (*32,67*). Film profile information should be considered one tool in search for useful mechanisms and rate forms.

D. Simulations Using Sandia's Mechanism

We now consider as a third model the mechanism proposed by researchers at Sandia National Laboratory, detailed in Sec. II of this chapter (*32,33*). The chemistry and reaction kinetic data at CVD temperatures and pressures were estimated using GCMS (gas chromatography–mass spectroscopy) and FTIR (Fourier transform infrared spectroscopy) in conjunction with isotopic labeling. The surface reactions proposed were based largely on interpretation of the work done by Crowell and coworkers. The reaction rate model parameters were fit to deposition rate data of Desu and coworkers. The stability of the surface species and the selection of the likely reactions from a series of possible reactions was calculated using *ab initio* thermochemical property calculations (*32*).

The gas phase homogeneous decomposition reaction was used to estimate the equilibrium concentrations of the reactive precursors at 998 K and TEOS at a pressure of 1 Torr. Using the reaction rate expressions provided for all the surface reactions (see Sec. II) the surface profile evolution was studied for 998 K and 1 Torr of TEOS in order to compare with the experimental profiles obtained in our laboratory. Figure 4 shows the film profile evolution for an ideal trench of aspect ratio 2.

The predicted sticking coefficients of TEOS and $Si(OH)E_3$ (see Sec. II), the heterogeneous reaction rates, and the deposition rates on the flats

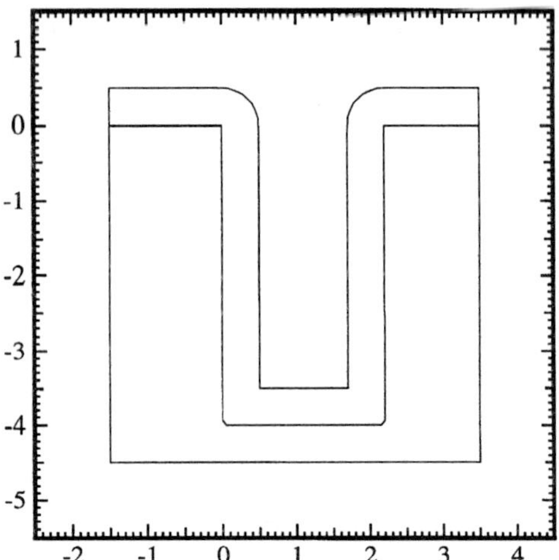

FIG. 4. EVOLVE simulation predicting almost perfect conformality for SiO_2 deposition using the mechanism developed at Sandia National Laboratory (see text) at 998 K and TEOS pressure of 1 Torr. Similar results are observed for higher temperatures. Scales on the predicted profile are in micrometers.

calculated using CHEMKIN/SURFACE CHEMKIN/AURORA, and EVOLVE were essentially identical. The decomposition rates obtained were identical to the results of Desu (*34*). However, under these conditions the simulated films were very conformal, in contrast with the experimental profiles shown in Figs. 1–3. Further refinement of the mechanism using topography simulators such as EVOLVE could allow better predictions.

VI. Plasma-Enhanced Chemical Vapor Deposition of Silicon Dioxide from TEOS

This section and the next discuss our studies of plasma-enhanced CVD of SiO_2 from TEOS, which have focused on identifying the role of oxygen atoms in the deposition process. This section covers the analysis of direct rf (PETEOS); the next section deals with remote microwave PETEOS (RμPETEOS). Essentially the same procedure discussed in the section on thermal TEOS has been used to validate proposed models. For details of the PETEOS study, see Refs. *60, 68*, and *69*.

A. Experimental

Experiments used for model development were performed in a research scale single wafer, cold wall parallel plate plasma reactor. The lower, grounded electrode contained a resistive heating element which allowed the wafers to be heated to 800 K. Wafer temperature was measured using a thermocouple located just below the surface of and near the center of the electrode/susceptor. Thermal SiO_2 layers were deposited on *p*-type silicon wafers and patterned to give long trenches with aspect ratios varying from 0.2 to 1.7. The wafers were coated with a thin layer of sputtered aluminum prior to the PETEOS experiments. The aluminum yielded features which are slightly re-entrant. SiO_2 was deposited on these test structures from mixtures of TEOS and O_2 using continuous 13.56-MHz rf excitation at applied power densities of 0.03 to 0.18 W/cm^2, at temperatures from 473 to 773 K, and at total pressures from 0.5 to 2.0 Torr. In most experiments TEOS and O_2 flow rates were set at 50 and 300 sccm, respectively. Assuming deposition occurs on both the working and counter electrodes, we estimate that overall TEOS fractional conversion was limited to a maximum of 16%, but was typically less than 8%. Film thicknesses were visually measured from cross-sectional SEM micrographs. Time-averaged deposition rates were determined by dividing the average film thickness by deposition time.

Our experimental results show that deposition rate increases with increasing rf power density, increasing total pressure, and decreasing wafer temperature. Step coverage, defined as the film thickness at the midpoint of the sidewall divided by the film thickness at the external surface near the feature, increases with decreasing rf power, increasing pressure, decreasing temperature, and decreasing trench aspect ratio. SEM micrographs of PETEOS sourced SiO_2 films indicate that there is a significant directionality in the deposition process. Under no conditions did step coverage exceed 85%, even for the lowest aspect ratio features. Most step coverages ranged from 40 to 65%. In all cases film thicknesses at the feature base exceeded those at the bottom of the feature sidewall, and approached those at the external surface near the feature. These results are consistent with those reported by Flamm and coworkers (*70,71*) and Chang *et al.* (*72*), and suggest that directional ion bombardment is responsible for a large portion of the observed film deposition.

B. Model Development

Simple models for the plasma volume, plasma sheath, and surface reaction kinetics of the process were formulated which together correctly predict trends in deposition rate with changes in power, pressure, temperature, and O_2 : TEOS flow ratios (*68,69*). The plasma chemistry submodel provides the oxygen atom flux to the wafer surface and the oxygen ion number density at the plasma-sheath boundary. The sheath submodel provides the ion flux to the wafer surface. The oxygen radicals were assumed to be cosine distributed; that is, to have a random velocity distribution. The angular distribution of the incoming ions was modeled as an exponential distribution, with a small adjustable standard deviation. In this section, we focus only on the surface chemistry model.

It is now well established that oxygen atoms are the dominant reactive neutral oxygen intermediate and that molecular oxygen ions (O_2^+) are the dominant positively charged ions in a high-frequency oxygen discharge (*73*). Our model therefore focuses on production and loss of these key plasma intermediates: The degree of directionality is controlled by competition between a highly directional ion-assisted pathway and a nondirectional oxygen radical induced pathway. We assume that the plasma sufficiently excites or dissociates TEOS such that the resulting intermediates or fragments readily adsorb on the growing SiO_2 surface. In fact, the surface is assumed to be covered by these species, which are referred to later as adsorbed TEOS. The rate determining step for film deposition is oxidative attack of the adsorbed TEOS fragments by oxygen atoms, perhaps enhanced by oxygen ions.

Within the model, the areal molar deposition rate of SiO_2 by the neutral pathway is written in terms of the molar concentrations as

$$R_1 = 0.9[\mathrm{O}]. \tag{65}$$

Assuming the reactive sticking factor for oxygen ions (to form SiO_2) is 1, the local SiO_2 deposition rate via the oxygen ion assisted pathway is equal to the local ion flux to the surface:

$$R_2 = \eta_{\mathrm{O}_2^+}. \tag{66}$$

As mentioned, the preceding two reactions do not depend on TEOS concentration, because it is assumed that the organosilane film precursors formed from TEOS completely saturate the surface under steady deposition conditions.

The recombination of oxygen atoms to form nonreactive oxygen molecules competes with the first reaction. The rate of consumption of oxygen radicals by this reaction is

$$R_3 = 2\gamma(T)\eta_O. \tag{67}$$

The recombination of oxygen atoms is activated; however, the temperature dependence cannot be represented by a single activation energy. Thus the "sticking factors" for oxygen atom recombination are read directly from the data of Greaves and Linnett (*74*, see also Sec. VII). The molar rate of generation of film is given by

$$R_{SiO_2} = R_1 + R_2 \tag{68}$$

and the local linear growth rate is obtained by multiplying by the molar volume of SiO_2.

C. Results

After establishing our model parameters, we were able to match film profiles observed in SEM cross-sections over a range of conditions using the standard deviation of the ion flux distribution as the single adjustable parameter. Figure 5a shows an SEM micrograph of a silicon dioxide film deposited at 10 W rf power, 1 Torr total pressure, and a wafer temperature

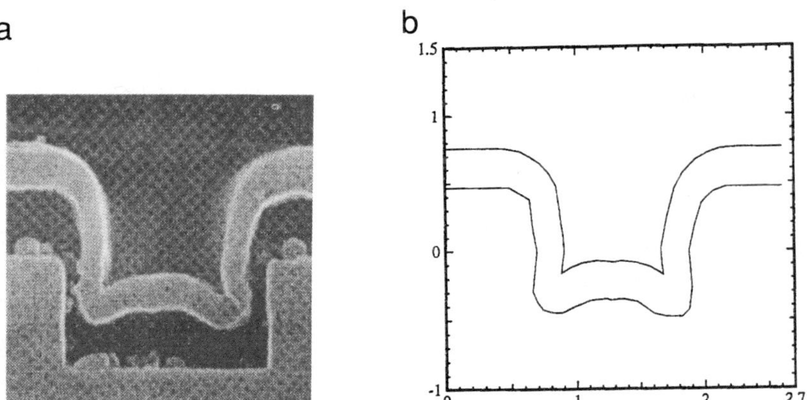

FIG. 5. (a) SEM of an SiO_2 film deposited using PETEOS, at 10 W, a wafer temperature of 573 K, and a pressure of 1 Torr. (b) EVOLVE simulation corresponding to part (a), using the submodels and parameters discussed in the text (*68*).

of 573 K to a thickness of 290 nm at a total rate of 72 nm/min. Figure 5b shows the film profile predicted by EVOLVE for these conditions using the constitutive models discussed earlier. Note that the original silicon surface is not shown in Fig. 5b. For the deposition conditions corresponding to Fig. 5a, the plasma model predicts an oxygen atom concentration of 3.6×10^{-9} gmol/cm^3 and an oxygen ion flux to the surface of 7.1×10^{14} ions/cm^2/s. Figures 6a and 6b are the experimental and predicted films for a 430-nm-thick silicon dioxide film deposited at 180 nm/min at 30 W, 1 Torr, and 573 K. The plasma model predicts an oxygen concentration in the plasma of 7.2×10^{-9} gmol/cm^3 and an oxygen ion flux to the surface of 2.7×10^{-1} ions/cm^2/s. The deposition rates determined by EVOLVE are essentially equal to the observed rates in both cases.

Recall that the only adjustable parameter in each case was the standard deviation of the ionic flux distribution. In these simulations we chose values (0.08 radian for Fig. 5b and 0.05 for Fig. 6b) which provided the best match for the ratio of the film thickness on the sidewall at a depth corresponding to the original silicon surfaces to the thickness on the flat area outside the trench. The predicted film profiles are only somewhat sensitive to changes in this value. The difference between these values for the standard deviation is in the direction predicted by the simple sheath model used, in that the predicted plasma sheath at 30 W is not as collisional as the 10-W sheath. In addition, these values are within the

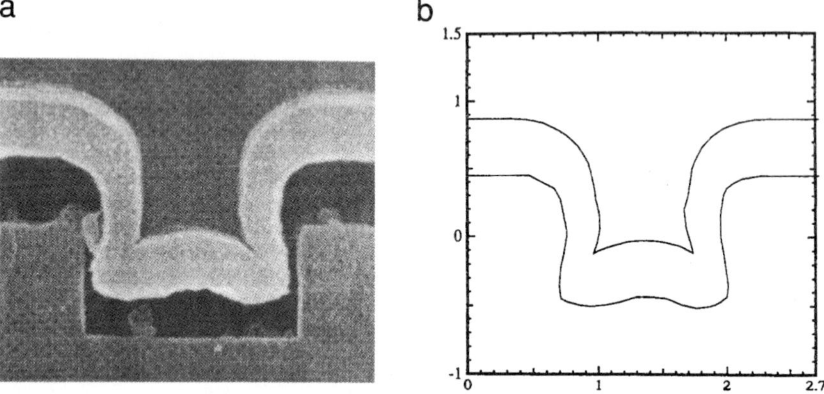

FIG. 6. (a) SEM of an SiO_2 film deposited using PETEOS, at 30 W, a wafer temperature of 573 K, and a pressure of 1 Torr. (b) EVOLVE simulation corresponding to part (a), using the submodels and parameters discussed in the text (*68*).

range of standard deviation values which could be used to characterize our Monte Carlo simulation results. The agreement between experiment and simulation is good for both cases, especially considering that the aluminum film profiles which serve as the original surfaces are rather rough and were digitized manually from the SEMs. Measurement of the simulated film thickness at the midpoint along the base reveals that the predicted ratio of ionic to radical induced deposition rates is a little higher than the actual ratio.

Significantly better agreement between simulated and experimental profiles can be obtained by using the sticking factor for oxygen in the atomic recombination reaction as a second adjustable parameter. The value used (0.0015 for 573 K) was measured for the wall of a quartz tube in a pure oxygen plasma (*74*). The value of this parameter is sensitive to the nature of the silicon dioxide surface (*74*). Thus it is conceivable that the growing surface exposed to the TEOS/O_2 plasma recombines atoms with a significantly different efficiency than that assumed. Nevertheless, we chose not to use the recombination coefficient as an adjustable parameter, since the constitutive models predict film profile trends qualitatively and at this stage of initial development are more credible if they include a minimum of adjustable parameters. For more quantitative film profile predictions, the models for both the ion flux distribution and the sticking factor for atomic recombination need further development.

Finally, it should be emphasized that the plasma submodels presented are valid only over the range of operating conditions for which they were developed. In addition, the surface reaction submodel may not be valid for BTRM simulations in which the feature is completely filled, since local reactant flux conditions within the feature change rapidly as feature closure is approached (*75*). On the other hand, PETEOS simulations outside the range for which the models were developed show that the model *structure* is useful. Silicon dioxide film profiles predicted using EVOLVE were in fairly good agreement with experimental films deposited at higher pressures, much higher power densities, and lower oxygen to TEOS partial pressure ratios (*76*).

VII. Remote Microwave PECVD of Silicon Dioxide from TEOS

This section focuses on our use of EVOLVE to understand the surface reactions involving oxygen atoms in the remote microwave plasma-enhanced CVD of SiO_2 using TEOS (RμPETEOS). A surface reaction model was proposed, then the unknown model parameter values were estimated using nonlinear regression to minimize the sum of squared errors between the computed profiles and the experimental profiles until the film profiles generated using EVOLVE compared well with the experimental film profiles. The extracted model parameters were then used to interpret the relative importance of various reaction steps within the mechanistic model framework. Although this procedure does not generate a validated kinetic model, this can be used as a tool to quantitatively predict film conformality over the range of experimental operating conditions employed to develop the model. Details of this study can be found in Refs. *61* and *77*.

A. Experimental

Depositions were carried out in an experimental microreactor which was coupled with a waveguide powered by a 300-W, 2.44-GHz power supply for the generation of the microwaves. Pure TEOS (Schumacher, electronic grade) vapor flowed into the deposition chamber from a vapor source mass flow controller. The TEOS flow rate was fixed at 35 sccm for all the depositions. Patterned dies of about 1 cm^2 were mounted on a resistive heater/holder arrangement in the deposition chamber. A thermocouple in contact with the top surface of the sample was used to measure the temperature. More complete descriptions of the experimental conditions can be found in Refs. *61, 77*, and *78*.

B. Model Development

The kinetic models used in the RμPECVD simulations were similar to those developed for a direct capacitively coupled rf PECVD of SiO_2 (*60,69* and Sec. VI]. In the direct PECVD model, the decomposition rate of SiO_2 was assumed to be limited by the rate of arrival of reactive oxygen ions and neutral atoms formed in the plasma to the surface of the film. TEOS and/or TEOS fragments, formed primarily due to collisions between the activated oxygen species and TEOS molecules in the plasma region, were assumed to saturate the surface of the film. This assumption is supported by experimental findings that the deposition rate is nearly independent of TEOS concentration and temperature, and proportional to the oxygen atom concentration over the range of conditions used. Film deposition rate was assumed to be the sum of the individual rates of oxygen-induced deposition and ion-enhanced deposition. The model also accounted for the pseudo-first-order recombination of atomic oxygen on the surface. This recombination reaction competes directly with the neutral induced deposition reaction by reducing the available atomic oxygen flux to the surface of the growing oxide film.

The model for the remote plasma system, however, assumes that there are no oxygen ions present in the deposition chamber volume. This is reasonable considering the relatively short lifetimes of the oxygen ions as compared to the lifetimes of the oxygen neutral atoms (*79*). As for the rf PETEOS, it is assumed that TEOS and/or TEOS fragments are formed primarily due to collisions between the activated oxygen neutral atoms and the TEOS molecules in the reactor volume, saturate the growing film surface, and therefore the SiO_2 deposition rate is assumed to be independent of TEOS concentration. Consistent with our treatment of deposition due to oxygen atoms in the rf PETEOS work, the deposition rate is assumed to be independent of temperature, and directly proportional to only the oxygen atom concentration. The reactions occurring on the surface of the growing film can then be listed as:

$$O_{(g)} + * \xrightarrow{k_{ads}} O^* , \tag{69}$$

$$O^* \xrightarrow{k_{des}} O_{(g)} + * , \tag{70}$$

$$O^* + O_{(g)} \xrightarrow{k_r} O_{2(g)} + * , \tag{71}$$

$$15O^* + TEOS^* \xrightarrow{k_D} SiO_2 + \text{desorbed by-products} , \tag{72}$$

where subscript (g) refers to the gas phase and * refers to an activated surface site. The first equation represents the adsorption of an oxygen atom on an unoccupied activated surface site with an adsorption rate constant k_{ad}. The second equation represents the desorption of the adsorbed oxygen species into the gas phase with a desorption rate constant k_{des}. The third reaction is an Eley–Rideal type reaction, in which a gas phase impinging oxygen atom combines directly with the adsorbed oxygen atom to form an oxygen molecule without first adsorbing on the growing film. This oxygen molecule subsequently desorbs from the surface. The fourth reaction is the assumed stoichiometric deposition reaction for SiO_2, based on the previous model for the deposition of SiO_2 in the direct rf PETEOS work (Sec. VI). We assume that the initial attack of adsorbed TEOS by a single oxygen atom is the rate determining step for SiO_2 film formation. Assuming Langmuir site exclusion adsorption behavior and a pseudo-steady-state situation for the adsorbed oxygen atom reaction rate, we can write

$$\frac{d[\mathrm{O}^*]}{dt} = R_{ads} - R_{des} - R_r - R_D = 0, \tag{73}$$

where R_{ads}, R_{des}, R_r, and R_D are the rates of adsorption, desorption, recombination, and deposition, respectively. The rate of recombination R_r of oxygen atoms can be written as

$$R_r = \frac{k_r\left[K_a - \dfrac{k_D}{k_{des}}\right]P_{\mathrm{O}}^2}{1 + \left[K_a + \dfrac{k_r}{k_{des}}\right]P_{\mathrm{O}}} \tag{74}$$

where P_{O} is the partial pressure of oxygen atoms at the surface of the wafer, K_a is the oxygen atom adsorption equilibrium constant (given by k_{ads}/k_{des}), and k_r is the Eley–Rideal reaction rate parameter. For a surface saturated with adsorbed oxygen atoms,

$$\left[K_a + \frac{k_r}{k_{des}}\right]P_{\mathrm{O}} \gg 1, \tag{75}$$

and the rate expression simplifies to a first-order dependence on partial pressure of oxygen:

$$R_r = k_1 P_{\mathrm{O}}, \tag{76}$$

where all the rate constant terms are lumped into the pseudo-first-order rate parameter k_1 for convenience. If the surface coverage by oxygen atoms is very low, then

$$\left[K_a + \frac{k_r}{k_{des}}\right]P_{\mathrm{O}} \ll 1 \tag{77}$$

and the recombination rate is second order in P_{O}:

$$R_r = k_2 P_{\mathrm{O}}^2, \tag{78}$$

where the rate constants have been lumped into one second-order recombination rate constant k_2.

C. RESULTS

The reactive sticking coefficient for oxygen atoms in the recombination reaction is given by the ratio of the rate of the reaction to the incident oxygen atom flux. For the pseudo-first-order situation, there is no dependence on the partial pressure of the oxygen atoms and the sticking coefficient takes a single value at a given temperature. In contrast, in the second-order situation, the sticking factor has an apparent first-order dependence on the oxygen atom partial pressure. It is therefore essential to explore the role of the oxygen atom partial pressure in determining the conformality of the deposited films.

The effect of substrate temperature on the conformality of the deposited films can be seen from the four experimental SEM micrographs of Figs. 7a, 8a, 9a, and 10a, with substrate temperatures varying from 250 to 400°C. It can be seen that an increase in the substrate temperature from 250 to 400°C results in a dramatic reduction of bottom and sidewall step coverages of the films. EVOLVE was used to simulate the observed trend

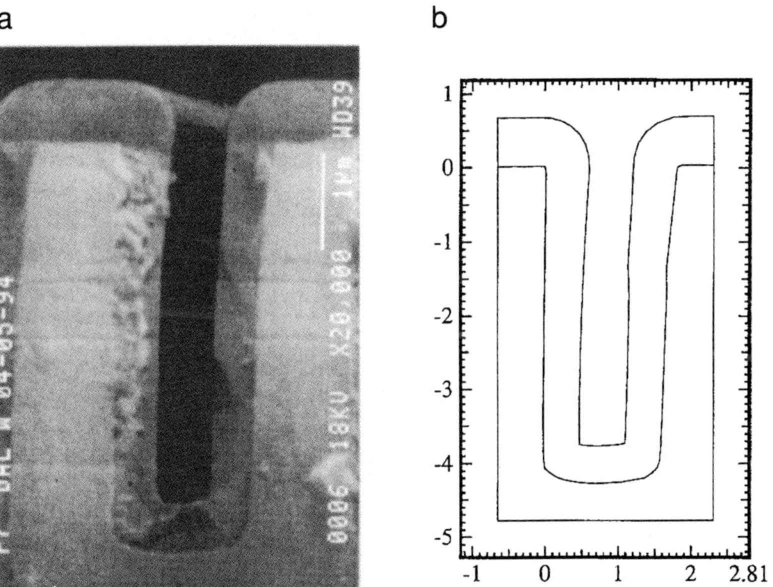

FIG. 7. (a) SEM cross-section of film deposited in trench with an aspect ratio of about 2.5. Substrate temperature is 250°C, total pressure of 0.4 Torr, and flow ratio of O_2 : TEOS is 3. (b) EVOLVE simulated profile at conditions of part (a) (*61*).

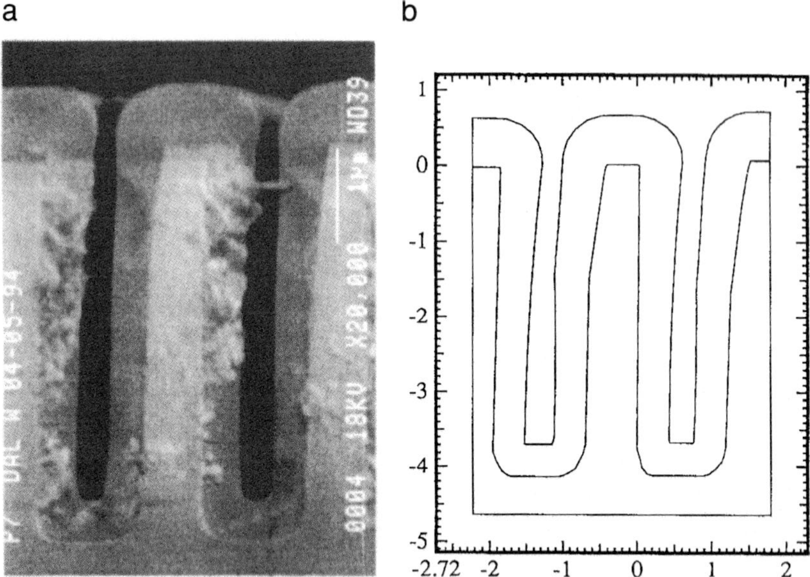

FIG. 8. (a) SEM cross-section of film deposited in trench with an aspect ratio of about 2.5. Substrate temperature is 250°C, total pressure of 0.4 Torr, and flow ratio of O_2 : TEOS is 3. (b) EVOLVE simulated profile at conditions of part (a) (*61*).

of decreasing film conformality of the deposited films with increasing temperature using the RμPECVD model with both the first-order oxygen atom recombination reaction and the second-order oxygen atom recombination reaction, with kinetic parameters chosen to obtain the best fit to the data (Figs 7b, 8b, 9b, and 10b).

For this purpose, three parameters need to be estimated for the RμPECVD model using the first-order recombination reaction: P_O, the surface partial pressure of oxygen atoms, k_D, the deposition rate constant (the deposition reaction is assumed to be not thermally activated), and k_1, the lumped pseudo-first-order recombination rate constant at the temperatures of interest. We used the value of k_D previously employed in simulations for the direct plasma process (*60,69*). The surface partial pressure of oxygen atoms, P_O, was chosen to match the experimental and simulated deposition rates on the flats. The values of P_O that were used in the simulations were at least an order of magnitude lower than the inlet oxygen gas partial pressures (experimental data, *77*). This outcome is reasonable since only a small fraction of the inlet oxygen gas is ionized by

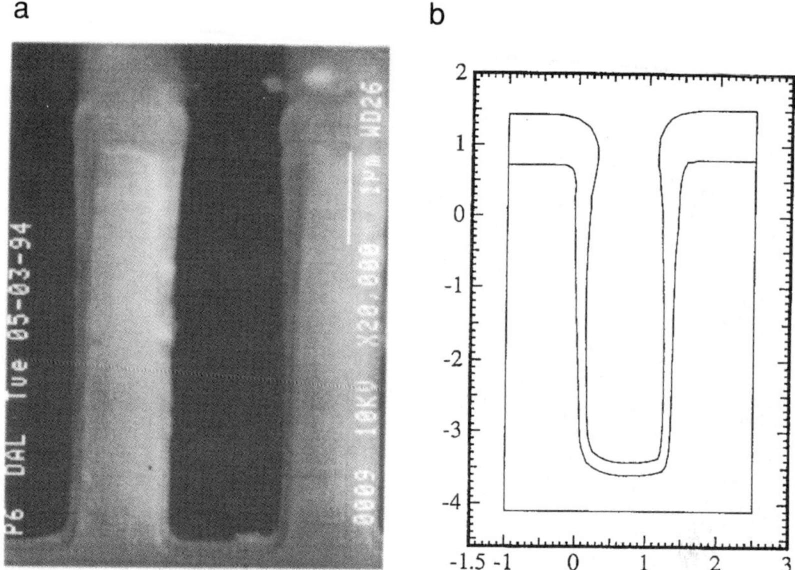

FIG. 9. (a) SEM cross-section of film deposited in trench with an aspect ratio of about 2.5. Substrate temperature is 400°C, total pressure of 1.0 Torr, and flow ratio of O_2 : TEOS is 3. (b) EVOLVE simulated profile at conditions of part (a) (*61*).

the plasma, and only a fraction of the excited gas, the neutral component, is assumed to reach the wafer surface. Thus the only adjustable parameter was the rate parameter k_1. The first guess at the values for k_1 were taken from experimentally determined values for the recombination coefficient over silica reported by Greaves and Linnett (*74*) in the temperature range of interest for this work. Using this approach, while the trends in film conformality computed by EVOLVE were similar to those obtained experimentally (Figs. 7a–10a), the values for the lumped recombination coefficient were found to be an order of magnitude higher than the initial guessed values. When this approach was applied to SEM micrographs with multiple trenches at 300 and 350°C (Figs. 11a and 12a), we found that although a good fit was obtained for the low-aspect-ratio trenches (i.e., the left trench and the right trench), the model underpredicted the high-aspect-ratio sidewall and bottom step coverages, as can be seen in Figs. 11b and 12b.

The second-order oxygen-atom-dependent recombination reaction was then used in the RμPECVD model. There are again three parameters which need to be estimated when using this recombination model: the

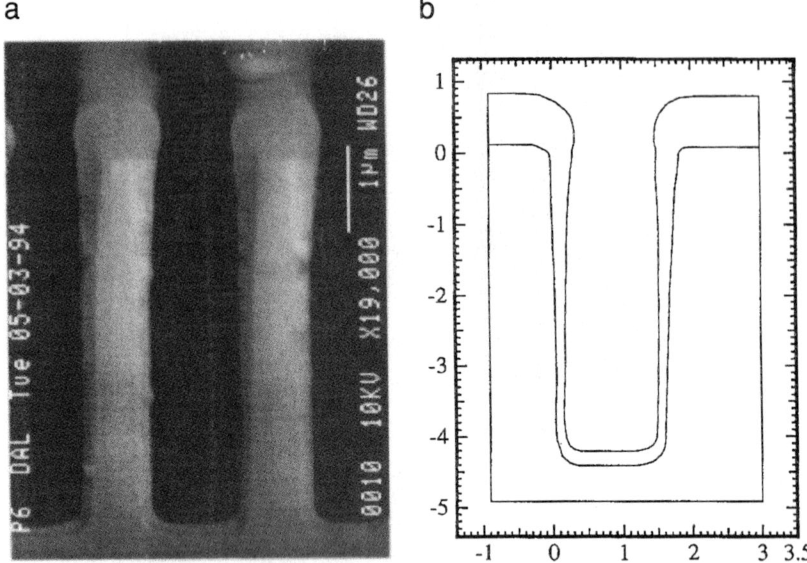

FIG. 10. (a) SEM cross-section of film deposited in trench with an aspect ratio of about 2.5. Substrate temperature is 400°C, total pressure of 1.0 Torr, and flow ratio of O_2 : TEOS is 3. (b) EVOLVE simulated profile at conditions of part (a) (*61*).

deposition rate constant k_D, the oxygen atom surface partial pressure P_O, and the second-order recombination rate constant k_2. The values for k_D and P_O were fixed as in the previous case. The value of k_2 was varied until the best match between the simulated and experimental profiles in single trenches was obtained. This approach was then applied to the multitrench SEM micrographs. This model gave a much better fit than that obtained by using the first-order recombination reaction model, for the low- as well as the high-aspect-ratio trenches, as can be seen in Figs. 11c and 12c.

The values of the recombination sticking factors obtained for the best fit in EVOLVE are plotted, along with values reported by Greaves and Linnett, in Fig 13. Both the first-order and the second-order recombination reaction reactive sticking coefficients were an order of magnitude higher than those determined by Greaves and Linnett (*74*). The discrepancy is attributed to the differences in the experiments. Greaves and Linnett studied oxygen atom recombination over SiO_2 surfaces in a pure oxygen plasma. Therefore the surfaces in their experiments were saturated with adsorbed oxygen atoms, leading to pseudo-first-order behavior. Moreover, the surface was chemically very different than the growing SiO_2

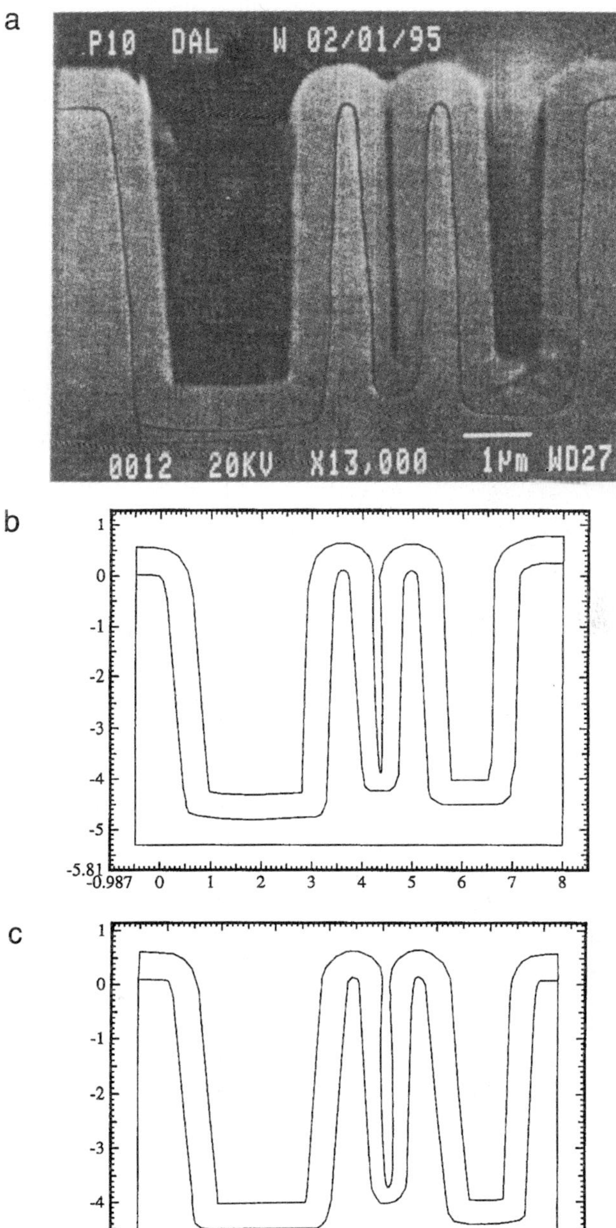

FIG. 11. (a) SEM cross-section of film deposited in multiple trenches with aspect ratios varying from about 1.5 to 4.5. Substrate temperature is 300°C, total pressure of 1.0 Torr, and flow ratio of O_2 : TEOS is 0.725. (b) EVOLVE simulated profile at conditions of part (a), using a first-order recombination reaction approach (see text). (c) EVOLVE simulated profile at conditions of part (a), using a second-order recombination reaction approach (see text) (*61*).

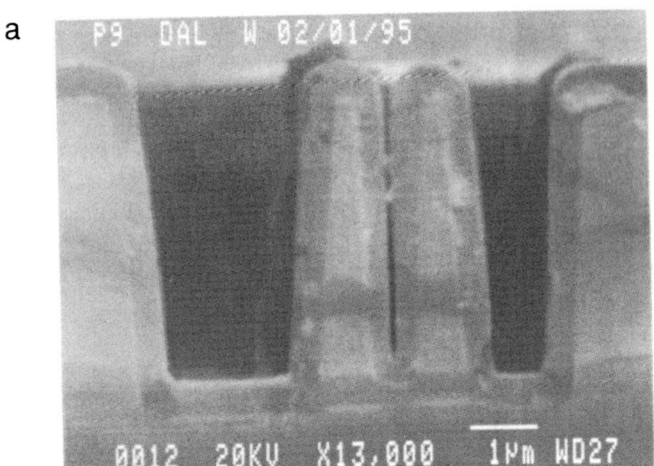

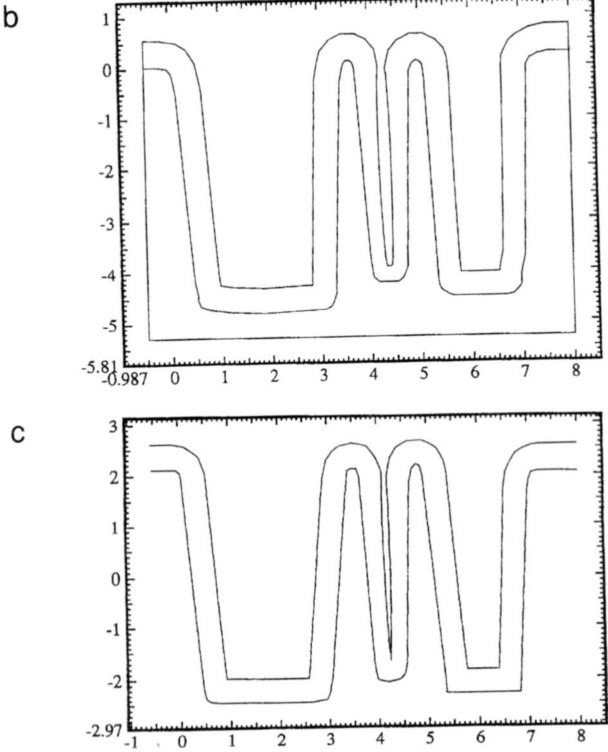

FIG. 12. (a) SEM cross-section of film deposited in multiple trenches with aspects ratios varying from about 1.5 to 4.5. Substrate temperatures is 300°C, total pressure of 1.0 Torr, and flow ratio of O_2 : TEOS is 0.725. (b) EVOLVE simulated profile at conditions of part (a), using a first-order recombination reaction approach (see text). (c) EVOLVE simulated profile at conditions of part (a), using a second-order recombination reaction approach (see text) (*61*).

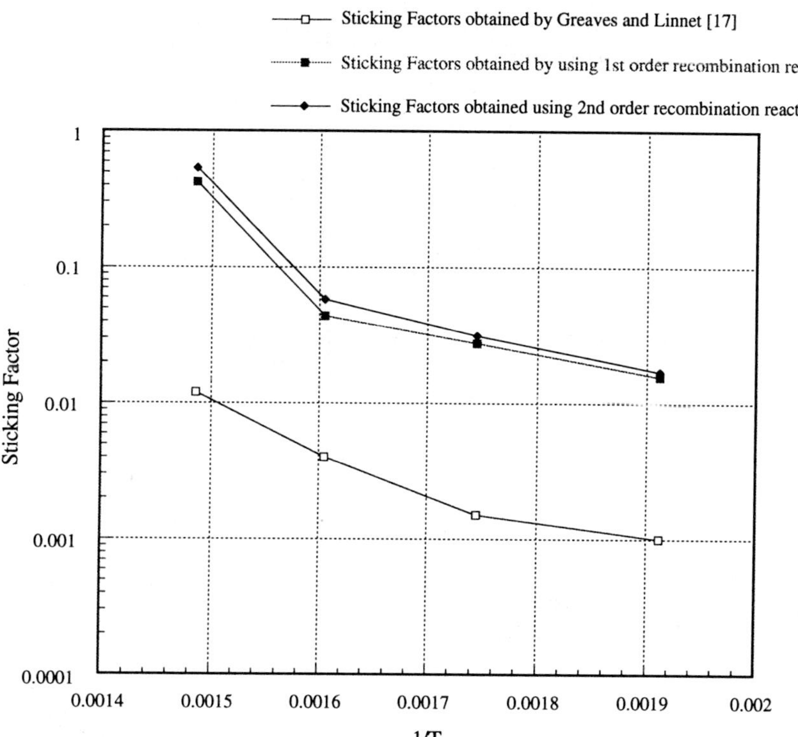

FIG. 13. Sticking factors obtained using a first-order and a second-order recombination reaction approach (see text), and those obtained by Greaves and Linnett (*74*) vs. 1/T (*61*).

surface explored in this study. According to Greaves and Linnett, the value of the reactive sticking coefficient is very sensitive to the chemical nature of the surface, as well as the water vapor content of the oxygen-containing atmosphere. It is therefore not surprising that higher reactive sticking coefficients are found for our surfaces.

Finally, Greaves and Linnett studied the oxygen atom recombination reaction with no other reactions occurring on the surface. In contrast, the adsorbed oxygen atoms are being continuously consumed by the deposition reaction in our case. This sink for surface-adsorbed oxygen atoms should shift the recombination reaction toward second-order behavior. An additional reason exists for the apparent second-order dependence. If the surface is indeed nearly saturated with TEOS fragments, then few sites are available for oxygen atom adsorption, and the oxygen atom coverage will be low. If this interpretation is correct, then a decrease in the O_2 : TEOS ratio in the feed should enhance step coverage, since the rate of oxygen atom recombination will be decreased. This trend was indeed observed in our previously reported conformality study (*68*).

VIII. Uncollimated and Collimated TiN PVD

This section summarizes our work toward developing an engineering model for TiN PVD with and without collimation. We do not deal with the collimator model in this section; details on it can be found in Refs. *48–51*. Flux distributions and deposition profiles on flat substrates determined using our Monte Carlo simulator for transport through the collimator and from the collimator exit to the substrate can be found in those references as well.

A. Experimental

Experimental data for Ti/TiN deposition were obtained using a Varian M2000/8 cluster sputter system (*51*). The collimators used in this study had hexagonal-shaped cells, with a diameter of 0.625 in. and cell heights of 0.625 and 0.9375 in., in order to obtain aspect ratios of 1.0 : 1 and 1.5 : 1. The TiN films were deposited in contacts with openings ranging from 0.40 to 0.75 μm and aspect ratios as high as 2.4. The contacts were formed in a deposited LPCVD TEOS/BPSG film using a standard plasma etch. The morphologies and thicknesses of the deposited films were determined by transmission electron microscopy (TEM). The TEM photographs were digitized to allow representation of the data in plots.

B. Uncollimated PVD of TiN

Examples of EVOLVE simulation results for uncollimated sputtered titanium nitride are compared to data in Figs. 14 and 15. A sticking factor of 0.7 and the parameter (ρ) of 2.0 in the generalized cosine distribution function (see Sec. III) are used in all simulation results presented in this chapter. In addition to the basic assumptions of the BTRM (see Sec. IV), we assume that species re-emit diffusely. In Fig. 14a, the left-hand side is the digitized TEM profile and the right-hand side is the EVOLVE predictions in a contact cross-section. Figure 14b compares the experimental and simulated film thicknesses of Fig. 14a as functions of arc length, starting at

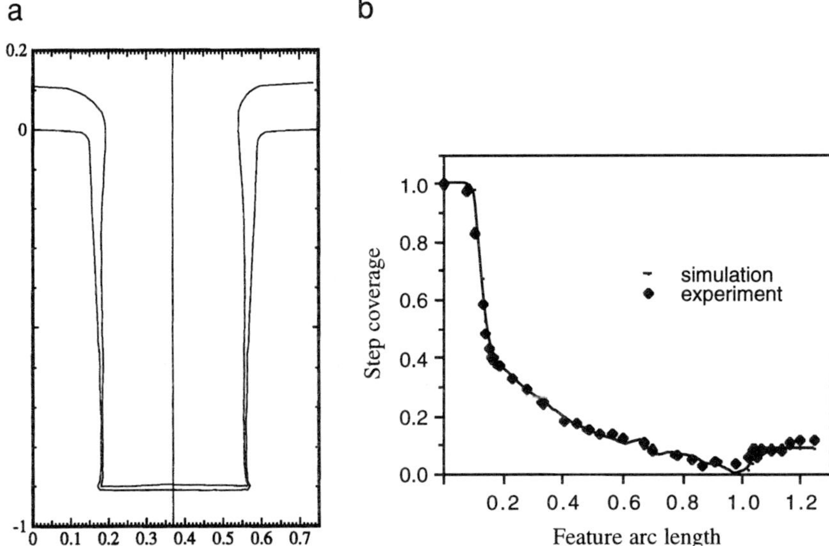

FIG. 14. (a) Experimental and computed TiN profiles in a contact. (b) Experimental and computed TiN thicknesses along the contact cross-section (normalized to top surface thickness) (*50,51*).

the top of the feature and traversing down along the side to the center of the feature's bottom. The deposited thickness is indicated by the data points and EVOLVE's predictions are shown as a solid line. The same is true for Fig. 15.

Other information is consistent with a subunity sticking factor in TiN PVD (*52,80*). The first is the apparent deposition of TiN in re-entrant features, which have no view to the source of incoming flux at the feature opening. The other evidence for a subunity sticking factor in related PVD processes comes from studies by Liu *et al.* (*81*) and Tsai *et al.* (*82*). In those studies, a subunity sticking factor was found for Ti PVD. Rogers and Cale (*53,83,84*) reported subunity sticking factors for both Ti and W in sputtered Ti–W films (also see Sec. IX).

C. Collimated PVD of TiN

Collimated TiN PVD is simulated using a sticking factor of 0.7 and a generalized cosine distribution parameter of 2.0 obtained from the simulations of uncollimated TiN PVD described in the previous section. Monte

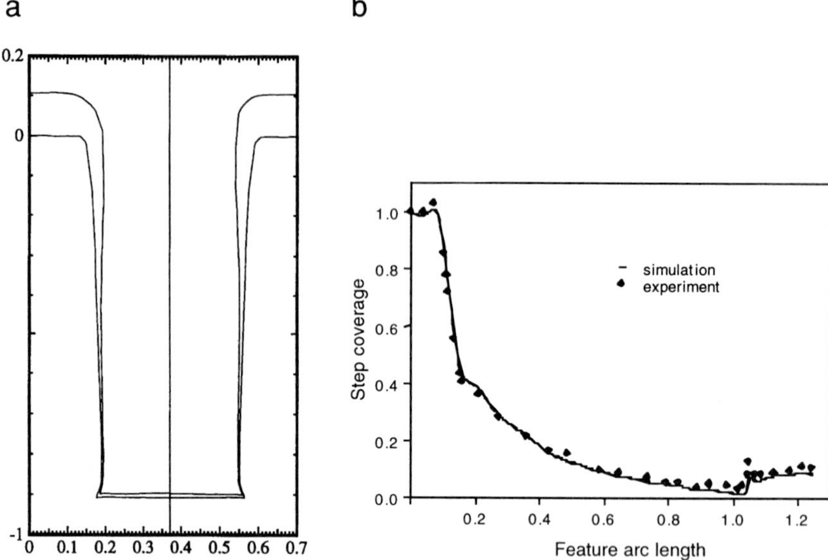

FIG. 15. (a) Experimental and computed TiN profiles in a contact. (b) Experimental and computed TiN thicknesses along the contact cross-section (normalized to top surface thickness) (*50*,*51*).

Carlo simulations are used to estimate flux distributions, as described in Ref. *48*. A sticking factor of 0.7 inside the collimator cell is assumed. This seems reasonable because the collimator is coated with TiN, as are the features. Collisionless transport from the collimator exit to the substrate is assumed.

Figures 16 and 17 compare the EVOLVE predictions of collimated TiN PVD with TEM experimental data obtained using Sematech's Varian M2000 using a 1 : 1 collimator. It is clear that EVOLVE, with the Monte Carlo collimator transport simulator, can be used to predict film profile evolution during collimated sputter deposition. In Fig. 18, EVOLVE predictions are compared with bottom step coverage data obtained at Sematech, Varian, and IBM (*52*), for vias with different aspect ratios using 1 : 1 collimators. The simulation results agree with experimental data fairly well, especially in view of the wide range of results by the three different research groups. In addition, the vias used in the simulations were idealized, whereas no details were given regarding the vias used in the reported experiments.

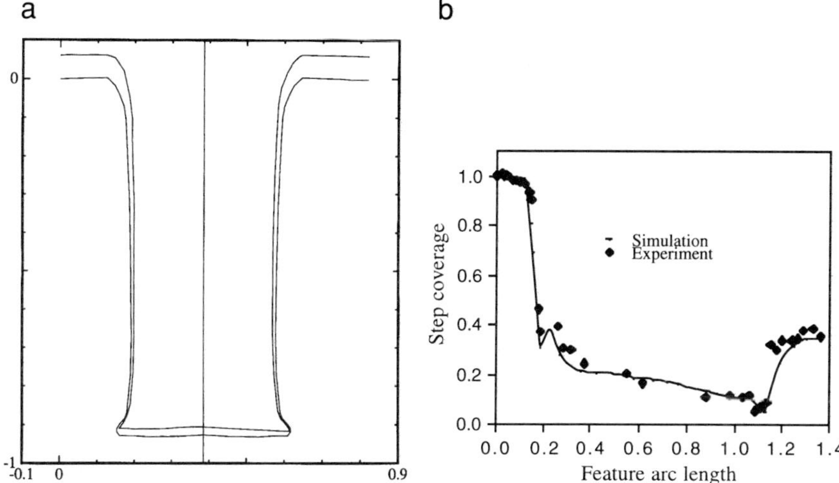

FIG. 16. (a) Experimental and computed TiN profiles in a contact. (b) Experimental and predicted TiN thicknesses along the contact cross-section (normalized to top surface thickness) (*50,51*).

The higher the aspect ratio of the collimator, the more "beaming" the collimated flux, and the better the bottom step coverage. By changing the input parameters to the collimator model, we get the flux distribution for a collimator aspect ratio of 1.5 : 1 ($\sigma = 0.7$, $\rho = 2$), and then we use this distribution as an input to EVOLVE. Figure 19 compares bottom step converge as a function of feature aspect ratio predicted by EVOLVE with experimental data (*52*) for aspect ratio 1.5 : 1 collimators. The IBM data extend to higher aspect ratios than those in the Sematech or Varian data sets. In addition, the IBM data show consistently higher bottom coverage than the Varian data, where these data sets overlap. Additional data for higher aspect ratio contacts and consistent equipment configuration and operation between research groups could explain this variation in the available published data, and allow more quantitative comparisons of simulation results with experimental data.

IX. Composition of PVD Titanium – Tungsten Films

This section discusses our work toward identifying and explaining the spatial variation in Ti–W film sputter deposited inside features. Perhaps the most important aspect of our work on composition profiles is that we

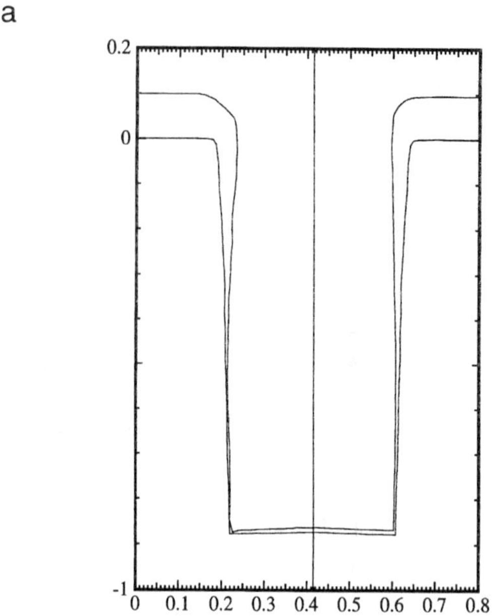

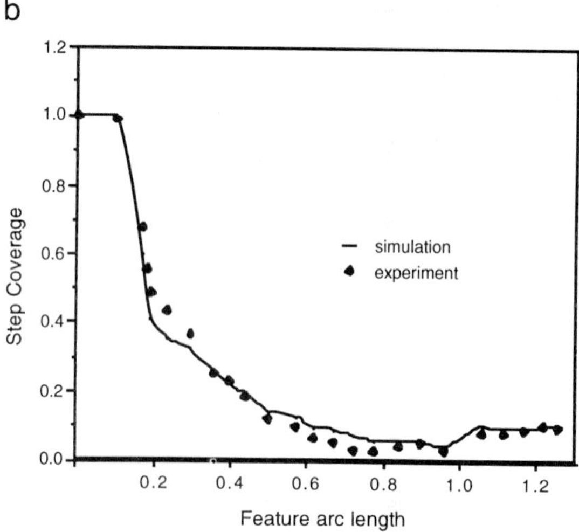

FIG. 17. (a) Experimental and computed TiN profiles in a contact. (b) Experimental and computed TiN thicknesses along the contact cross-section (normalized to top surface thickness) (*50,51*).

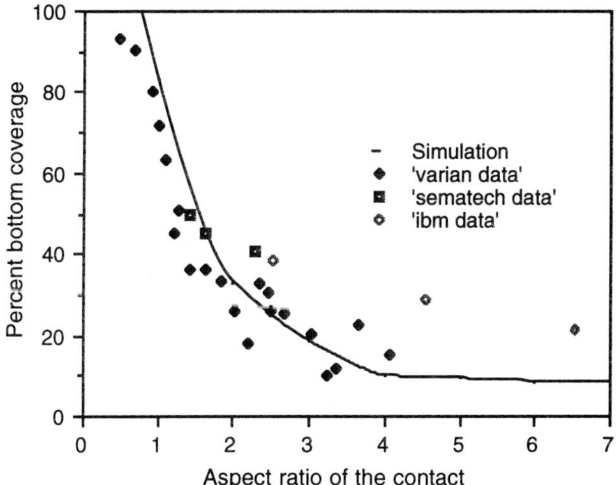

FIG. 18. EVOLVE predictions and data from 1.0 : 1 collimators (*50,51*).

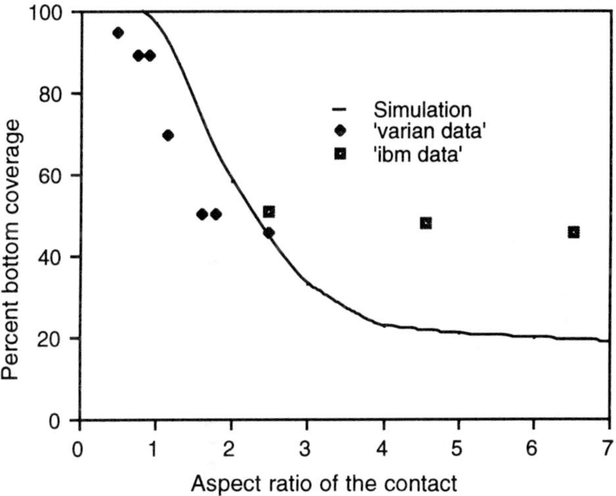

FIG. 19. EVOLVE predictions and data from 1.5 : 1 collimators (*50,51*).

predicted that composition profiles can occur, before they were observed experimentally. Thus, this demonstrates one of the roles which modeling can play in the development of process understanding. Our work, discussed in Refs. *53, 83*, and *84* and that of Liu *et al.* (*85*) show that Ti–W films deposited into trenches and holes of various aspect ratios become Ti rich as a function of depth into the features. Our work involved two phases, which are discussed separately.

A. Composition Profiles

Qualitative understanding of the possible causes of spatial composition variations can be gained by studying the equations which represent the BTRM. Film composition gradients are due to differences in the rates at which different elements are incorporated into the growing film, which could well be caused by local variations in the species fluxes to the local surfaces in features. Such gradients in species fluxes are common in deposition processes. For chemical vapor deposition systems, compositional variations can be caused by the competition between reactions, each of which in general is affected to a different extent by the gradients in the component fluxes along the feature surfaces. Indeed, composition profiles were predicted using modeling before they were found experimentally. Cale *et al.* (*28,29,75*) used this to explain the position dependence of the silicon-to-tungsten ratio inside features, when dichlorosilane was used to reduce tungsten hexafluoride. For PVD systems in which atomic species are incorporated into the growing film, compositional variations inside features can be caused by differing flux distributions from the source, different effective sticking factors of the depositing species, and/or different flux distributions of re-emitted species. The effective sticking factor is the fraction of atoms of a particular species which is permanently incorporated into the growing film after striking a surface. It accounts for re-emission (diffusive, specular, or some combination) as well as resputtering of deposited material. Differences in these properties between species can cause spatial variations in films sputtered into surfaces, but not in films sputtered onto flat substrates.

B. First Experiments

In the first set of experiments, samples were prepared using conventional photoresist patterning and reactive ion etching techniques to define various aspect ratio features with a nominal depth of 6 μm. Ti–W films were then deposited in a single wafer, dc magnetron sputtering system. The film has a nominal composition of 31 at.% Ti. Auger electron spectroscopy (AES) was then used to determine composition profiles along the Ti–W film inside the patterned features. AES spectra were acquired at specific points down a feature sidewall which had been exposed by cleaving the sample. Experimental details can be found in Ref. *53*. The composition profile for a trench nominally the same as that shown in Fig. 20a is shown in Fig. 20b. The film clearly becomes enriched in Ti as the base of the trench is approached, with a sharp drop in the Ti fraction on the base.

C. Model Development

EVOLVE was used to study the cause of the observed spatial compositional variations in Ti–W films (*53*). Two models were used for Ti re-emission: (1) a diffuse re-emission model and (2) a simple specular re-emission model. The assumptions appropriate to this study are (in addition to the four assumption basic to the BTRM given in Sec. IV.A):

5. Species interact with the surface in one of two modes: (a) they accommodate and re-emit diffusely or (b) they do not interact and re-emit specularly.
6. The fraction of incoming species which accommodate the surface and have a chance to become part of the evolving film depends on the angle of the incoming atom relative to the surface normal.

The sticking factor for tungsten is assumed to be unity in the simulations used for the analysis of this set of experiments. Figure 20b compares the Ti atomic fractions as a function of normalized depth into a trench which is nominally the same as the one shown in Fig. 20a, predicted using two re-emission models in EVOLVE. In the diffuse re-emission model, a sticking factor of 0.38 provides a good fit of the experimental data near the bottom of the feature. For specular re-emission, the Ti atoms which arrive at low enough angles relative to the local surface normal are assumed to equilibrate with the surface and become incorporated into the film. Atoms which arrive at larger angles with respect to the surface normal are more likely to reflect specularly. To account for this angular dependence on the sticking factor, the probability of being re-emitted specularly is modeled simply as the sine of the angle (in three-dimensional space) of the arriving species with respect to the local surface normal. Atoms which arrive normal to the local surface stick, whereas those arriving at a grazing angle

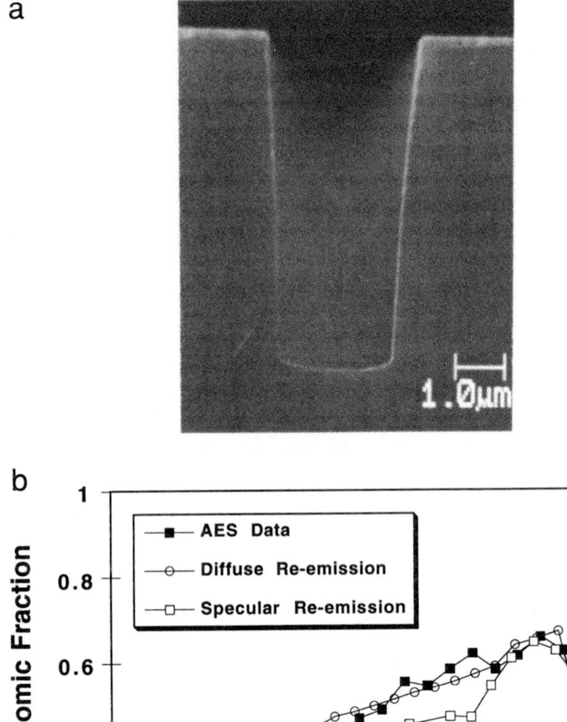

FIG. 20. (a) SEM of representative trench after Ti–W sputter deposition. (b) Ti atomic fraction of a Ti–W film versus normalized depth into a 2.4 aspect ratio trench as determined experimentally using AES, and as calculated using EVOLVE simulations modeling the re-emission as diffuse or specular (see text) (*83,84*).

are re-emitted. Thus, the average sticking factor depends on location in the feature in this model, because of the directionality of the net transport due to deposition. As in the case of diffuse re-emission, the predicted composition profile agrees qualitatively with the experimental data, with better agreement near the feature mouth. In the simulations, the total fluxes of titanium and tungsten were adjusted to match the experimental growth rate and the film composition on the flat surface away from features.

The simulation results described above might be taken to imply that both specular and diffuse re-emission may play roles in determining film composition profiles. The comparison between simulated and experimental composition profiles can indeed be improved by using more complex (e.g., combined re-emission) models. Because of the number of adjustable parameters introduced by such an approach, and a lack of understanding regarding resputtering of Ti, a large amount of information is needed. Contributing to the difficulty in extracting model parameter values from experiments are (1) the details of the features impact the distribution of the film properties, particularly for PVD films; that is, the experimental (and simulated) composition profiles depend significantly on the details of the feature profiles, even for features which are nominally the same, and (2) the computed profiles are insensitive to assumptions regarding the details of the proposed models and their parameters. We cannot simulate deposition in trenches which have cross-sections identical to those used to obtain experimental composition profiles; simulations require the trench to be cross-sectioned, while AES analyses require the trenches to be cleaved lengthwise. To refine our model of how Ti interacts with the surface, a number of experiments should be performed and average response should be computed. In addition, a model for the role of resputtering of Ti would be needed for completeness.

D. Second Experiments

A second set of experiments was run to test our conclusions from the preceding study. Ti–W films were deposited onto test structures which had large areas with no direct line-of-sight to the target (*86*). A detailed description of the test structure fabrication process and analyses of the resulting samples have been previously reported (*83,84*); however, a brief summary is presented here for completeness. The test structures were fabricated by first growing a 2-μm-thick layer of SiO_2 on a silicon substrate. Then a 4000-Å-thick polysilicon layer was deposited onto the SiO_2.

The polysilicon layer was patterned into lines using standard lithography and reactive ion etching techniques. The underlying SiO_2 was etched in a solution of 10 : 1 water : HF for a time sufficient to undercut substantially the polysilicon lines. Layers of Ti–W or Ti, nominally 600 nm thick, were then sputtered onto the structures. TEM and AES were used to identify material in the non-line-of-sight areas.

Figures 21a and 22a are scanning electron micrograph (SEM) cross-sections of structures with Ti–W and Ti metal layers, respectively. A

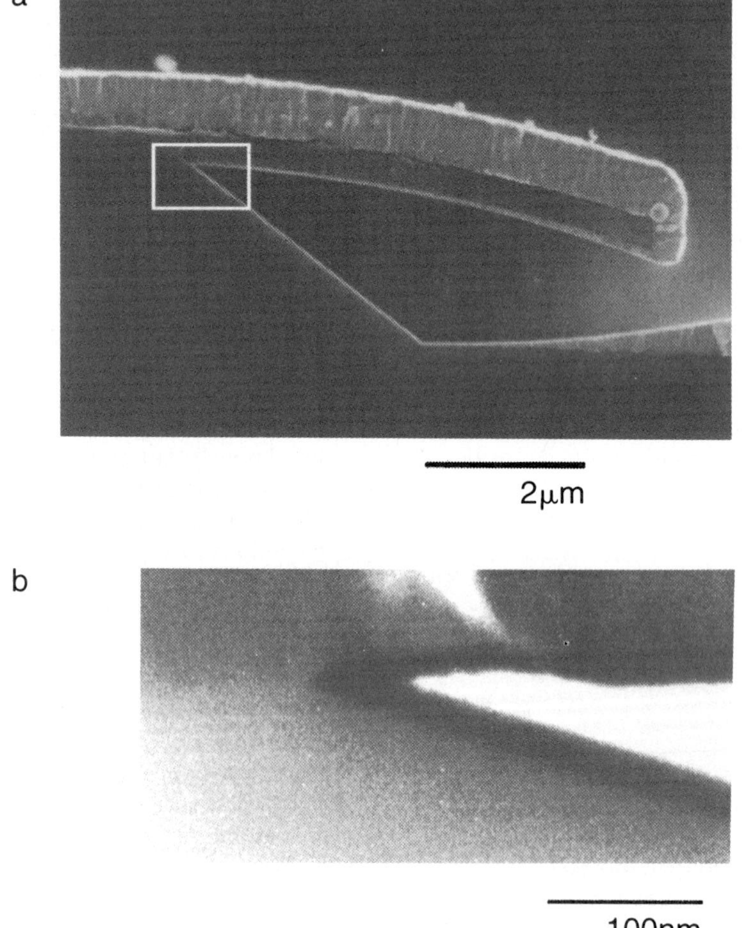

FIG. 21. (a) SEM cross-section of a test structure after Ti–W deposition. (b) TEM cross-section of a test structure with Ti–W in an area similar to that highlighted in part (a) (*83,84*).

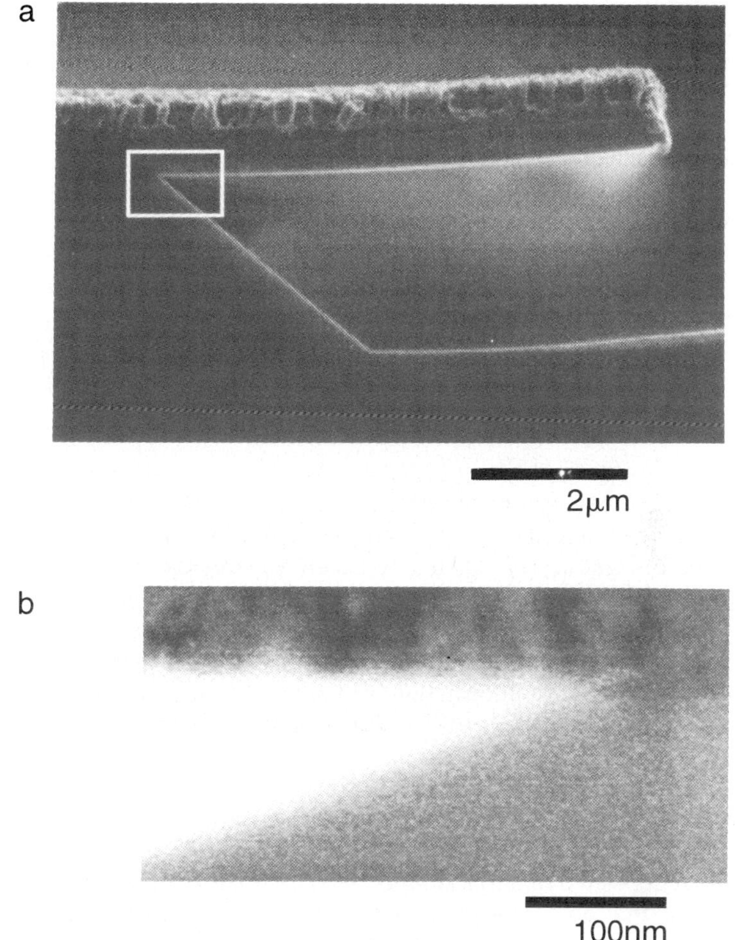

FIG. 22. (a) SEM cross-section of a test structure after Ti deposition. (b) TEM cross-section of a test structure with Ti in an area similar to that highlighted in part (a) (*83,84*).

continuous layer of redeposited material is clearly evident in the SEM cross-section of the Ti–W sample, even though the compressive stress of the Ti–W film has bowed the polysilicon beam downward. However, there is no redeposited layer evident in the SEM cross-section of the Ti sample, even though the tensile stress of the Ti film has bowed the polysilicon beam slightly upward, allowing entering atoms to strike more of the feature's interior surface. TEM images were used to measure the thickness of the redeposited material at a number of positions on both samples.

Figure 21b and 22b are TEM micrographs of the areas highlighted on the SEMs in Figs. 21a and 22a, respectively. These micrographs show that there is a measurable amount of redeposited material present on the Ti–W sample but not on the Ti sample. AES analyses of the Ti–W on top of the polysilicon beam and on the sidewall near the corner where it meets the beam (tape was used to tear off the beam, exposing the sidewall) showed that the redeposited material contained approximately 81 at.% Ti compared to 31 at.% Ti in the film on top of the beam. Similar AES analyses of the Ti sample showed Ti was present in the top corner of the structure, but the layer was either very thin or discontinuous.

E. Simulations

EVOLVE was used to simulate the deposition of Ti–W and Ti onto the test structures described. The composition and thickness of material in the non-line-of-sight area predicted by EVOLVE simulations of sputter deposition onto the test structures are compared to AES and TEM data. For Ti–W deposition simulations, a major uncertainty is introduced because the polysilicon beams bow under the stress of the growing Ti–W film. Since we do not have information on the time evolution of the beam shape during deposition, the process was simulated with two very different starting profiles. First, the bow was assumed to form early in the process, so the starting profile was digitized from the SEM in Fig. 21a. Second, the bow was assumed to form late in the process, so the starting profile was digitized from an SEM of a straight structure before metal deposition.

The redeposited material on the Ti–W sample contained both Ti and W, indicating that both Ti and W have subunity effective sticking factors. The flux distributions from the source are assumed to be cosine, as in previous simulations (*53*). There is little information available regarding the flux distributions of re-emitted material. In this paper we assume cosine distributions; that is, diffuse re-emission was used in the analysis of these experiments. The Ti and W effective sticking factors were varied until the calculated compositions qualitatively agreed with the experimental data at selected points. The total Ti and W fluxes to the wafer surface were set after Ti and W effective sticking factors were determined in order to match the experimental growth rate on the top surface of the structures. Each simulation was run until the thickness of the film on top of the polysilicon beam reached the film thickness observed experimentally, approximately 600 nm.

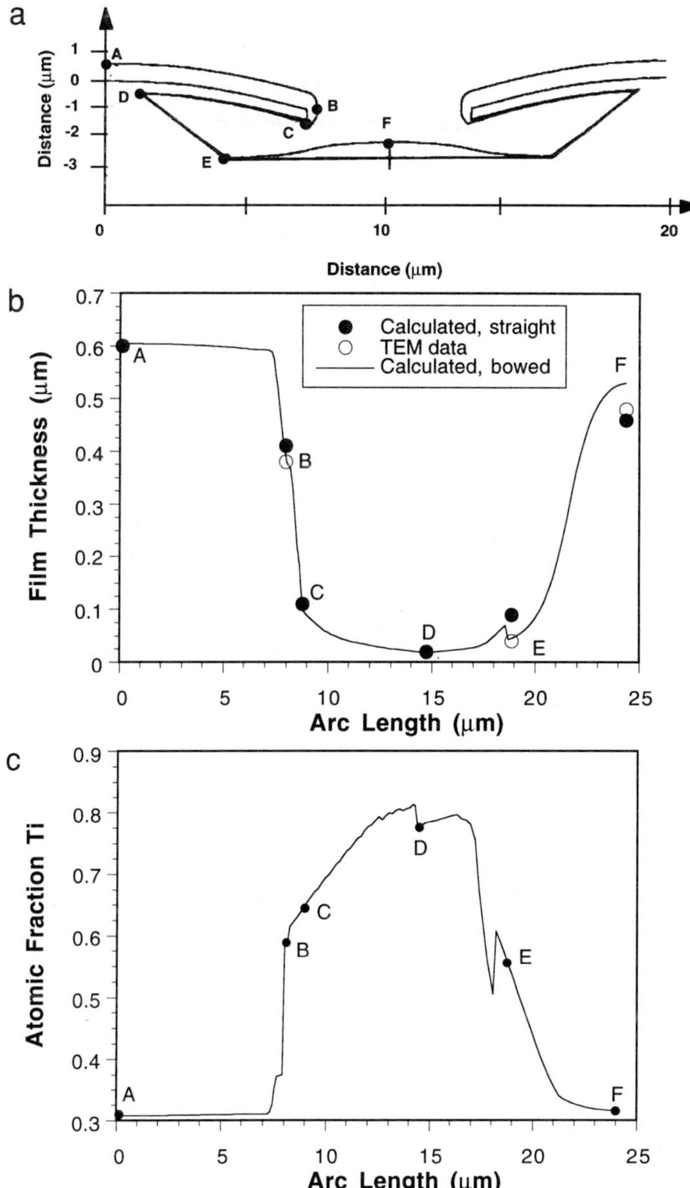

FIG. 23. (a) Ti–W thickness as determined by an EVOLVE simulation starting with a bowed structure. (b) Ti–W thickness profile versus arc length along the EVOLVE calculated surface and Ti–W thicknesses at points A through F in part (a), as determined by TEM measurements, and an EVOLVE simulation starting with a straight structure. (c) Titanium fraction as a function of arc length (*83,84*).

Figure 23a is the Ti–W thickness profile calculated by EVOLVE using a starting profile with the beam bowed. Ti and W effective sticking factors were estimated to be 0.30 and 0.74, respectively. Figure 23b plots the predicted film thickness as a function of arc length calculated by simulations starting with a bowed profile. Arc length is defined to be the distance along the predicted surface, starting from a reference point on the upper left corner of the structure. TEM measurements and calculated results determined at points A through F, corresponding to locations A through F in Fig. 23a, are also plotted in Fig. 23b. The predictions arrived at starting with the straight profile are higher at all points, except point F, than those arrived from starting with a bowed profile. Intuitively, this makes sense. Compared to a bowed beam, a straight beam would allow more re-emitted atoms to strike the interior of the feature, while blocking more incoming source volume atoms from striking point F. Note that the calculated thicknesses agree fairly well with the experimental data. Figure 23c plots the atomic fraction of Ti in the Ti–W film as a function of the arc length. Notice that the fraction of Ti predicted by simulation is maximum around the corner where the polysilicon beam meets the SiO_2 sidewall. The composition in this area is predicted to be approximately 78 at.% Ti, in good agreement with the AES determined composition of 81 at.% Ti.

Simulations of Ti sputter deposition were also performed. The incoming flux distribution of Ti atoms from the target and the flux distribution of re-emitted atoms were assumed to be cosine. The Ti effective sticking factor was varied to fit the experimentally determined thickness profile. Figure 24a is a plot of the Ti profile predicted using a unity Ti effective sticking factor. Figure 24b plots the predicted film thickness as a function of arc length along the surface and the TEM determined thicknesses at points A through F. As in the case of the Ti–W simulations, the calculated thickness in the Ti case fit well to the TEM distributions.

The large difference in Ti effective sticking factors in Ti–W versus Ti sputter deposition may be due to either (1) a difference in sticking factors of Ti atoms on a Ti–W surface and Ti atoms on a Ti surface or (2) enhanced resputtering effects during Ti–W deposition compared to Ti deposition compared to Ti deposition. However, neither the experimental data nor the simulation results obtained in this study are sufficient to determine the exact mechanism. Work presented by Bergstrom *et al.* (*87*) suggests that the Ti depletion is due to preferential resputtering of the deposited Ti atoms. It was proposed that energetic Ar particles backscattered from W atoms on the target cause the resputtering. Their

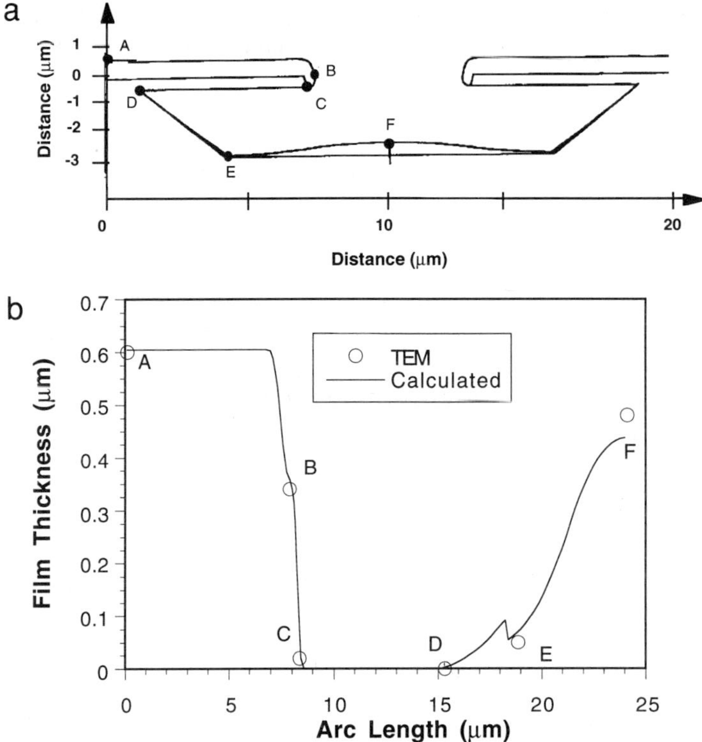

FIG. 24. (a) Ti thickness as determined by an EVOLVE simulation. (b) Ti thickness profile versus arc length along the EVOLVE calculated surface in part (a), compared with TEM data (*83,84*).

experiments showed that the depletion could be minimized if Xe were used as the sputtering gas instead of Ar.

X. Simultaneous Deposition and Sputter Etch

This section discusses our simulation studies of a CVD technique and a PVD technique with increased directionality of the depositing species which have been proposed to help achieve better filling of high-aspect-ratio features (*88*). Directionality is achieved using ion fluxes in the systems used in the studies discussed later: an electron cyclotron resonance (ECR) plasma reactor system, and the ionized magnetron sputter (IMS) deposition system. In addition to increasing the directionality of the depositing species, these systems utilize the sputter etching capability of the energetic ions to modify the evolving film profile.

A. IMS Deposition of Aluminum

Rossnagel and Hopwood (*89,90*) have shown that trenches and vias can be filled by ionizing a large fraction of the metal using a radio-frequency induction (RFI) coil placed between the target and substrate. The ion fluxes to the wafer surface are almost unidirectional (and perpendicular to the wafer surface) in a IMS deposition system, however, the metal atoms have wider angular flux distributions similar to typical magnetron sputter tools. The directionality of the ions is due to their acceleration in the plasma sheath region close to the wafer. On the other hand, the neutral metal atoms arriving at the surface either are not ionized by the RFI coil or have been neutralized by collisions. A small dc bias voltage (about 20 V) is sufficient for high directionality of the ions. However, most of the ions do not have sufficient energy to (re)sputter deposited material at these dc bias voltages. The sputtering yield of these ions can be enhanced by increasing the bias voltage, and this feature can be effectively utilized to fill high-aspect-ratio trenches and vias.

In general, the ratio of ion flux to neutral flux can be controlled by changing the power applied to the RFI coil. The topography of the deposited films is highly dependent on this ratio. While higher ion fluxes lead to a more directional deposition, higher neutral fluxes lead to formation of overhangs at the mouth of the feature, which lead to void formation. If the energy of the ions is large enough, they will preferentially sputter away the overhangs, allowing for further deposition into the features. If the ion flux and energies are large enough, void formation can be avoided in small-aspect-ratio features. On the other hand, void formation is enhanced in high-aspect-ratio features by increasing the energy of the sputtering ions.

Hamaguchi and Rossnagel (*91*) used simulations to explain the control obtained in filling trenches and vias by varying ion energies (as a result of changing dc bias). While the expected trends were observed in small-aspect-ratio features, the opposite results were seen in the case of high-aspect-ratio features and large film thicknesses (i.e., when the film thickness is larger than the trench width). This was explained using the argument that the sputtered material collects on the opposite sidewall, because the sticking factor of this resputtered material has a sticking factor near 1.0. This leads to lateral buildup of the resputtered material and closure of the feature. The flux distribution of the incoming ions was modeled as exponential, while the neutral species was modeled using a generalized cosine distribution with a collimation cutoff angle of 52°.

B. ECR Deposition of SiO_2

The ECR system used by Bose *et al.* (*92*) to study planarized deposition of SiO_2 interlevel dielectric layers also relies on ion fluxes. They observed trends in feature fill opposite to those of Rossnagel *et al.* in their IMS depositions (*89*), with changes in etch-to-deposition-rate ratio. While the ECR process is inherently different from the ionized magnetron deposition process, there are several similarities. Ions are accelerated across a plasma sheath making them very directional, and the neutral species exhibit more random flux distributions. Labun (*93*) used a pair of simulation packages sequentially: EVOLVE for the deposition and SAMPLE for the sputter process, to show the role of sputtering during deposition in trenches of varying aspect ratio.

C. Simulations

To better understand the processes involved, we used EVOLVE to simulate the evolving surface profiles during simultaneous deposition and sputter etch processes. Table I lists some of the model parameters (or models) for these simulations. We assume that the resputtered material leaves the surface with a cosine distribution, and has a unity sticking factor. Then the flux of sputtered material leaving a segment can be written in terms of the incoming fluxes of sputtering ions as

$$\eta_s^{\text{out}} = E_{p^+} + E_{g^+}, \tag{79}$$

where s stands for the sputtered material, p for precursor, g for plasma gas, η for flux, and E for etch rate (in moles per area per time). The etch rates in turn depend on the ion fluxes and their sputter yields, and can be written as

$$E_i(\theta) = Y_i(\theta)\eta_i^{\text{in}}(\theta), \tag{80}$$

where θ is the angle measured relative to the local surface normal, $Y(\theta)$ is

TABLE 1
SUMMARY OF THE PARAMETERS USED IN THE SIMULATIONS

(a) Parameters used in the ECR simulations		EVOLVE v4.1a	Ref. *93*
Incident species	Sticking coefficient	0.4	0.4
	Flux distribution of ions	Exponential	Exponential
	Flux distribution of neutrals	Cosine	Cosine
Resputtered species	Sticking coefficient	1.0	—
	Flux distribution	Cosine	—

(b) Parameters used in the IMS simulations		EVOLVE v4.1a	Ref. *91*
Incident species	Sticking coefficient	1.0	1.0
	Flux distribution of ions	Exponential	Unidirectional
	Flux distribution of neutrals	Over-cosine (Parameter = 12) see Ref. *35*	cosine (Cutoff of 52°)
Resputtered species	Sticking coefficient	1.0	1.0
	Flux distribution	Cosine	Cosine

the sputter yield of deposited material due to the incoming depositing species ions or working gas ions impinging at θ, and the total sputter rate at each point is obtained by integrating over all angles in the global coordinate system [see Eq. (39) in Sec. IV]. The dependence of the sputter yield on the angle θ is modeled as

$$Y(\theta) = A\cos\theta + B\cos^2\theta + C\cos^4\theta, \tag{81}$$

where $Y(\theta)$ is defined as the number of atoms that are removed from the surface for every impinging ion.

The variation of the sputter yield as a function of the incident angle has been studied for several systems by Lee (*94*), and can be explained as follows. As the angle of incidence is increased from normal incidence ($\theta = 0$), the etch rate of most materials increases at first, reaches a maximum for angles in the range of 45 to 60°, and then decreases to 0 as the angle of incidence approaches 90° (*94,95*). Due to the lack of information, we use the same yield curves for SiO_2 sputtering by oxygen and argon ions in the ECR system, and for sputtering of aluminum by aluminum and argon ions in the IMS system. The energy dependence on the shape of the sputter yield curve was ignored for the same reason. However, a constant multiplier was used to account for the ion energies by matching the sputter

TABLE II
PARAMETERS DEFINING THE NORMALIZED SPUTTER YIELD CURVES FOR Ar SPUTTERING OF SiO_2 AND Al USED IN THE SIMULATIONS

	A	B	C
Sputtering of SiO_2	10.7	−11.7	2.0
Sputtering of Al	1.63	1.96	−2.59

rates on the wafer flats with the experimental results. The parameters of the normalized sputter yield curves used in this work for Ar sputtering of SiO_2 and Al are given in Table II.

For the ECR simulations, the deposition kinetics are assumed to be controlled by the oxygen ion flux to the surface. Further, the surface is assumed to have a complete layer of adsorbed silane and/or other precursors, similar to the TEOS case as reported by Cale and coworkers (*60,61,68,77*). Following the work of Labun (*93*), we used a constant sticking factor of 0.4 for the depositing species, and assumed an exponential distribution. ECR simulation results obtained using EVOLVE, for three sets of ideal trenches with aspect ratios of 1.5, 2.5, and 3.5, are shown in Fig. 25. SEM micrographs of trenches of similar aspect ratio for varying etch to deposition ratios (E/D), obtained by varying the bias on the substrate, are shown in Fig. 26 (*93*). An increase in E/D from 0.30 to 0.45 affects the deposition profile considerably, changing it from a profile with a well-defined void to one where no void is present. This can be explained as follows: For a pure deposition system (no substrate biasing or low E/D), a high deposition rate on the flats and shoulders relative to the deposition rate inside the feature leads to overhang formation, poorer step coverage, and eventual pinch-off. As the bias on the substrate is increased (significant etch rate or high E/D), high-energy ions bombard the growing film and preferentially etch away the film deposited on the shoulders. This preferential etching on the shoulders is due to the fact that the maximum sputter yield (for Ar ions with respect to SiO_2) occurs between ~ 45 and $\sim 60°$, which is the range of angles that the normal to the film deposited on the features' shoulders makes with the impinging ions. This leads to increasing feature fill by reducing the overhang (as it is etched at the maximum rate), keeping the feature mouth wider and allowing more deposition inside the feature. For high enough values of E/D, no void is formed (E/D is ~ 0.45 in this case). In general, the elimination of the void by sputtering is dependent on the ion energy and the sputter yield of

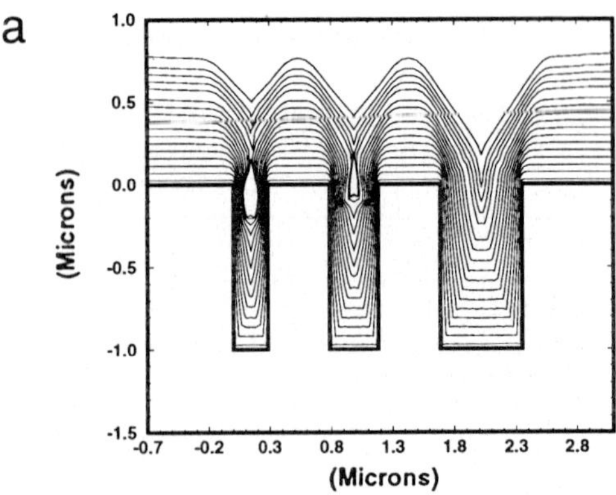

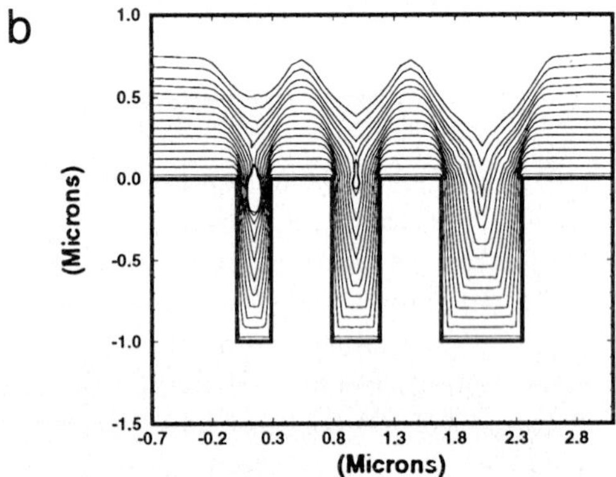

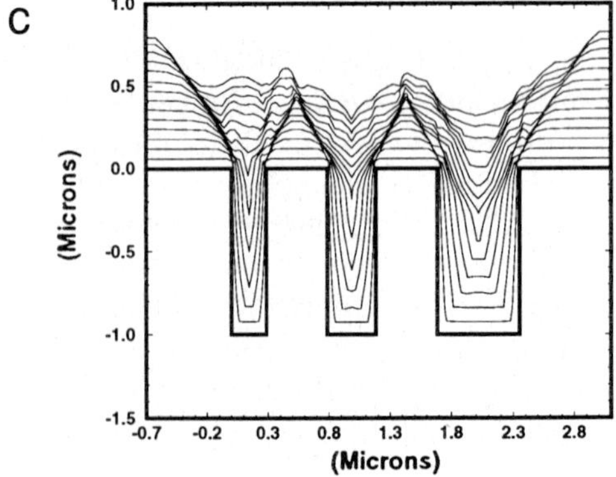

FIG. 25. EVOLVE simulation results showing the decreasing void size with increasing etch to deposition rate ratio for three ideal trenches of 1.5, 2.5, and 3.5 in the ECR deposition of SiO_2. (a) $E/D = 0.3$, (b) $E/D = 0.35$, and (c) $E/D = 0.5$.

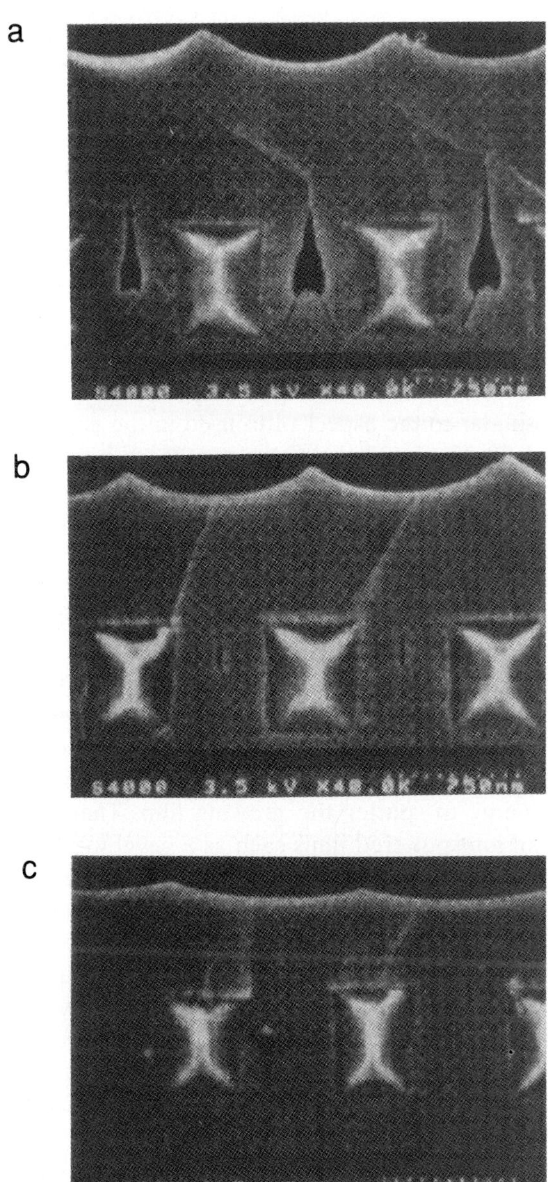

FIG. 26. SEMs showing the decreasing void size with increasing etch to deposition rate ratio in the ECR deposition of SiO_2 (*93*). (a) $E/D = 0.3$, (b) $E/D = 0.35$, and (c) $E/D = 0.5$ (*93*).

the ions. The simulation results shown in Fig. 25 follow the trend seen in the SEMs for the various E/D.

The results presented in Figs. 25a–c show that for a fixed E/D ratio (say, 0.30), moving from a high-aspect-ratio trench (aspect ratio = 3.5) to a low-aspect-ratio trench (aspect ratio = 1.5) reduces the void information from a well-defined one to almost no void. The interiors of low-aspect-ratio features receive more depositing species flux compared to their high-aspect-ratio counterparts. This leads to a relatively smaller difference between the deposition rates on the flats and the insides of low-aspect-ratio features. Hence, more of the low-aspect-ratio feature gets filled, leading to smaller void formation.

EVOLVE was used to simulate IMS depositions in ideal trenches of aspect ratio 2, similar to the aspect ratio used in the experiments reported in Refs. *89* and *90*. Figure 27 shows the results, with an Al ion to neutral flux ratio of 1 : 1, and varying ion energies striking the sample surface (by adjusting the bias). The void size, defined as the fractional filling, follows a trend opposite to that seen in the ECR reactor system. Because the IMS deposition system is a PVD system (the sticking factor of the impinging species is assumed to be 1 in these simulations), an inherent overhang buildup occurs. Coupled with this effect, the sputtered material is redeposited with a cosine distribution and a sticking factor of 1. Hence though the ions sputter the overhang, they are not able to completely etch away the overhang, leading eventually to the formation of a void; that is, at low ions energies ($\sim$20 eV) at the sample surface, the impinging ions do not have enough energy to sputter the growing film. Therefore none of the characteristics of ion sputtered films such as a bevel formation is seen (see Fig. 27d). As the energy of the ions increases to 80 eV, the ions sputter the growing film. This reflects in the formation of the bevel angles, but the profile comes close to closure (see Figs. 27e and 27f). As the energy is further increased to 120 eV, a large amount of sputtering of the growing film takes place. This leads to large amounts of redeposition on the sidewalls, eventually causing pinch-off. Though no direct relation between the ion energy at the surface and etch rate has been obtained, the simulations follow a similar trend (Figs. 27a–c). The effect of aspect ratio is shown in Fig. 28.

Our simulation results yield the same trends as those used in the previous studies of these systems. The contrasting simulation results lead us to propose that sticking factors and flux distributions play important roles, in addition to redeposition, to explain the larger voids at higher E/D

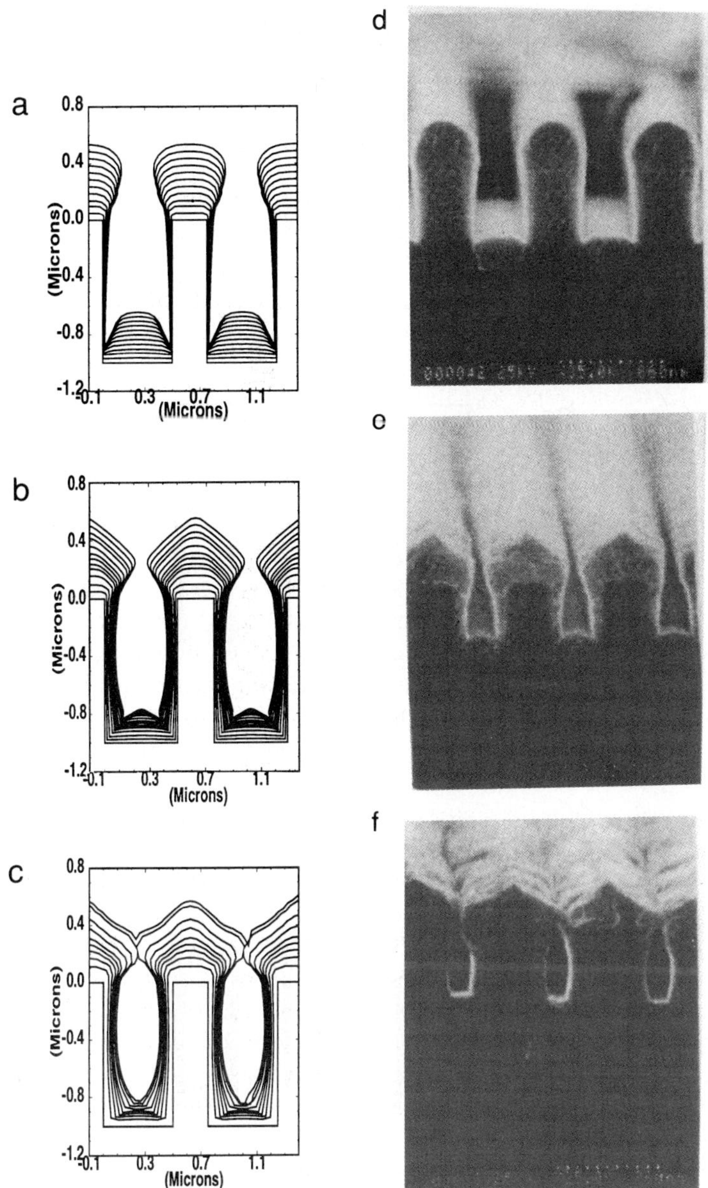

FIG. 27. (a–c) EVOLVE simulations results showing decreasing filling with increasing etch to deposition rate ratio for an ideal trench of aspect ratio 2.0. (d–f) SEMs showing the decreasing filling with increasing etch to deposition ratio in the IMS deposition of aluminum (*89,90,91*).

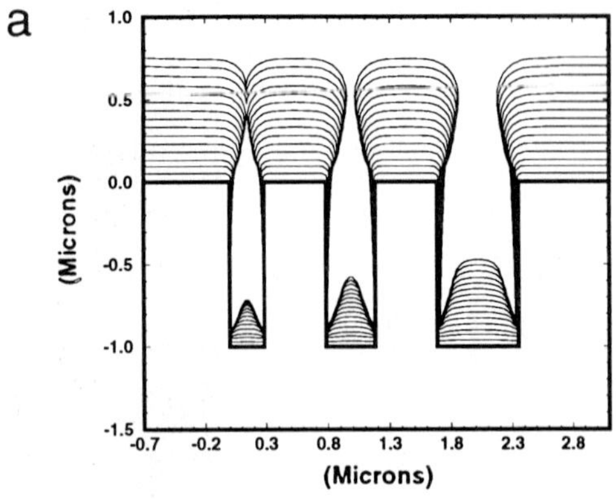

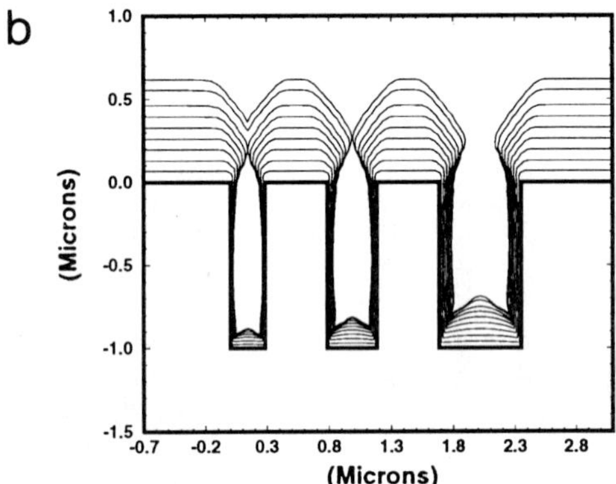

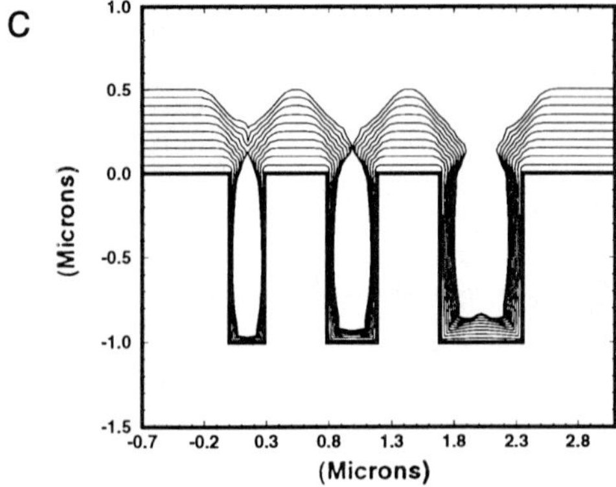

FIG. 28. EVOLVE simulation results showing the decreasing amounts of filling in the trenches with increasing etch to deposition rate ratios in the case of IMS deposition of aluminum for three ideal trenches of aspect ratios 1.5, 2.5, and 3.5. (a) $E/D = 0.0$, (b) $E/D = 0.38$, and (c) $E/D = 0.5$.

for the IMS process. To demonstrate this, we have studied the effects of varying the sticking parameter and the flux distributions of the neutrals. Sticking coefficient values of 0.4 (value used in the ECR simulations) and 1.0 (value used in IMS simulations) were used. The flux distribution on the other hand was either over-cosine as in the case of IMS, or exponential as in the case of ECR simulations. These simulations were performed with and without redeposition of the sputtered material. Figure 29 shows the results of using an (a) overcosine and (b) exponential flux distribution (with the same parameters given earlier) with a sticking factor of 1 and 0.4, in an otherwise unaltered IMS input file. The sputter etched material is assumed

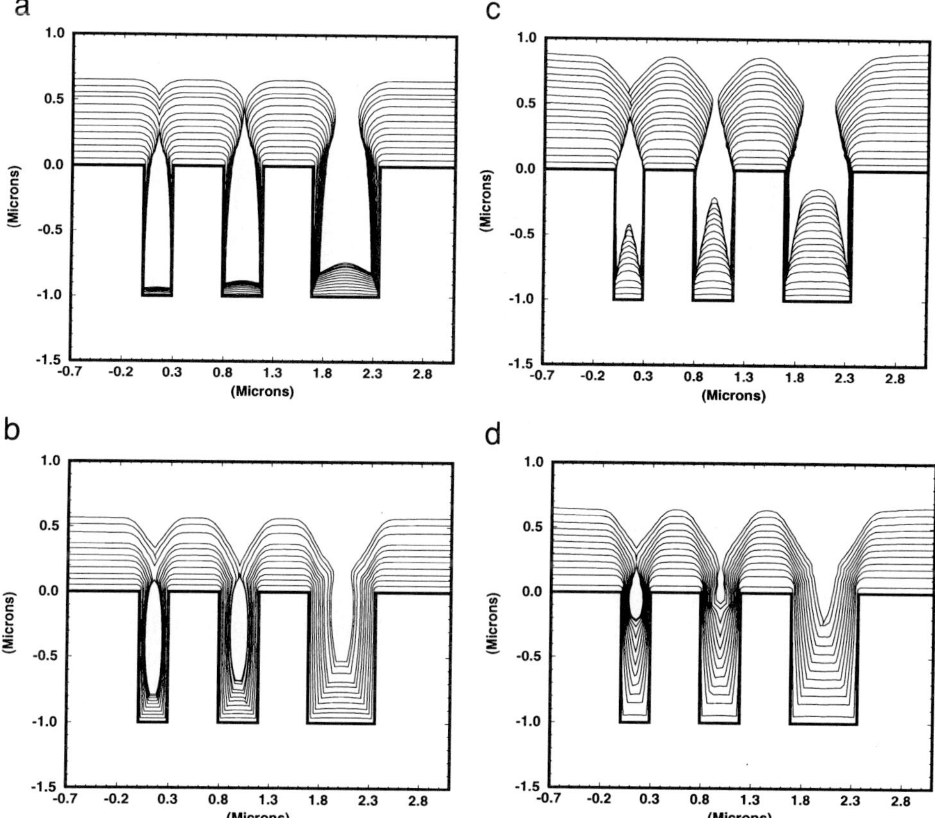

FIG. 29. EVOLVE simulations showing the effect of changing the sticking factor and flux distribution in a standard IMS file. Results include complete redeposition of the sputtered species.

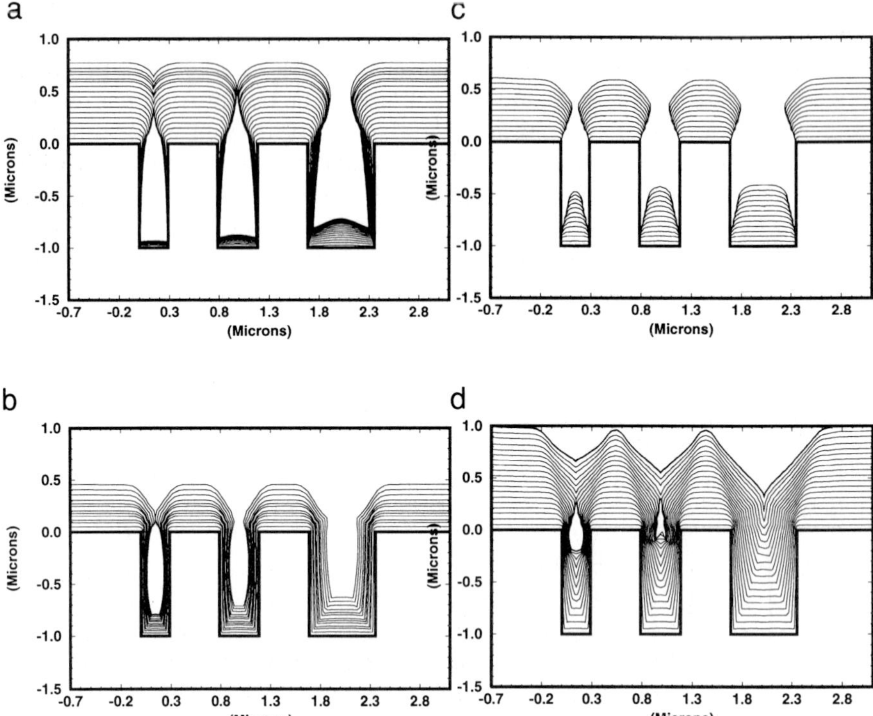

FIG. 30. EVOLVE simulations showing the effects of the two factors—sticking coefficient and the flux distribution of the depositing in the absence of redeposition of the sputtered species. Results indicate that the redeposition does not play a very significant role.

to have a cosine distribution with unity sticking factor. Figure 30 shows the same results without any redeposition (sticking factor = 0). It can be seen that we can qualitatively move from the ECR trend to the IMS trend by changing these aspects of the two models. Also, the results indicate that the sticking coefficient and flux distributions are the most significant parameters in determining the shape of the void and that the redeposition plays only a minor role.

XI. Curvature-Driven Surface Diffusion

In this section we focus on surface diffusion. Surface diffusion is accepted to play a role in physical vapor deposition (PVD) processes and in some CVD processes. It is suspected to play a role in the other CVD and plasma-enhanced CVD processes. We study surface diffusion during the sputter deposition (PVD) of aluminum–(1.5%)copper films. By comparing film profiles computed using EVOLVE with experimental profiles, we show how to develop an expression for surface diffusivity as a function of temperature. The method we demonstrate should prove useful for process design, because it uses data obtained from the films as deposited in the sputter equipment. We then demonstrate both a feature size dependence on conformality and a deposition rate dependence on conformality, consistent with the model equations. More details of this work can be found in Refs. *55* and *56*.

A. Experiment

In the experiments, silicon dioxide films were grown thermally on 4-in. silicon wafers to a thickness corresponding to the desired trench depths (0.5, 1, and 2 μm). Trenches with a range of widths were etched into the silicon dioxide to each silicon substrate. A thin layer (about 100 nm) of thermal oxide was then grown on the exposed Si surface to prevent deposited metal from contacting the silicon. Al–(1.5%)Cu films were then deposited onto these wafers at a target deposition rate of 10 nm/s in a sputtering system (using a conical magnetron electrode), at temperatures of 303, 423, 523, and 623 K. Deposition temperatures were measured by a thermocouple at the backside of the wafers. The sputtering gas (Ar) pressure was 4 m Torr and the target to substrate distance was 2.54 cm. No substrate bias was applied in order to minimize resputtering of deposited material. After each film was deposited, a low-temperature plasma oxide was deposited to aid in imaging the metal films. We have shown that the deposition of these delineating plasma oxides does not affect the metal film profiles.

B. Model Development

When surface diffusion and/or reaction mechanisms that involve surface species are considered, material balances on ballistic species as well as on adsorbed species must be formulated to determine the distributions of fluxes and surface concentrations. Cale *et al.* (*97*) have presented the analytical procedure and have demonstrated a feature size dependence of step coverage in the presence of surface diffusion. In that study, surface diffusion was modeled as due to gradients in the surface concentration of aluminum passing through an adsorbed precursor state before becoming incorporated into the growing film. That mode of surface diffusion plays a role in epitaxial growth processes, and will occur to some extent in PVD; however, the predicted film profiles did not match experimental profiles very well at the intersections of the trench wall and its base. In addition, the diffusivity parameters required to match the experimental trends in step coverage or "fill factor" (*96*) for sputter deposited aluminum films give diffusivity values which are unreasonably large (*96,98*). In this chapter, we consider surface diffusion driven by gradients in surface energy caused by gradients in surface curvature. This problem has been treated thoroughly for nondeposition systems (*99,100*).

The infinite trench problem is inherently two dimensional, since there is no variation in deposition rate along the length of the trench. Any surface diffusion is in the plane of the cross-section. As in other sections of this chapter, similar equations hold for features of circular horizontal cross-section. Since the problem is expressed in two dimensions, we deal with lengths along the film surface rather than differential areas. The arc length s runs from the left to right along the trench surface. When measured relative to a fixed point on the perimeter of the trench, arc length alone specifies the position of a point on the surface.

We ignore the copper in the films in the submodels used in EVOLVE to simulate the deposition of the Al–(1.5%)Cu films; that is, we treat the film as though it were a single component material for purposes of simulating profile evolution. To date, the detailed mechanism for curvature-driven surface diffusion has not been resolved (*98–105*). In fact, it is probable that more than one mechanism contributes to the observed surface diffusivity, particularly at temperatures just below the melting point of the film (*101*). One approach is to assume that surface diffusion occurs due to surface concentration gradients in diffusing species before they become part of the growing film or are excited out of the solid film (*97,101–103*) The most common approach used in the analysis of surface diffusion driven by gradients in film curvature is to ignore the details of the actual mechanism (*98,103*). This phenomenological approach is used in this analysis. The highest temperature used in this work is below 0.75 times the melting point of the film, where a sharp increase in surface diffusivity and its activation energy is observed for many metals (*101*).

The increase in chemical potential (per atom) of a curved surface, relative to that on a flat surface, is (*101*)

$$\mu(K) = K\gamma\Omega, \tag{82}$$

where K is the signed curvature expressed in terms of arc length along the surface, γ is the surface free energy of the film, and Ω is the volume per atom in the film. The surface diffusive flux due to this mode is expressed in terms of an effective diffusion coefficient, as might be seen from observations of grain boundary grooves or probe tips (*101*). The flux of aluminum via this mode can be written as

$$J_c = -\left[\frac{D\gamma\Omega\chi}{k_B T}\right]\frac{\partial K}{\partial s}, \tag{83}$$

where χ is the density of aluminum atoms at the surface of the film and k_B is Boltzmann's constant. The curvature of the film is expressed as

$$K = \frac{\dfrac{dy}{ds}\dfrac{d^2x}{ds^2} - \dfrac{dy}{ds}\dfrac{d^2y}{ds^2}}{\left[\left(\dfrac{dx}{ds}\right)^2 + \left(\dfrac{dy}{ds}\right)^2\right]^{3/2}} \tag{84}$$

and K is positive for surfaces concave inward and negative for surfaces concave outward. According to Eqs. 83 and 84, curvature-driven surface diffusion results in a net flux of surface atoms from points on the surface which are concave inward to points which are concave outward.

The reactions used for the sputter deposition of aluminum are

$$\mathrm{Al}(b) \underset{k_{-1}}{\overset{k_1}{\longleftrightarrow}} \mathrm{Al}(p) \underset{k_{-2}}{\overset{k_2}{\longleftrightarrow}} \mathrm{Al}(s), \tag{85}$$

where b means ballistic, p means precursor and s means aluminum atoms which are part of the growing solid film. We assign numbers 1, 2, and 3 to Al(b), Al(p), and Al(s), respectively. It is assumed that the sticking factor of the incoming ballistic species is unity and does not depend on their incident angle and there is no resputtering of deposited material. The two reaction rate expressions used are

$$R_1 = \eta_1 \tag{86}$$

and

$$R_2 = k_2\chi_2 - k_{-2}\chi_3. \tag{87}$$

Equation 86 reflects our assumption that arriving atoms stick to the surface with unity sticking factor and no material leaves the surface ($k_1 = 1$ and $k_{-1} = 0$). Equation (87) gives the net rate at which precursor atoms become part of the growing film [see the discussion after Eq. (90)].

The rate of change in position of the local film surface at the position represented by s is given in terms of the rate of the second reaction as

$$\frac{\partial \Gamma(s,t)}{\partial t} = R_2 \upsilon_3, \tag{88}$$

where υ_3 is the molar volume of aluminum and R_2 is its molar rate of generation per area of surface as given by Eq. (87). Time has been added to the argument list in this equation, since the local deposition rates depend on the extent of feature fill. We assume that the surface position changes slowly relative to the redistribution of fluxes to the walls caused by the changes in geometry during deposition processes; that is, we can determine the flux distribution to the evolving film surface by considering a "snapshot" of the feature at any time.

Material balances for the two mobile species on a differential area of the trench lead to

$$R_1 = \eta_1 \tag{89}$$

for species 1 and

$$0 = R_1 - R_2 + \left[\frac{D\gamma\Omega\chi}{k_B T}\right]\frac{\partial^2 K}{\partial s^2} \tag{90}$$

for species 2. The surface concentration of solid aluminum χ_3 is equal to the number of atoms per square centimeter χ. The diffusivity values determined in this work do not depend on the values used for k_2 and k_{-2}, because the diffusion of species 2 is driven by curvature in our model. The film profiles depend only on the value of the coefficient of the second derivative of curvature in Eq. (90), which in turn depends on $\chi = \chi_3$ and not on χ_2. The latter surface concentration is determined by the values of k_2 and k_{-2}. If there is no ballistic flux at a given film position, the equilibrium surface concentration of mobile precursor adatoms χ_2 is calculated directly by setting R_2 to zero. The kinetic model [Eqs. (86) and (87)] is given in the interest of completeness. As our understanding of the mechanism of curvature driven surface diffusion improves, perhaps the diffusive flux will be written in terms of the surface concentration of adatoms, which is a function of local film curvature (*101,103*). In that case, the values of the kinetic parameters k_2 and k_{-2} will impact the film profiles and hence the predicted diffusivities.

Dividing Eqs. (89) and (90) by the deposition rate on the flat areas away from the trench, which is also the flux to those areas because there is no curvature, we obtain equations expressed in terms of dimensionless variables (overbar):

$$\bar{\eta}_1 = \bar{R}_1 = \frac{\eta_1}{\eta_{1v}} = \frac{\eta_1}{R_{1v}} = \frac{\eta_1}{R_{2v}} = \frac{R_1}{R_{1v}} \tag{91}$$

and

$$0 = \bar{R}_1 - \bar{R}_2 + \Delta \frac{\partial^2 \bar{K}}{\partial \bar{s}^2} = \frac{R_1}{R_{1v}} - \frac{R_2}{R_{1v}} + \left[\frac{D\gamma\Omega\chi}{R_{1v} s_r^3 k_B T} \right] \frac{\partial^2 (K s_r)}{\partial (s/s_r)^2}, \tag{92}$$

where s_r is a reference length and Δ is defined implicitly.

Thc integral represented by the first term of Eq. (92) can be evaluated directly to determine the local incoming ballistic flux of aluminum atoms from the source volume as a function of position on the surface, after the flux distribution from the source volume has been specified. The interior surface of the trench is discretized and the second derivative is replaced by second-order finite difference approximations. The two boundary conditions for the curvature are that it is constant at the endpoints of the deposition region on the flat surfaces on either side of the trench.

Figure 31 shows cross-sections of selected films from the experimental matrix; four profiles in low-aspect-ratio trenches deposited at nominal (setpoint) temperatures of 303 K (Fig. 31a), 423 K (Fig. 31b), 523 K (Fig. 31c), and 623 K (Fig. 31d), and two cross-sections of films deposited in higher aspect ratio trenches at 523 K (Figs. 31e and 31f). Focus on the clearly delineated interface between the bright oxide (top) layer and the dark Al (middle) layer in these cross sections.

Figure 32 shows film profiles predicted by EVOLVE, using Eq. (92) and corresponding to those shown in Fig. 31. The predicted film profiles match the trends in the experimental profiles well. In particular, the degree of rounding of the corners at the base of the trenches is predicted well. The shape of the base is not predicted as well; however, the correct trends in the thickness of the base with temperature and aspect ratio are predicted. The lack of quantitative comparison between predicted and experimental base profiles can be attributed to several factors:

1. The profiles shown in Fig. 31 are examples of the many SEM cross-sections taken for each feature size. There is considerable variation between the SEMs for a given feature size and aspect ratio, even from the same wafer. This variability is partially explained by differences in the grain structure in the different cross-sections analyzed.

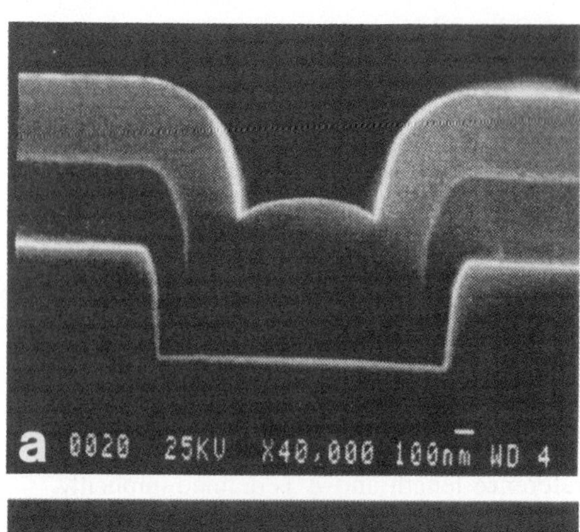

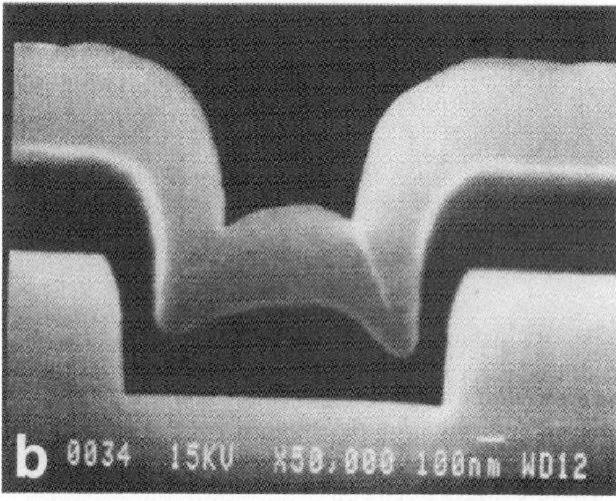

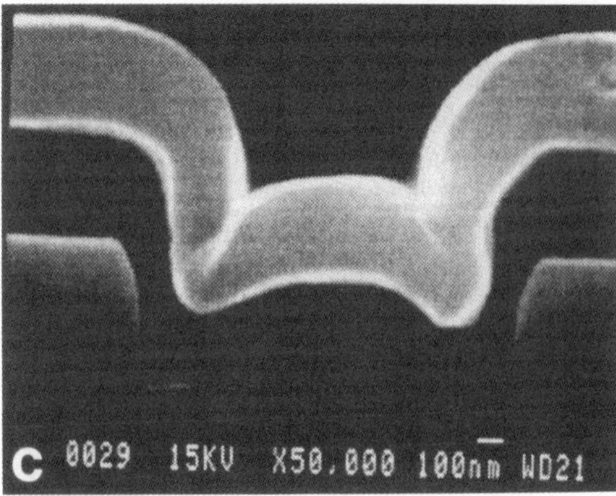

FIG. 31. Experimental film profiles: (a) 303 K, (b) 423 K, (c) 523 K, (d) 623 K, (e) 523 K, and (f) 523 K (*55*).

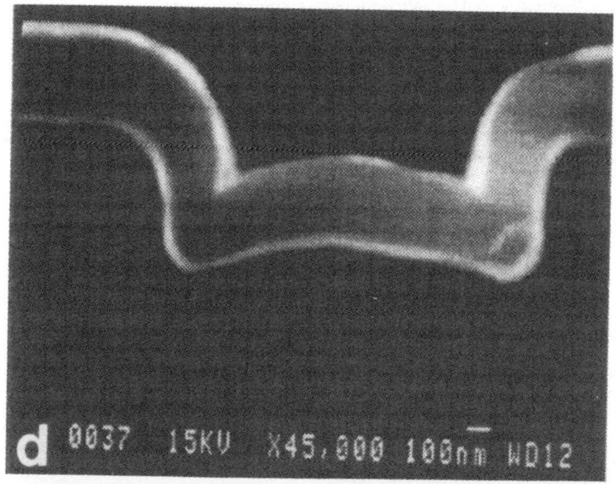

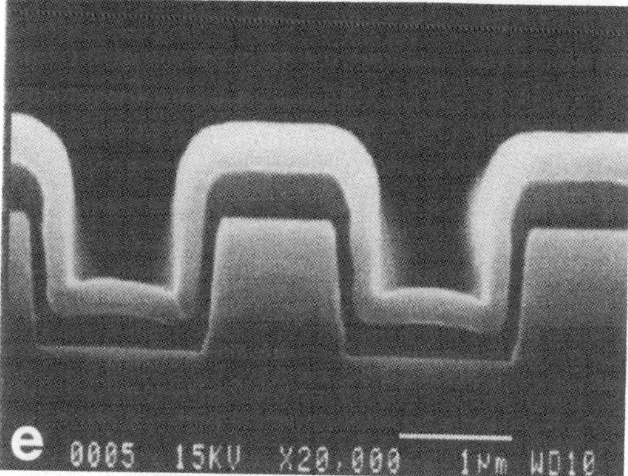

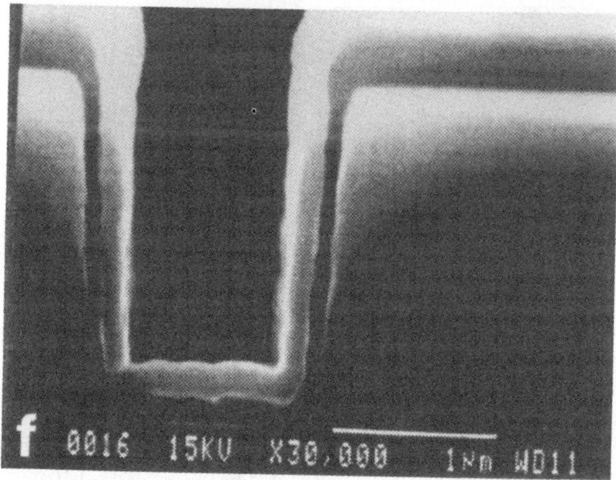

FIG. 31. Continued

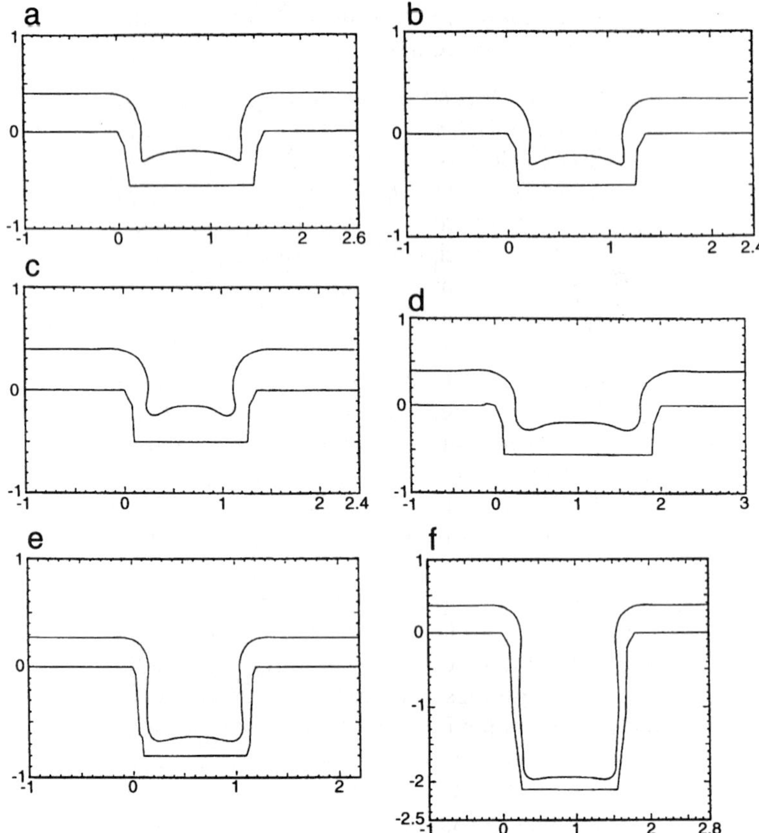

FIG. 32. Calculated film profiles corresponding to those shown in Fig. 31, using the parameter values discussed in the text (*55*).

2. The use of the *generalized cosine distribution* or GCD to express the angular dependence of the flux to the surface (see Sec. III). The value of the GCD parameter ρ used in all simulations is 3. The range of conditions (aspects ratios and film thicknesses) did not allow a more precise estimate of this parameter; however, it should be the same for all experiments in this study since all depositions were performed at the same rate.
3. The effective deposition temperature (the temperature of the surface) is not known, because there is considerable energy flux associated with the incoming species. We did not attempt to predict

deposition temperatures, but instead have used the temperatures measured by the thermocouple.

The asymmetry of the deposited film profiles in a couple of the figures (e.g., Figs. 31b and 31c) might well be caused by some combination of the first and second factors. Because of the variability of the experimental film profiles, as well as the other two factors, we can only expect semiquantitative agreement between simulated and specific experimental profiles. The impact of item 1 could in principle be addressed by comparing a simulated profile with the average of a number of aluminum film profiles in essentially identical trenches. By considering a number of cross-sections, a model which represents the average response can be developed. However, even small variations in the starting trench profiles will cause significant variations in predicted deposited film profiles. Thus, trends in film conformality can be determined using our model; however, exact comparison with individual experiments cannot be expected.

In the model represented by Eq. (92), the evolution of film profiles depends on the dimensionless diffusion parameter Δ and the flux distribution of incoming moieties. In an attempt to isolate the effect of surface diffusion on film profiles, the parameter of the flux distribution model is estimated by comparing simulated and experimental profiles for depositions done at a temperature of 323 K, where surface diffusion does not appear to affect film profiles ($\Delta \approx 0.0$). The value estimated for the generalized cosine parameter is 3, and this value was used for all simulations in this analysis. Once the flux distribution parameter is specified, Δ is estimated by comparing simulated and experimental profiles. To estimate a diffusivity value for a given experiment, it is necessary to estimate values for the other parameters which define Δ. Both the surface energy and the surface diffusivity are unknown and they are multiplied in Eq. (90). The values used for the atomic volume and the surface site density are $1.7 \times 10^{-23}\ \mathrm{cm^3/atom}$ and $1.5 \times 10^{15}\ \mathrm{sites/cm^2}$, respectively. In the absence of better information, the surface energy is assumed to to be temperature independent and to have a value of $1100\ \mathrm{erg/cm^2}$ (*98,106*). The remaining unknown in the coefficient of the derivative of Eq. (90) is the surface diffusivity, which we model as an activated process. The preexponential factor and activation energy in the diffusivity relationship estimated by comparing simulated and experimental film profiles is

$$D = 6 \times 10^{-4} \exp(-5800/T), \tag{93}$$

where temperature is in Kelvin and the diffusivity is in $\mathrm{cm^2/s}$.

Comparison of diffusivity values obtained using this method with previously reported experimental results should be approached with caution, because surface diffusivity depends on crystal orientation and surface composition (*101*). Neither surface compositions nor preferred orientations were determined in this study. Nevertheless, it is interesting to note that the activation energy given in Eq. (93) is in good agreement with experimental information on Al surface self-diffusion (*98*) and aluminum grain boundary diffusion (*107*) as well as the activation energy estimated from electromigration data (*108*). Surprisingly, there are apparently no published data for the preexponential factor for aluminum or aluminum–copper surface self-diffusion obtained from experiments dealing directly with surface diffusion. The preexponential factor is much lower (two orders of magnitude) than that estimated from electromigration data (*108*). It is likely that other phenomena are lumped together with surface diffusion in the electromigration based estimates. It is reassuring to note that the value for the preexponential factor in Eq. (93) falls in the range of the majority of reported preexponential factors for surface self-diffusion in metals (*98*).

As discussed in relation to the kinetic model [see Eq. (90)], film profiles are governed in our model by the dimensionless diffusion parameter Δ [see Eq. (92)], for a specified flux of incoming sputtered metal atoms and a specified feature. A similar result for concentration driven surface diffusion was pointed out by Cale *et al.* (*97*). Although the surface diffusion model used in Ref. *97* is different from the one used here, the governing equations have a similar structure. Thus, our model predicts a feature size dependence on step coverage; step coverage increases with decreasing feature size for the same feature shape. Heuristically, this is because the nominal diffusion length is fixed for a given diffusivity value and deposition rate in the model used in this analysis. Thus, surface diffusivity has more impact in smaller features for the same deposition conditions.

Figure 33a shows a cross-section of a trench with an initial aspect ratio of 0.20, which is close to the aspect ratio (0.24) of the trench in Fig. 31d. The feature shown in Fig. 33 is four times the size of the feature shown in Fig. 31d. Both films were deposited at the same set of conditions. The step coverage, computed as the ratio of the thinnest film thickness on the sidewall divided by the film thickness on the flat surface is 27% in Fig. 33 and 34% in Fig. 31d. This is despite the fact that the aspect ratio of the trench in Fig. 33 is somewhat smaller than that in Fig. 31d and the film thickness (0.36 μm) in Fig. 33 is smaller than the film thickness (0.38 μm) in Fig. 31d. Both of these factors serve to decrease step coverage, whereas the step coverage for the smaller feature size is higher. The model

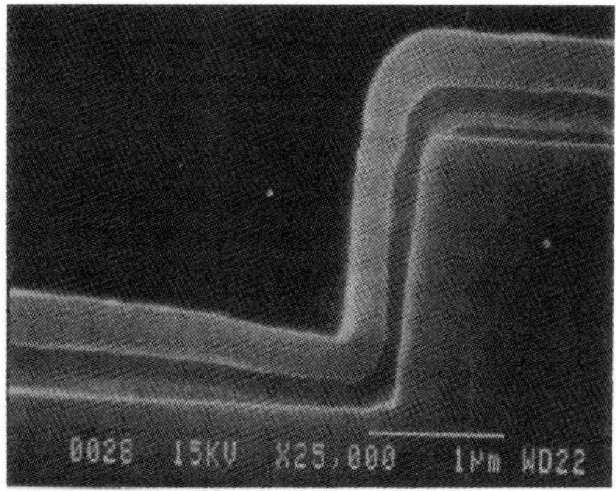

FIG. 33. Experimental film profile for a trench with about the same aspect ratio as that of Fig. 31d, demonstrating a feature size effect on step coverage in the presence of surface diffusion (*55*).

predictions of the feature size dependence on step coverage is not quite this dramatic; however, some of the difference in experimental step coverage could well be due to experimental variability. Model predictions of a step coverage dependence on feature size is certainly validated by these two cross-sections, as well as by a number of other profiles not shown.

We used the same experimental procedure to show that deposition rate can effect the conformality of sputter deposited Al–(1.5%)Cu films. The deposition rate was varied to demonstrate that surface diffusion is a rate process and the conformality realized depends on deposition rate as well as temperature. Figures 34 and 35 show cross-sections of two selected experimental films (right side of each figure) compared with film profiles predicted by EVOLVE using Eq. (90). The film shown in Fig. 34 was deposited at 523 K at a rate of 22 nm/s and the one shown in Fig. 35 was deposited at 523 K at a rate of 2 nm/s. Figure 34 displays less effect of surface diffusion than the film shown in Fig. 35. These figures clearly show that the effect of surface diffusion decreases with increasing deposition rate at the same thermocouple determined temperature. Surface diffusion plays a larger role at lower deposition rates, because it has more time to affect film profiles, particularly for the same film thickness. It is important to note that the sizes of the features are similar.

Our results also indicate that diffusivity increases with deposition rate at a given measured temperature (*56*). This can be attributed to an increase

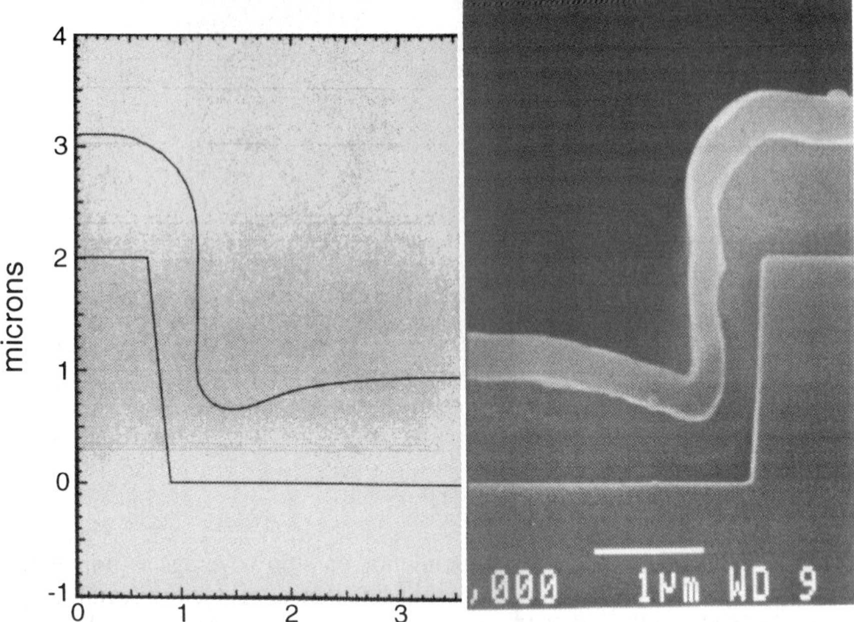

FIG. 34. Simulated and experimental aluminum film deposited at 523 K at 22 nm/s.

in surface temperature with deposition rate, because of the substantial energy flux associated with incoming moieties. Because of this dependence of diffusivity on deposition rate, any model developed for use in fabrication should be determined at specific deposition rates. This effect supports our conclusion presented in the previous paragraph that the effect of surface diffusion decreases with increasing rate at a constant temperature —even with a higher diffusivity at the higher rate, there is less effect.

Though we have demonstrated how engineering estimates of diffusivity values can be made, it is clear that any reported diffusivity results should be approached with caution, because surface diffusivity depends on many factors: alloy composition, surface orientation, and sputtering conditions. Our results indicate that surface diffusion is a critical rate process in determining film conformality. However, it is well known in practice that trying to use curvature-driven surface diffusion can lead to separation of the metal; that is, opening a circuit. Models which account for grain structure may lead us to a better understanding of this phenomenon (*109*).

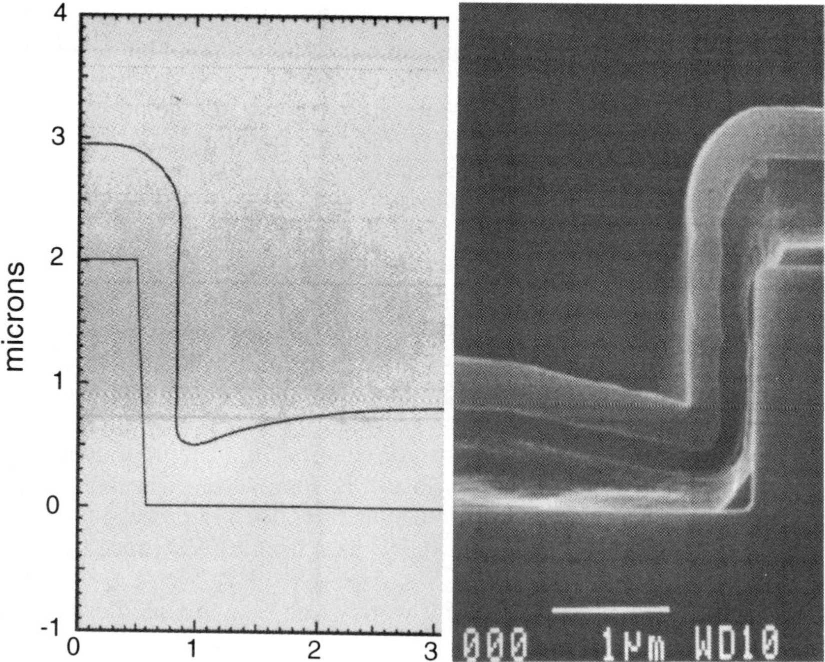

FIG. 35. Simulated and experimental aluminum film deposited at 523 K at 2 nm/s.

XII. Programmed Rate CVD of Tungsten

Increased wafer throughput in the unit processes used to fabricate devices is a universal goal. For deposition processes, the goal is to maximize throughput while maintaining acceptable film properties. Higher deposition rates in general lead to poorer step coverages due to reactant depletion caused by the higher reaction rates, or due to depletion along the depth of the features. There may well not be enough degrees of freedom in conventional *constant rate CVD* (CRCVD) processes to increase throughput significantly while maintaining acceptable film properties (but see Refs. *110* and *111*). This section differs from the previous ones in the sense that it does not contain simulations; however, the PRCVD protocol resulted from modeling and simulation work. Thus, it demonstrates one of the roles which modeling can play in process development: testing new concepts before they can be tested experimentally.

A. Programmed Rate CVD

Cale and coworkers introduced the concept of *programmed rate CVD* (PRCVD), in which additional degrees of freedom are added by ramping one or more deposition parameters during deposition (*112,113*). In that work, we showed theoretically that substantial increases in throughput can be obtained by ramping the deposition temperature down during deposition. The analysis was based on feature scale simulations. As deposition inside features proceeds, aspect ratios naturally increase. To achieve good step coverage using CRCVD processes, the deposition conditions must be chosen such that good conformality is achieved even towards feature closure. This limits the deposition rate. The PRCVD protocol takes advantage of this increasing aspect ratio; that is, deposition conditions at the start of the deposition are set such that the rate is much higher than for CRCVD processes. Conditions are changed in a preprogrammed way to decrease the rate as feature closure is approached. The final rate is about the same as that used throughout the CRCVD process. Thus, the average rate is much higher for PRCVD than for CRCVD processes which yield the same step coverage.

This section discusses our recent work which demonstrated experimentally that the concept is valid for LPCVD of tungsten: Throughput can be increased at the same step coverage, and other film properties are as good or better than those obtained by the CRCVD processes considered. The same concept can be applied to other processes and using other deposition parameters. The PRCVD process described next focuses on improving throughput for the same step coverage and acceptable film properties, and follows the protocol described by Cale *et al.* (*112,113*). Our recent experimental studies are reported in Refs. *114–116*.

B. Experimental

Our CVD W films are produced using a Spectrum 202, cold wall, single-wafer LPCVD reactor with a load lock chamber and a radiant heat source. The wafer temperature is measured at the wafer center on the backside of the wafer using a thermocouple. The blanket tungsten process we use is the hydrogen reduction of tungsten hexafluoride. Depositions have been performed over a range of pressures, hydrogen to WF_6 flow rate ratios, and temperature (CRCVD) or temperature ramps (PRCVD). Before the hydrogen reduction, a 40-nm nucleation layer of CVD W based on the silane reduction of WF_6 is deposited. The nucleation layer and blanket

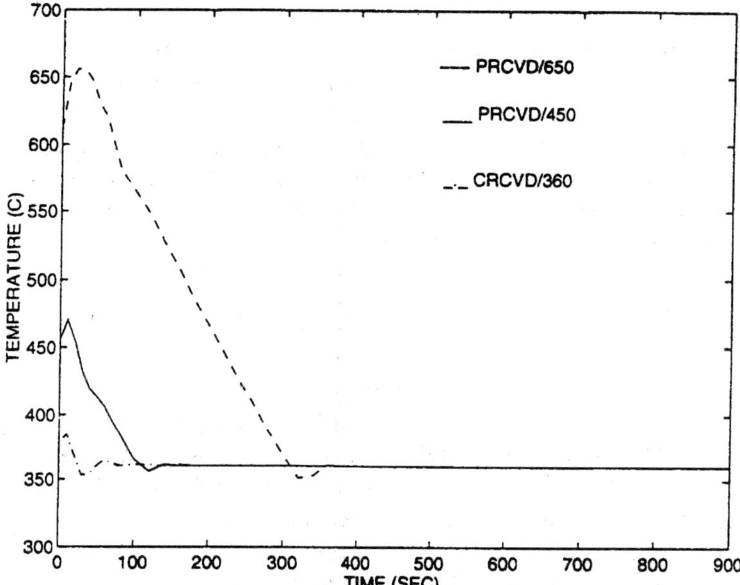

FIG. 36. Representative group temperature trajectories (*116*).

tungsten films are deposited in succession; that is, the wafer remains in the process chamber for both processes and is subsequently removed. The same nucleation process is used for both CRCVD and PRCVD processes.

For all depositions, the reactor temperature is allowed to stabilize in a flow of hydrogen, then the coreactants are introduced. For CRCVD processes, the deposition conditions are held essentially constant for the duration of the depositions. For PRCVD experiments, the temperature ramps are started after a 12-s delay at the initial temperature. After reaching 360°C (the lower temperature for the PRCVD experiments), the temperature is held constant until an expected film thickness is achieved. Figure 36 shows examples of temperature trajectories achieved in practice.

Experimental responses measured on unpatterned wafers are sheet resistance (four-point probe), film thickness (SEM and weight gain), density (RBS), composition (SIMS), reflectivity (nanospec), grain structure (TEM), and residual stress (Flexus wafer curvature measurements) for unpatterned wafers. The residual film stress (tensile) was found to decrease with increasing temperatures, as predicted by Leusink *et al.* (*117*). Compositional analysis of selected samples indicate the PRCVD wafers can have significantly lower fluorine concentration in the bulk of the film compared to the CRCVD wafers. Resistivity results indicate that PRCVD wafers have resistivities as low as or lower than CRCVD wafers. From these results, we conclude that critical properties of PRCVD films can be as good as, or better, than those of CRCVD films.

For patterned wafers, experimental responses are step coverage and film thickness, both of which are determined from SEM cross-sections. The time saved is calculated on the deposition rate from the CRCVD case and is calculated in the following manner:

$$\text{time\_saved}(\%) = [(t_{\text{CRCVD}} - t_{\text{PRCVD}})/t_{\text{CRCVD}}] \times 100. \tag{94}$$

Here, t_{PRCVD} is the time required to deposit a given thickness using PRCVD, and t_{CRCVD} is the time required for a CRCVD process to deposit the same thickness as the PRCVD case. To calculate the percent time saved, the average deposition rate is determined from a CRCVD wafer. The time required to deposit the thickness of a PRCVD film using this average deposition rate (i.e., t_{CRCVD}) is then calculated, and compared to actual time required for the PRCVD film (i.e., t_{PRCVD}) using Eq. (94).

Figures 37a and 37b show films in features of about the same aspect ratio, deposited at a total pressure of 2 Torr, a total flow of 600 sccm, and a $H_2 : WF_6$ flow ratio of 15. The starting temperature for the PRCVD process was 450°C, and it was decreased at a rate of 3°C/s to 360°C. The deposition time was 90 s. The CRCVD temperature was 360°C and the deposition time was 450 s. The conformality for both films is excellent. Adjusting the CRCVD time (using the observed average deposition rate) so that the same film thickness would be deposited, yields a time saving of about 75%. At higher fractional feature closures, the time savings are expected to be even larger; however, for the purposes of this work, feature closure was not desired.

Results from these experiments show that PRCVD blanket tungsten can be used to improve reactor throughput significantly relative to CRCVD processes. Film properties such as residual stress and composition are improved for PRCVD W films, and good resistivities and densities can be achieved. On the other hand, Fig. 37 highlights one of the difficulties with implementing the PRCVD concept. The film thickness is not known as a function of time during the deposition. In the absence of direct film thickness measurements, PRCVD processes would be much more difficult to develop than CRCVD processes because of the increased degrees of freedom. For example, the timing of the ramp and the ramp rate must be carefully chosen so that it occurs as feature aspect ratios are increasing.

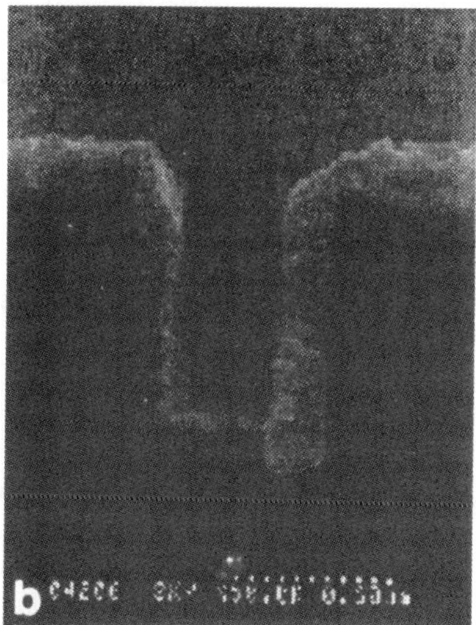

FIG. 37. (a) SEM of a tungsten film deposited using PRCVD. (b) SEM of a tungsten film deposited using CRCVD. For both films, the pressure was 2 Torr, the flow rate ratio was 15 : 1 ($H_2 : WF_6$), the flow rate was 600 sccm. For the PRCVD film, the starting temperature was 450°C, the final temperature was 360°C, and the temperature ramp was −3°C/s. The deposition time was 90 s. For the CRCVD case, the temperature was 360°C, and the deposition time was 450 s (*116*).

The PRCVD concept was developed by considering the desired trajectories in deposition conditions: the temperature and fluxes to the wafer surface (*113,114*). In our initial work, the temperature trajectories were *ad hoc*. In addition, we did not account for the changes on the reactor scale which naturally occur as deposition rate decreases. Conversion of reactants decreases, leading to increased reactant partial pressures at the wafer surface. Recently, we have theoretically demonstrated how to determine optimal temperature trajectories using feature scale simulations, and to use reactor scale simulations to develop model based methods to control the reactant partial pressures at the wafer surface (*118–120*). Thus, modeling and simulation are providing the framework for model-based process design and control.

XIII. Conclusions

We have presented a selected subset of our studies over the years, which were aimed at elucidating the process fundamentals critical to processes of interest to the microelectronics industry. We limited our attention to feature scale simulation studies, although an exciting new area of interest is that of coupling reactor scale and feature scale models and simulators (see, e.g., Refs. *121–124*). In addition, we dealt only with our work on low-pressure deposition processes, neglecting high-pressure depositions, etches, and thin film flow processes. While we demonstrated how we combine experimental information with simulation efforts, we did not thoroughly discuss our approach to estimating model parameters and to designing experiments (see Refs. *125* and *126*, as well as many of the referenced works throughout the chapter). We also did not report on several studies which used simulations to help answer processing questions; instead we concentrated on projects designed to improve our process understanding. A particularly interesting application of feature scale simulations to process development is illustrated in Ref. *127*.

Nor did we consider how surfaces are moved during simulations, which is an area of major interest today. We use a front tracking approach for multiple materials, which uses a conservation law formulation in an attempt to conserve matter (*128–130*). A new approach to moving surfaces, which shows great promise, is the level set formulation which has been made popular by Sethian and coworkers (e.g., see Refs. *131* and *132*). This approach is particularly promising for evolution of completely 3-D surfaces.

We consider the major growth front in the area of deposition simulation to be the prediction and tracking of grain structure evolution during processing. This will require an extension of current atomistic approaches to simulating nucleation and cluster evolution during the early stages of deposition to complete coverage and grain boundary formation. These are issues to be resolved before we can develop first principles simulation codes for these problems.

Acknowledgments

We gratefully acknowledge the financial support from a number of sources: the Semiconductor Research Corporation (and its member companies), the National Science Foundation, Motorola, Sematech, the Materials Research Corporation, the Air Force Office of Scientific Research, the Office of Naval Research, and the Advanced Research Projects Agency.

A number of students and postdoctoral associates participated in the work discussed in this chapter. Their names appear in the author lists of the referenced material. Greg Raupp, a colleague at ASU, was central to several of the studies. A number of industrial collaborators participated in the work, or provided good problems and suggestions; in particular, we have benefitted significantly through technical interactions with Tom Zirkle, Erik Egan, and

Len Borucki of Motorola; Andrew Labun of Digital Equipment Corporation, and Anthony Toprac of SEMATECH. Of course, there have been a number of people who have helped in nontechnical ways, most notably Mike Witty, Bill Lynch, Bill Holton, and Bill Atkins at the Semiconductor Research Corporation, and Ed Hall at Motorola.

References

1. A. Challa, T. S. Cale, and J. Drucker, *in* "Evolution of Epitaxial Structure and Morphology" (A. Zangwill, D. Jesson, D. Chambliss, and R. Clarke, eds.), MRS Symp. Proc. Vol. 399. MRS, Pittsburgh, Pennsylvania, 1996.
2. A. Challa, T. S. Cale, and J. Drucker, *in* "Evolution of Thin Film and Surface Structure and Morphology" (B. G. Demczyk, E. Garfunkel, B. M. Clemens, E. D. Williams, and J. J. Cuomo, eds.), MRS Symp. Proc., Vol. 355, p. 101. MRS, Pittsburgh, Pennsylvania, 1995.
3. A. Challa, "Microstructural Evolution During the Early Stages of Deposition: Monte Carlo Simulations." Ph.D. Dissertation, Arizona State University, Tempe, Arizona (May, 1996).
4. H. Liao and T. S. Cale, *J. Vac. Sci. Technol. B*, in press.
5. G. Thallikar, H. Liao, T. S. Cale, and F. R. Myers, *J. Vac. Sci. Technol. B* **13**, 1875 (1995).
6. H. Liao and T. S. Cale, *Thin Solid Films* **253**, 419 (1994).
7. H. Liao, "High Pressure Chemical Vapor Deposition and Thin Film Thermal Flow Process Simulations." Ph.D. Dissertation, Arizona State University, Tempe, Arizona (December, 1995).
8. T. S. Cale and G. B. Raupp, *J. Vac. Sci. Technol. B* **8**, 1242 (1990).
9. T. S. Cale, T. H. Gandy, and G. B. Raupp, *J. Appl. Phys.* **68**, 3645 (1990).
10. T. S. Cale and G. B. Raupp, *J. Vac. Sci. Technol. B* **8**, 649 (1990).
11. H. Liao, J. J. Hsieh, A. J. Toprac, J. N. Musher, R. G. Gordon, B. H. Weiller, and T. S. Cale, *in* "Advanced Metallization for ULSI Applications 1994" (R. Blumenthal and G. C. A. M. Janssen, eds.), p. 231. MRS, Pittsburgh, Pennsylvania, 1995.
12. H. Liao and T. S. Cale, *J. Vac, Sci. Technol. A* **12**, 1020 (1994).
13. D. G. Coronell and K. F. Jensen, *J. Electrochem. Soc.* **139**, 2264 (1992).
14. G. J. Parker, W. N. G. Hitchon, and D. J. Koch, *Phys. Rev.* **E51**, 3694 (1995).
15. A. Kersch, W. Morokoff, and C. Werner, *J. Appl. Phys.* **75**, 2278 (1994).
16. E. Meeks, H. K. Moffat, J. F. Grcar, and R. J. Kee, AURORA: A FORTRAN Package for Modeling Well Stirred Plasma and Thermal Reactors with Gas and Surface Reactions, Sandia Report-SAND96-8218.UC-405, February 1996.
17. T. S. Cale, Users' Guide for EVOLVE, a low pressure deposition and etch process simulator developed with funding from the Semiconductor Research Corporation, the National Science Foundation and Motorola. EVOLVE Vol. 2.0 was released in 1991, Vol. 3.0 was released in 1992, Vol. 4.0 was released in 1994 and Vol. 4.1 was released in 1995.
18. M. Fujinaga, I. Torrori, T. Kunikiyo, T. Uchida, N. Kotani, and K. Tsukamoto, *IEEE Trans. Comp. Aided Des. Int. Cir. Syst.* **14**, 631 (1995).
19. E. W. Scheckler and A. R. Neureuther, *IEEE Transactions of Computer Aided Design of IC's and Systems* **13**, 219 (1994).
20. J. Lorenz, (ed.), "3-Dimensional Process Simulation." Springer-Verlag, Berlin, 1995.

21. H. Liao and T. S. Cale, *Thin Solid Films* **236**, 352 (1993).
22. T. S. Cale, M. B. Chaara, H. Liao, and G. B. Raupp, *in* "Advanced Metallization for ULSI Applications in 1993" (Y. Shacham, D. Favreau, and Y. Horiike, eds.), p. 159. MRS, Pittsburgh, Pennsylvania, 1994.
23. C. G. Hill, "An Introduction to Chemical Engineering Kinetics and Reactor Design," Wiley, New York, 1977.
24. A. Yuuki, Y. Matsui, and K. Tachibana, *Jpn. J. Appl. Phys.* **28**, 212 (1989).
25. M. M. IslamRaja, M. A. Capelli, J. P. McVittie, and K. C. Saraswat, *J. Appl. Phys.* **70**, 7137 (1991).
26. D. G. Coronell and K. F. Jensen, *in* "Evolution of Surface and Thin Film Microstructure" (H. A. Atwater, E. Chason, M. H. Grabow, and M. G. Lagally, eds.), MRS Symp. Proc. Vol. 280, p. 169. MRS, Pittsburgh, Pennsylvania, 1993.
27. T. T. Kodas and M. J. Hampden-Smith (eds.), "The Chemistry of Metal CVD." VCH Verlag, Weinhelm, 1994.
28. T. S. Cale, G. B. Raupp, J.-H. Park, M. K. Jain, and B. R. Rogers, *in* "First International Conference on Transport Phenomena in Processing" (S. Guceri, ed.), p. 127. Technomic Publ., 1993.
29. G. B. Raupp, T. S. Cale, M. K. Jain, D. Srivinas, and B. Rogers, *Thin Solid Films* **193**, 234 (1990).
30. R. J. Kee, F. M. Rupley, and J. A. Miller, "CHEMKIN-II: A FORTRAN Chemical Kinetics Package for the Analysis of Gas Phase Chemical Kinetics," Sandia Report-SAND89-8009B.UC-706, 1994.
31. M. E. Coltrin, R. K. Kee, and F. M. Rupley, "SURFACE CHEMKIN (Version 4.0): A FORTRAN Package for Analyzing Heterogeneous Chemical Kinetics at a Solid Surface-Gas-Phase Interface," Sandia Report-SAND90-8003C.UC-706, 1994.
32. P. Ho, R. J. Buss, M. E. Coltrin, H. K. Moffat, and C. F. Melius, *in* "Proceedings of the Seventh Annual Dielectrics and CVD Metallization Symposium," p. 297. San Diego, 1995.
33. J. E. Crowell, L. L. Tedder, H. C. Cho, F. M. Cascarano, and M. A. Logan, *J. Electron Spectrosc. Relat. Phenom.* **54/55**, 1097 (1990).
34. S. B. Desu, *J. Am. Ceram. Soc.* **72**, 1615 (1989).
35. T. S. Cale, *J. Vac. Sci. Technol. B* **9**, 2551 (1991).
36. T. S. Cale, *in* "Modeling and Simulation of Thin-Film Processing" (D. J. Srolovitz, C. A. Volkert, M. J. Fluss, and R. J. Kee, eds.), MRS Symp. Proc., Vol. 389, p. 95. MRS, Pittsburth, Pennsylvania, 1995.
37. V. K. Singh, E. S. G. Shaqfeh, and J. P. McVittie, *J. Vac. Sci. Technol. B* **10**, 1091 (1992).
38. J. J. Hsieh and R. V. Joshi, *in* "Advanced Metallization for ULSI Applications 1991" (V. V. S. Rana, R. V. Joshi, and I. Ohdomari, eds.), p. 77. MRS, Pittsburgh, Pennsylvania, 1992.
39. M. Dalvie, R. T. Faruoki, and S. Hamaguchi, *IEEE Transactions on Electron Devices* **39**, 1090 (1992).
40. Y. Akiyama, S. Matsumura, and N. Imaishi, *Jpn. J. Appl. Phys.* **34**, 6171 (1995).
41. C. Cercignani, "The Boltzmann Equation and Its Applications." Springer-Verlag, Berlin, 1988.
42. G. N. Patterson, "Introduction to the Kinetic Theory of Gas Flows." Univ. of Toronto Press, Toronto, 1971.
43. T. S. Cale, G. B. Raupp, M. B. Chaara, and F. A. Shemansky, *Thin Solid Films* **220**, 66 (1992).
44. F. R. Myers and T. S. Cale, *J. Electrochem. Soc.* **139**, 3587 (1992).

45. F. R. Myers, M. Ramaswami, and T. S. Cale, *J. Electrochem. Soc.* **141**, 1313 (1994).
46. F. R. Myers, M. W. Peters, M. Ramaswami, and T. S. Cale, *Thin Solid Films* **253**, 522 (1994).
47. M. Ramaswami, "Monte Carlo Simulation of Plasma Sheaths in Single and Dual Frequency Plasma Systems." M.S. Thesis, Arizona State University, Tempe, Arizona (May, 1994).
48. Z. Lin and T. S. Cale, *Thin Solid Films* **270**, 627 (1995).
49. Z. Lin and T. S. Cale, *J. Vac. Sci. Technol. A* **13**, 2183 (1995).
50. Z. Lin, "Flux Distributions and Deposition Profiles from Hexagonal Collimators During Sputter Deposition." M.S. Thesis, Arizona State University, Tempe, Arizona (August, 1995).
51. Z. Lin, T. S. Cale, A. J. Toprac, S.-Q. Wang, and J. Schlueter, *in* "Advanced Metallization for ULSI Applications 1994" (R. Blumenthal and G. C. A. M. Janssen, eds.), p. 333. MRS, Pittsburgh, Pennsylvania, 1995.
52. A. J. Toprac, B. P. Jones, J. Schlueter, and T. S. Cale, *in* "Evolution of Thin Film and Surface Structure and Morphology" (B. G. Demczyk, E. Garfunkel, B. M. Clemens, E. D. Williams, and J. J. Cuomo, eds.), MRS Symp. Proc., Vol. 355, p. 355. MRS, Pittsburgh, Pennsylvania, 1995.
53. B. R. Rogers and T. S. Cale, *Thin Solid Films* **236**, 334 (1993).
54. C. C. Fang, V. Prasad, and F. Jones, *J. Vac. Sci. Technol. A* **11**, 2778 (1993).
55. T. S. Cale, M. K. Jain, R. Duffin, and C. J. Tracy, *J. Vac. Sci. Technol. B* **11**, 311 (1993).
56. D. S. Taylor, "Conformality of Deposited Interconnect and Contact Materials." M.S. Thesis, Arizona State University, Tempe, Arizona (May, 1994).
57. G. Sewell, "The Numerical Solution of Ordinary and Partial Differential Equations." Academic Press, San Diego, 1988.
58. G. B. Raupp, F. A. Shemansky, and T. S. Cale, *J. Vac. Sci. Technol. B* **10**, 2422 (1992).
59. T. S. Cale, M. B. Chaara, G. B. Raupp, and I. J. Raaijmakers, *Thin Solid Films* **236**, 294 (1993).
60. G. B. Raupp, T. S. Cale, and H. P. W. Hey, *J. Vac. Sci. Technol. B* **10**, 37 (1992).
61. M. Virmani, D. A. Levedakis, G. B. Raupp, and T. S. Cale, *J. Vac. Sci. Technol. A*, **14**, 977 (1996).
62. F. A. Shemansky, "Kinetics and Mechanism of Silicon Dioxide Deposition through Thermal Pyrolysis of Tetraethoxysilane." Ph.D. Dissertation, Arizona State University, Tempe, Arizona (December, 1991).
63. A. C. Adams and C. D. Capio, *J. Electrochem. Soc.* **126**, 1042 (1979).
64. J. Schlote, K.-W. Schroder, and K. Drescher, *J. Electrochem. Soc.* **138**, 2393 (1991).
65. M. M. IslamRaja, C. Chang, and J. P. McVittie, *J. Vac. Sci. Technol.* **11**, 720 (1993).
66. E. A. Haupfear, E. C. Olson, and L. D. Schmidt, *J. Electrochem. Soc.* **141**, 1943 (1994).
67. M. E. Bartram and H. K. Moffat, *in* "Proceedings of the Thirteenth International Conference on Chemical Vapor Deposition," (T. M. Besmann, M. D. Allendorf, McD. Robinson, and R. K. Ulrich, eds.), Electrochemical Society, PV96-5, p. 842, 1996.
68. T. S. Cale, G. B. Raupp, and T. H. Gandy, *J. Vac. Sci. Technol. A* **10**, 1128 (1992).
69. T. S. Cale and G. B. Raupp, *in* "Plasma Properties, Deposition and Etching" (J. J. Pouch and S. A. Alterovitz, eds.), p. 1. Trans. Tech. Pub., Aedermannsdorf, Switzerland, 1993.
70. D. E. Ibbotson, J. J. Hsieh, D. L. Flamm, and J. A. Mucha, Spring Meeting of the Electrochemical Society 1990, Extended Abstract 131.

71. J. J. Hsieh, D. E. Ibbotson, J. A. Mucha, and D. L. Flamm, *in* "Proceedings of the Sixth International IEEE VLSI Multilevel Interconnection Conference" (T. E. Wade, ed.), p. 411. IEEE, 1989.
72. C.-P. Chang, C. S. Pai, and J. J. Hsieh, *J. Appl. Phys.* **67**, 2119 (1990).
73. N. Selamoglu, J. A. Mucha, D. E. Ibbotson, and D. L. Flamm, *J. Vac. Sci. Technol. B* **7**, 1345 (1989).
74. J. C. Greaves and J. W. Linnett, *Trans. Far. Soc.* **55**, 1355 (1959).
75. T. S. Cale, T. H. Gandy, M. K. Jain, G. B. Raupp, M. Govil, and A. Hasper, *in* "Advanced Metallizations for ULSI Applications" (V. V. S. Rana, R. V. Joshi, and I. Ohdomari, eds.), p. 93. MRS, Pittsburgh, Pennsylvania, 1992.
76. T. E. Zirkle, C. I. Drowley, W. G. Cowden, and T. S. Cale, *Thin Solid Films* **220**, 45 (1992).
77. G. B. Raupp, D. A. Levedakis, and T. S. Cale, *J. Vac. Sci. Technol. A* **13**, 676 (1995).
78. M. Virmani, "Experimental Study of Remote Microwave Deposition of Fluorinated Silicon Dioxide from FASi-4." M.S. Thesis, Arizona State University, Tempe, Arizona (May 1996).
79. C. H. Chou and J. Phillips, *J. Appl. Phys.* **68**, 2415 (1990).
80. C. F. Hoener, E. Pylant, E. G. Boden, and S.-Q. Wang, *J. Vac. Sci. Technol. B* **12**, 1394 (1994).
81. D. Liu, S. K. Dew, and M. J. Brett, *J. Appl. Phys.* **72**, 1339 (1993).
82. W. Tsai, M. L. Brett, and R. N. Tait, *Thin Solid Films* **253**, 386 (1994).
83. B. R. Rogers, C. J. Tracy, and T. S. Cale, *J. Vac. Sci. Technol. B* **12**, 2980 (1994).
84. B. R. Rogers, Y.-K. Chang, and T. S. Cale, *J. Vac. Sci. Technol. A* **A14**, 1142 (1996).
85. D. Liu, S. K. Dew, M. J. Brett, T. Smy, and W. Tsai, *in* "Proceedings of the Eleventh International VLSI Multilevel Interconnection Conference" (T. Wade, ed.), p. 478. VMIC, 1995.
86. L.-Y. Cheng, J. P. McVittie, and K. C. Saraswat, *Appl. Phys. Lett.* **58**, 2147 (1991).
87. D. B. Bergstrom, F. Tian, I. Petrov, J. Moser, and J. E. Greene, *Appl. Phys. Lett.* **67**, 3102 (1995).
88. M. Virmani, V. Madadev, and T. S. Cale, *in* "Proceedings of the Second International Dielectrics for VLSI/ULSI Multilevel Interconnection Conference," p. 139. VMIC, 1996.
89. S. M. Rossnagel and J. Hopwood, *J. Vac. Sci. Technol. B* **12**, 449 (1994).
90. S. M. Rossnagel and J. Hopwood, *Appl. Phys. Lett.* **63**, 3285 (1993).
91. S. Hamaguchi and S. M. Rossnagel, *J. Vac. Sci. Technol. B* **13**, 183 (1995).
92. A. Bose, M. M. Garver, and R. A. Spencer, *in* "Proceedings of the Tenth International VLSI Multilevel Interconnect Conference" (T. Wade, ed.), p. 89. VMIC, 1993.
93. A. H. Labun, *J. Vac. Sci. Technol. B* **12**, 3138 (1994).
94. R. E. Lee, *J. Vac. Sci. Technol.* **16**, 164 (1979).
95. R. V. Stuart and G. K. Wehner, *J. Appl. Phys.* **33**, 2345 (1962).
96. A. Aronson, I. Wagner, and M. G. Ward, *in* "Transactions of the 49th and 50th Schools on Thin Film Technology." Materials Research Corporation, 1991.
97. T. S. Cale, T. H. Gandy, G. B. Raupp, and B. Ramaswami, *Thin Solid Films* **206**, 54 (1991).
98. C. Liu, J. M. Cohen, J. B. Adams, and A. F. Voter, *Surf. Sci.* **253**, 334 (1991).
99. W. W. Mullins, *J. Appl. Phys.* **28**, 333 (1957).

100. P. G. Shewmon, "Diffusion in Solids." J. Williams Book, 1983.
101. G. Newmann and G. M. Newmann, "Surface Self-Diffusion of Metals." Diffusion Monograph Series-1, 1972.
102. N. A. Gjostein and J. P. Hirth, *ACTA Metallurgica* **13**, 991 (1965).
103. N. A. Gjostein, *Transactions of the Metallurgical Society of AIME* **221**, 1039 (1961).
104. P. J. Feibelman, *Phys. Rev. Lett.* **58**, 2766 (1987).
105. P. J. Feibelman, *Phys. Rev. Lett.* **65**, 729 (1990).
106. W. R. Tyson and W. A. Miller, *Surf. Sci.* **62**, 267 (1977).
107. L. L. Levenson, *Appl. Phys. Lett.* **55**, 2617 (1989).
108. I. J. Raaijmakers, R. A. A. Hack, and A. G. Dirks, *in* "Proceedings of the International Symposium on Trends and New Applications in Thin Films," p. 479. Strasbourg, France, 1987.
109. R. Brain, Ph.D. Dissertation, California Institute of Technology, Pasadena (1995).
110. M. K. Jain, "Maximizing Step Coverage at High Throughput During Low Pressure Deposition Processes." Ph.D. Dissertation, Arizona State University, Tempe, Arizona (December, 1992).
111. T. S. Cale, G. B. Raupp, and M. K. Jain, *Thin Solid Films* **193**, 51 (1990).
112. T. S. Cale, M. K. Jain, and G. B. Raupp, *J. Electrochem. Soc.* **137**, 1526 (1990).
113. T. S. Cale, M. K. Jain, and G. B. Raupp, *in* "Tungsten and Other Advanced Metals for ULSI/VLSI Applications V" (S. S. Wong and S. Furukawa, eds.), p. 179. MRS, Pittsburgh, Pennsylvania, 1990.
114. K. M. Tracy, S. Bolnedi, and T. S. Cale, *in* "Proceedings of the 12-th International VLSI Multilevel Interconnection Conference" (T. Wade, ed.), p. 643. VMIC, 1995.
115. K. M. Tracy, S. Bolnedi, G. J. Leusink, and T. S. Cale, *in* "Advanced Metallization and Interconnect Systems for ULSI Applications in 1995," in press. MRS, Pittsburgh, Pennsylvania, 1995.
116. K. M. Tracy, "Programmed Rate Chemical Vapor Deposition: Blanket Tungsten Film Characterization," M.S. Thesis, Arizona State University, Tempe, Arizona (May, 1996).
117. G. J. Leusink, T. G. M. Oosterlaken, G. C. A. M. Janssen, and S. Radelaar, *in* "Advanced Metallization for ULSI Applications in 1993" (D. P. Favreau, Y. Shacham-Diamond, and Y. Horiike, eds.), p. 335. MRS, Pittsburgh, Pennsylvania, 1994.
118. P. E. Crouch, L. Song, K. S. Tsakalis, and T. S. Cale, *in* "Proceedings of the Fourth IEEE/UCS/SEMI International Symposium on Semiconductor Manufacturing," p. 233. IEEE, 1995.
119. T. S. Cale, P. E. Crouch, L. Song, and K. S. Tsakalis, *in* "Proceedings of the 1995 American Control Conference," p. 1289. American Control Council, IEEE, 1995.
120. T. S. Cale, P. E. Crouch, L. Song, and K. S. Tsakalis, *in* "Proceedings of the Symposium on Process Control, Diagnostics and Modeling in Semiconductor Manufacturing" (M. Meyyappan, D. J. Economou, and S. W. Butler, eds.), p. 97. Electrochemical Society, Proc. Vol. 95-2, 1995.
121. T. S. Cale, J.-H. Park, T. H. Gandy, G. B. Raupp, and M. K. Jain, *Chem. Eng. Commun.* **119**, 197 (1993).
122. M. K. Gobbert, T. S. Cale, and C. A. Ringhofer, *in* "Proceedings of the Symposium on Process Control; Diagnostics and Modeling in Semiconductor Manufacturing" (M. Mayyappan, D. J. Economou, and S. W. Butler, eds.), p. 553. Electrochemical Society, Proc. Vol. 95-2, 1995.
123. M. K. Gobbert, C. A. Ringhofer, and T. S. Cale, *J. Electrochem. Soc.* **143**(8), in press (1996).
124. M. Gobbert, T. S. Cale, and C. Ringhofer, *VLSI Design*, in press.

125. S. Bolnedi, D. Wang, G. B. Raupp, and T. S. Cale, *in* "Advanced Metalization and Interconnect Systems for ULSI Applications in 1995," p. 523. MRS, Pittsburgh, Pennsylvania, 1995.
126. M. B. Chaara and T. S. Cale, *Thin Solid Films* **220**, 19 (1992).
127. T. E. Zirkle, S. R. Wilson, S. Sundaram, T. S. Cale, and G. B. Raupp, *J. Vac. Sci. Technol. A* **11**, 905 (1993).
128. D. S. Ross, *J. Electrochem. Soc.* **135**, 1260 (1988).
129. J. C. Arnold, H. H. Sawin, M. Dalvie, and S. Hamaguchi, *J. Vac. Sci. Technol. A* **12**, 620 (1994).
130. G. Rajagopalan, V. Mahadev, and T. S. Cale, *VLSI Design*, in press.
131. D. Adalsteinsson and J. A. Sethian, *J. Comput. Phys.* **120**, 128 (1995).
132. Z. Hsiau, E. C. Kan, J. P. McVittie, and R. W. Dutton, *in* "Proceedings of the International Electron Devices Meeting," p. 101. IEEE, 1995.

Author Index

Numbers in parentheses are reference numbers; the complete citations can be found at the end of each chapter.

D

E

F

G

H

I

J

K

L

M

N

O

P

R

S

T

Subject Index

I

K

L

M

T

V

W

Recent Volumes in this Series

Maurice H. Francombe and John L. Vossen, *Physics of Thin Films*, Volume 16, 1992.

Maurice H. Francombe and John L. Vossen, *Physics of Thin Films*, Volume 17, 1993.

Maurice H. Francombe and John L. Vossen, *Physics of Thin Films, Advances in Research and Development, Plasma Sources for Thin Film Deposition and Etching*, Volume 18, 1994.

K. Vedam (guest editor), *Physics of Thin Films, Advances in Research and Development, Optical Characterization of Real Surfaces and Films*, Volume 19, 1994.

Abraham Ulman, *Thin Films, Organic Thin Films and Surfaces: Directions for the Nineties*, Volume 20, 1995.

Maurice H. Francombe and John L. Vossen, *Homojunction and Quantum-Well Infrared Detectors*, Volume 21, 1995.

ISBN 0-12-533022-7